Understanding Aeronautical Charts

Understanding Aeronautical Charts

Terry T. Lankford

FIRST EDITION
FIRST PRINTING

TAB Books is a division of McGraw-Hill, Inc.

Library of Congress Cataloging-in-Publication Data

Lankford, Terry T.
Understanding aeronautical charts / by Terry T. Lankford.
p. cm.
Includes bibliographical references and index.
ISBN 0-8306-3912-8 ISBN 0-8306-3911-X (p)
1. Aeronautical charts. I. Title.
TL587.L36 1992
629.132′54—dc20 91-33019
CIP

TAB Books offers software for sale. For information and a catalog, please contact TAB Software Department, Blue Ridge Summit, PA 17294-0850.

Acquisitions Editor: Jeff Worsinger
Editor: Norval G. Kennedy
Director of Production: Katherine G. Brown
Book Design: Jaclyn J. Boone
Cover photograph by Thompson Photography, Baltimore, MD. TPFS

Contents

Introduction

AN ESSENTIAL PART OF FLIGHT PREPARATION IS THE ACQUISITION, interpretation, and application of aeronautical charts and related publications. Any map user seeks information; a pilot needs specific details on terrain, airspace, landing areas, and navigational aids. The cartographer's task is to design a map with the least distortion for the intended purpose. The success of a map is dependent upon the cartographer and user's knowledge and on their joint realization of the purpose and limitations of the map.

No matter how short or simple the flight, Federal Aviation Regulations place the responsibility for flight preparation on the pilot. To effectively use available resources during preflight planning, a pilot must understand what information is available, how it is distributed, and how it can be obtained and applied. During the flight, a pilot must constantly use charted information for navigation and communications.

Almost since the beginning of commercial aviation it became apparent that charts could not contain all of the vast and varied information necessary for a safe and efficient flight. Because of the expense of chart production, charts could not economically be updated at every change. To this end aeronautical publications were developed. Publications primarily provide planning information, data to update charted information, or data that remains relatively unchanged.

In *Understanding Aeronautical Charts*, technical concepts and terms are explained in nontechnical terms, progressing beyond decoding and translating, to interpreting and applying information to actual flight situations. Discussions include applying chart information to visual flight rule (VFR) and instrument flight rule (IFR) operations, plus low level and high level flights.

Chart limitations are discussed. There is an examination of chart information and publication usage while flight planning using direct user access terminals (DUAT). This is a sound foundation for the novice and a practical review for the experienced pilot; a thorough knowledge of aeronautical charts and their relationship to the air traffic control system is essential to a safe, efficient operation.

The strictly VFR pilot should not overlook the information in the chapters on instrument charts. The VFR pilot might want to utilize products designed for instrument flying. For example, some pilots supplement visual charts with enroute low altitude charts for the enroute chart information (airway radials, minimum enroute altitudes, ATC frequencies). Instrument approach procedure charts feature all related communication and navigation frequencies with an airport sketch or detailed diagram for larger airports, which can be very valuable to the VFR pilot. Any pilot planning a flight into a major airport should obtain a copy of the airport sketch or airport diagram.

The importance of chapters explaining aeronautical publications cannot be over stressed because these publications are chart extensions, as important as the charts themselves. Supplemental publications provide information on how to use the air traffic control system, airspace, and procedures. Various commercial supplements are available that might go beyond flying data to include airport diagrams, plus commercial information about airport services, transportation, and lodging.

This book provides information that is required to pass written and practical examinations, which prepares the dispatcher as well as the pilot—from student through airline transport, from recreational airplane to business jet—to operate safer and more efficiently within an ever increasingly complex environment.

Chapters 1 and 2 provide the reader with a general background on the development of maps and charts, especially aeronautical charts. I've always felt that to really understand a subject it was essential to understand its history, development, and how it has evolved to the present.

We begin with a brief history of maps and aeronautical charts. Dedicated aerial maps were not produced until World War I. It took commercial aviation, beginning with the airmail routes, to spawn any active interest in aeronautical charts in the United States. It soon became evident that charts were needed to cover the entire country. And, World War II required charts for the world. The increase in commercial aviation along with the new radio navigation aids caused the development of radio facility and airport letdown charts. With the wealth of information available, charts could no longer display all the information needed by the pilot. To supplement charts, a series of publications evolved. Today a pilot has access to visual and instrument charts, and publications for the world.

Cartographer limitations are examined. How do we transfer a globe onto a flat surface, then locate a specific point? The various chart projections, each with its advantages and limitations, are discussed, along with limitations common to all charts; methods and criteria used by the cartographer are explained. The problems of scale, simplification, and classifications are considered. Another section discusses the problem of currency. In an ever changing environment, how does a pilot obtain the latest information required for a safe and efficient flight? A pilot must be fully aware of the system used to update aeronautical information. This includes the information available from flight service stations and DUAT weather briefings, the services available through commercial chart producers, and their limitations.

Chapters 3 through 5 cover visual charts. A pilot has access to charts covering the world. These charts are of little value without the knowledge and understanding to apply their vast wealth of information. The chapters begin with the terminology and symbols used on visual charts, then go on to explore visual charts used in routine flying, and supplemental charts available for other parts of the world and those used for special purposes. Dozens of different charts have been developed for aviation. Each has its own criteria, purpose, and limitations; each is analyzed. The pilot can then apply the array of charts available, with respect to regulations, type of operation, and pilot ability, to efficiently operate within today's aviation system.

Visual chart terminology and symbols are discussed. Topographical features of terrain, hydrography—water and drainage features, culture—manmade objects, and obstructions, are defined and chart symbology is explained. Next, aeronautical terminology and symbols for airports, NAVAIDs, and airspace are considered. From this analysis, and a practical application of terminology and symbols, the reader should be able to define and explain any visual chart symbol, and apply their meaning to various flight situations.

A thorough analysis of standard visual charts includes visual planning, sectional, terminal area, and world aeronautical charts, which are used most often for VFR flying. A pilot should be able to choose the best product for the planned flight; inappropriate chart selection has often led to pilot disorientation, and, unfortunately, at times to fatal accidents.

Defense Mapping Agency (DMA) and National Ocean Service (NOS) charts use similar format, and provide visual chart coverage for most areas of the world. DMA products are of limited value in the U.S. because of NOS chart series; however, some DMA charts might be helpful for planning, or other specialized missions. Special NOS charts are also presented, such as the Grand Canyon VFR aeronautical chart, which provides guidance through canyon airspace affected by a special Federal Aviation Regulation. Charts and related material published by foreign countries and one private publisher, Jeppesen Sanderson, are explored. This should give the reader a sense of the products available, but sometimes overlooked.

Chapters 6 through 8 examine instrument procedure charts. Since the late 1960s, more and more pilots are obtaining an instrument rating. General aviation aircraft of today are approaching and exceeding commercial aircraft performance of 1940s and 1950s. Each chart series is analyzed with respect to use, regulations, and limitations. Jeppesen Sanderson is a primary supplier of nongovernment charts; therefore, its charts and services are examined.

This book is not intended to make recommendations regarding chart publishers, merely to provide the reader with information for educated choices. It is the pilot's responsibility to select the chart and information publisher that best suits his or her needs.

A number of points can be inferred from these chapters. The FAA and other Government Agencies have gone to great lengths to establish a safe airway system. But, the pilot's safety, and that of the passengers and those on the ground, ultimately rests with a pilot's chart knowledge. The pilot must know what equipment is available and operational on the aircraft. Certain routes and procedures can only be flown with specific equipment, such as ADF or DME. The pilot must then correctly interpret the chart, refusing any ATC clearance that cannot be complied with due to equipment deficiencies.

The instrument chart chapters contain interrelated material; therefore, the reader is occasionally asked to refer to a previous chapter, usually a specific figure number. I apologize for any inconvenience, but this seemed the best way to integrate the material.

The IFR-specific chapters begin with enroute instrument charts. This might seem like starting in the middle, with subsequent chapters on departure and arrival procedures, and approaches; however, because most terminology and symbols on enroute charts also apply to the other chart series it was apparently a logical starting point. A solid understanding of enroute products is a foundation for examining specific charts used for departure, arrival, and approach.

Standard instrument departure (SID) and standard terminal arrival route (STAR) charts are discussed. Once the domain of air carrier, corporate, and military pilots, and contained in separate publications, SIDs and STARs are an integral part of today's IFR system, and now published along with their associated instrument approach procedures. A pilot's acceptance of a SID or STAR is an agreement to comply with the requirements of the procedure. In addition to terminology and symbols, procedural requirements are explored. For example, what is the pilot expected to do when radio communications are lost during a radar vector STAR?

Instrument approach procedure (IAP) charts are dissected. Each item of information is decoded, defined, and explained. A half dozen different instrument approach procedures, including RNAV and loran RNAV approaches, are analyzed with respect to information available, pilot procedures and requirements, and lost communications. NOS and Jeppesen products are compared, with an explanation of the advantages and disadvantages of each service. A thorough understanding of approach plates should provide the pilot with the knowledge to apply any approach procedure to a flight situation.

The final chapters discuss and analyze publications that support aeronautical charts. Recall that aviation complexities forced selected information off the chart and into supplemental publications. Aeronautical publications that support NOS VFR and IFR charts are discussed. These publications, which are direct extensions of charts, provide the detailed information beyond the scope of charts, and serve as interim documents to update charts between publication cycles or during periods of temporary, but extended, outages. Details on how to obtain NOS aeronautical charts and related publications are provided.

Additional supplementary information available to the pilot is discussed. These documents should not be overlooked. They include DMA's flight information publications, Canadian supplements, additional documents available through NOS, and those produced by private vendors. Included are addresses and telephone numbers for supplement product providers, many of which offer free catalogs.

The following chapters, hopefully told with a little humor and practical examples of application, explain how to use, translate, interpret, and apply aeronautical charts.

Navigation principles and techniques are not within the scope of this book; therefore, they are not included. The text is not intended to decode, translate, and interpret Notices to Airmen or provide a detailed explanation of the ins and outs of controlled airspace. The reader should remember that, especially in aviation, the only thing that doesn't change is change itself. Everything possible has been done to ensure that the information in this text is current and accurate at the time of writing.

Finally, I must thank Jeppesen Sanderson for its extraordinary helpfulness and cooperation providing information about the company's history and also providing permission to reproduce selected charts and approach plates.

1
Charts history

THE MAP OR CHART IS A UNIQUE MEANS OF STORING AND COMMUNICATING geographic or other information. It allows the user to reduce the world to a size within the normal range of vision. It is the most effective means of relating the relative location, size, direction, and distance of locations and objects on the earth.

For our purposes we will define *map* as a graphic representation of the physical features, natural, artificial, or both, of the earth's surface, by means of signs and symbols, at an established scale, on a specified projection, and with a means of orientation. A *chart* is a special purpose map designed for navigation. And specifically an aeronautical chart is a chart designed to meet the requirements of aerial navigation.

Although the origin of the map has been lost, it appears it was developed independently among many peoples. A Babylonian map of the sixth or fifth century B.C. is the earliest surviving world map. At about the same time came the first known references to maps of Greek origin. The Romans produced the first known road maps.

Ptolemy, a second century A.D. Egyptian mathematician, astronomer, and geographer, developed principles for map making, including divisions and coordinates. His reference system used lines parallel with and at equal intervals from the equator to the poles, and lines north and south at right angles to the parallels, equally spaced along the equator: *latitude* and *longitude*. Ptolemy acknowledged his work was based on Hipparchus (second century B.C.) who was the first geographer to use parallels of latitude and meridians of longitude. Interestingly, Columbus, more than 1,000 years later, based his reasoning that if you sailed west you would reach the east, on Ptolemy's assumption that the world was round.

In the 16th century, Gerardus Mercator began working on the problem of transferring a sphere, the earth, onto a flat map. He reintroduced latitude and longitude, which had only been sporadically used. His maps allowed the navigator to draw a straight line between locations and determine a constant course. In 1569, he introduced his first map of the world using the *Mercator projection*.

By the 18th century, the German mathematician, astronomer, and physicist Johann Heinrich Lambert improved the cone projection derived from Ptolemy to conform with two standard parallels. He developed a half dozen projections, one of which was *Lambert conformal conic*; however, it was more than a century before cartographers fully appreciated the value of these projections, and not until World War I was it adopted by the allies for military maps.

With commerce growing, it became apparent that map makers needed a common reference for coordinates. To this end, in October 1884 the first International Meridian Conference adopted the Observatory of Greenwich, England, as the prime meridian. Longitude was reckoned in two directions, east and west of the Greenwich meridian.

According to Walter W. Ristow in his book *Aviation Cartography*, the Prussian artillery officer and balloonist Hermann Moedebeck was the "self-styled 'spiritual godfather' of aeronautical charts . . . ." In 1888, Moedebeck called attention to the difficulty of distinguishing types of terrain from the air and observed that "good charts would, of course, overcome this disadvantage . . ."; however, it was almost two decades before anything further was written about air maps. Moedebeck retired from the army in 1908.

The third Congress of the International Aeronautical Federation met in Brussels in 1907, established an International Commission for Aeronautical Charts, and chose Hermann Moedebeck chairman. Each member country was to select existing maps and superimpose aeronautical symbols. Moedebeck organized the German chart effort and by 1909 published the first sheet chart series, considered the first true aeronautical charts.

There was little interest in aeronautical charts in the United States before World War I; however, the Europeans continued development, although, at the beginning of the war aeronautical charts were still in the experimental and development stage.

World War I stimulated interest in aeronautical charts in the United States. Lawrence Sperry, Sperry Gyroscope Company, had prepared an aeronautic map of Long Island in 1917. The map was presented to Rear Admiral Robert E. Peary, chairman of the Committee on Aeronautic Maps and Landing Places of the Aero Club of America.

Commercial aviation in the United States was launched on August 12, 1918, with the initiation of regular airmail service between Washington and New York. Pilots were forced to use railroad and road maps, or pages from atlases. As late as 1921, with transcontinental airmail operations day and night, no aeronautical charts existed. Pilots noted times and course between prominent landmarks. If they were lucky they flew two trips behind veteran pilots; if not, just one trip. Notes from various pilots were assembled and published by the Post Office Department. These *Pilot's Directions* contained distances, landmarks, compass courses, and emergency landing fields, with services and communications facilities at principal points along the route in a narrative form:

Hazelhurst Field, Long Island.—Follow the tracks of the Long Island Railroad past Belmont Park racetrack, keeping Jamaica on the left. Cross New York over the lower end of Central Park.

Newark, N.J.—Heller field is located in Newark and may be identified as follows: The field is 1¼ miles west of the Passaic River and lies in the V formed by the Greenwood

Lake Division and Orange branch of the New York, Lake Erie and Western Railroad. The Morris Canal bounds the western edge of the field. The roof of the large steel hanger is painted an orange color.

These narrative checkpoints covered the routes at 10 to 25 mile intervals.

With the passage of the Air Commerce Act in 1926, the Department of Commerce became responsible for the production of aeronautical charts for the nation's airways. The first chart published in 1927 was a strip map that covered the air route from Kansas City to Moline, Ill. These early charts depicted prominent topographical features for visual flying and contained the locations of the newly installed airway lighted beacon system for night operations. The strip map concept was extended to other lighted airways between major airports throughout the late twenties.

The original airway beacon was a 24-inch rotating searchlight containing a parabolic mirror. It was powered by a 110-V, 1,000-W lamp that produced approximately 1,000,000 candlepower. Beacons were established at intervals of 10 to 15 miles along the airway. Rotating at six rpm, they produced a clear flash every 10 seconds. Beacons were supplemented by green or red coded flashes; green for beacons at landing fields (sites were numbered from west to east, or south to north, depending upon the direction of the airway); red or green course lights pointed in the direction of the airway.

Lights flashed a Morse code letter that identified the site. For simplicity, letters that contained fewer elements—dots and dashes—than numbers, were used. A sequence of letters evolved: W-U-V-H-R-K-D-B-G-M. These letters formed the popular mnemonic, "When Undertaking Very Hard Routes, Keep Direction By Good Methods." This sequence was repeated every 100 miles. Figure 1-1 shows a strip map for the route Los Angeles to Las Vegas, March 1931, with the airway beacons.

The strip map has a scale of 1:500,000, the same as today's sectional charts. Topography was shown as contours and color tints. Culture features included railroads, highways, cities and towns, and prominent electric transmission lines. Airports were shown by type, military or civilian. The lighted airways and new low frequency radio ranges were depicted.

Supplemental aeronautical information was published in the *Domestic Air News*, until 1929 when it was replaced by *Air Commerce Bulletins*. These publications contained official aviation information assembled and distributed by the Department of Commerce. These were the forerunners of today's *Airport/Facility Directory* and *Notice to Airmen* publication. In 1932, the free bulletin series described airports, intermediate landing fields, and meteorological conditions in the various states, along with the low frequency A/N radio ranges for air navigation. By 1940, bulletins contained notices to airmen, air navigation radio aids, danger areas in air navigation, and a directory of airports. They were prepunched notebook size and included airway radio facility charts.

With more flying conducted away from established airways, it became apparent that a system of charts was needed to provide complete coverage. Recommendations of the Committee on Aerial Navigation Maps in 1929 prompted the Coast and Geodetic Survey to develop a series of 92 sectional aeronautical charts for the United States. Sectionals were perhaps the best topographic maps; however, strip maps continued to be published until 1932 when the initial series of 31 sectionals was completed.

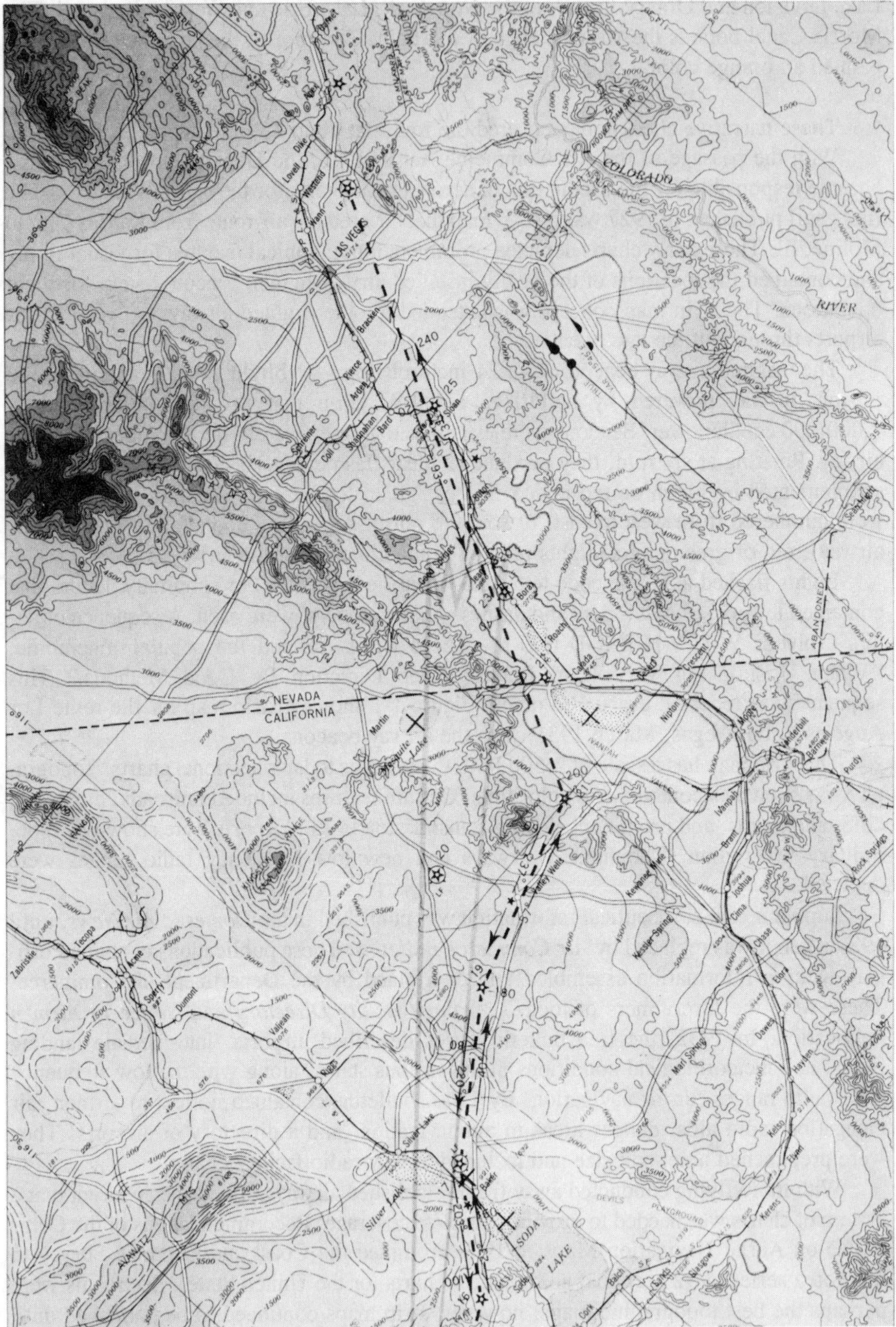

Fig. 1-1. This Los Angeles to Las Vegas strip map, dated March 1931, shows one of the lighted airways of the period.

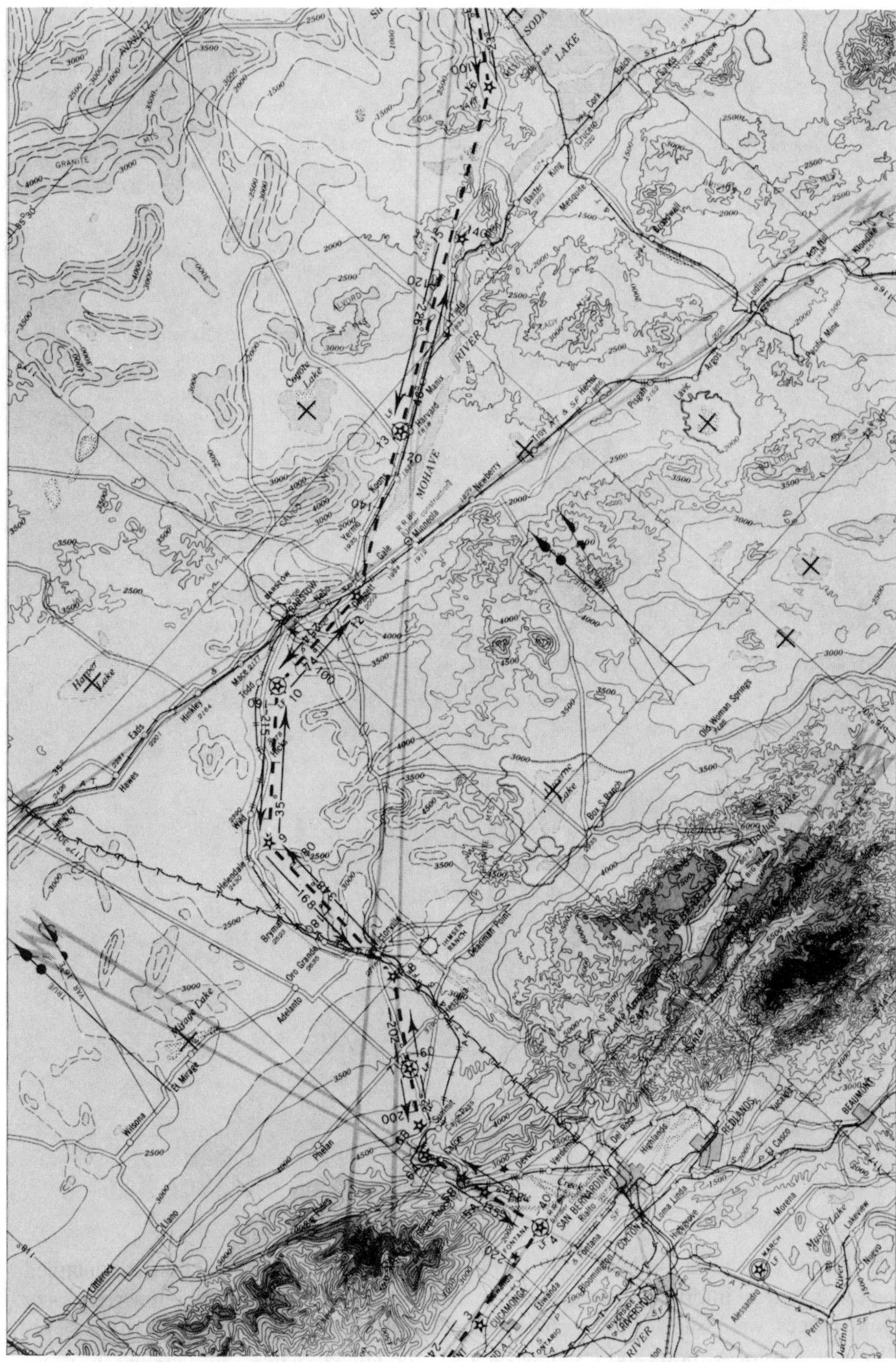

Fig. 1-1. Continued.

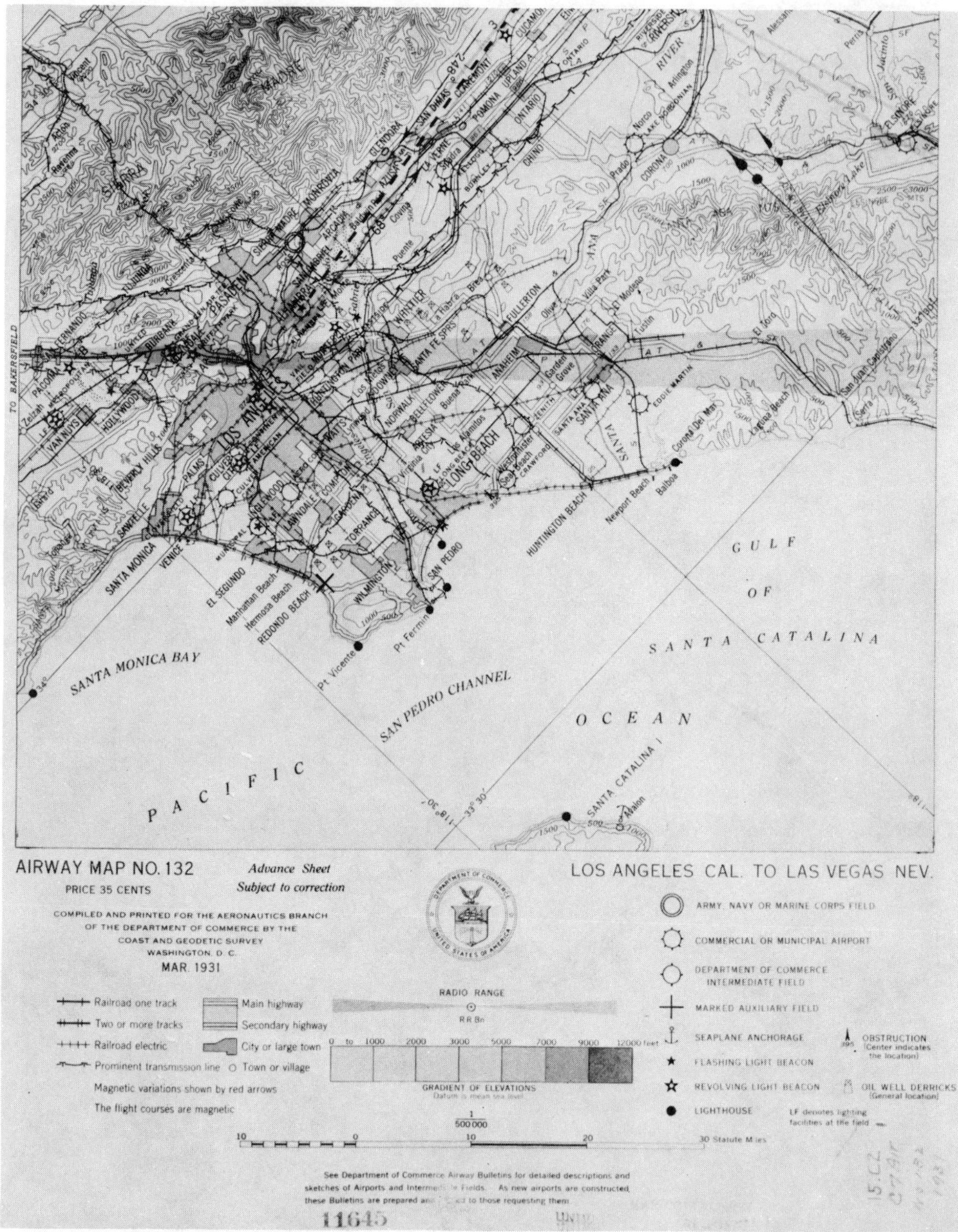

Fig. 1-1. Continued.

Figure 1-2 shows the June 1932 Los Angeles Sectional Chart. Sectionals contained the same general information as strip maps. Lighted airways as well as the low frequency radio ranges were shown. This chart refers the user to see Airway Bulletin No. 2, "Descriptions of Airports and Landing Fields in the United States," for detailed information on airports and landing fields. The Air Corps used strip maps until 1935 when the

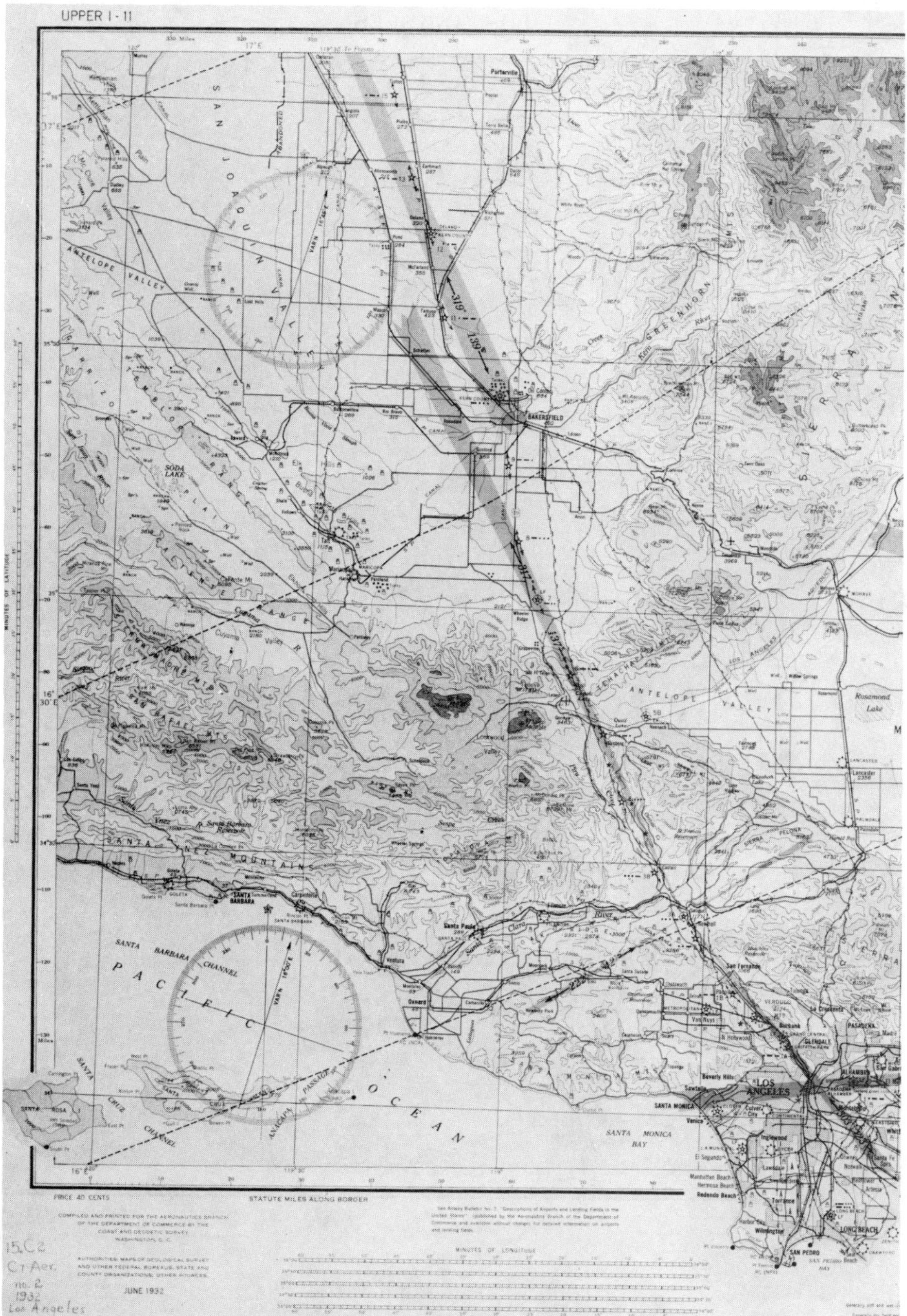

Fig. 1-2. This June 1932 Los Angeles sectional was one of the first in the second generation of aeronautical charts.

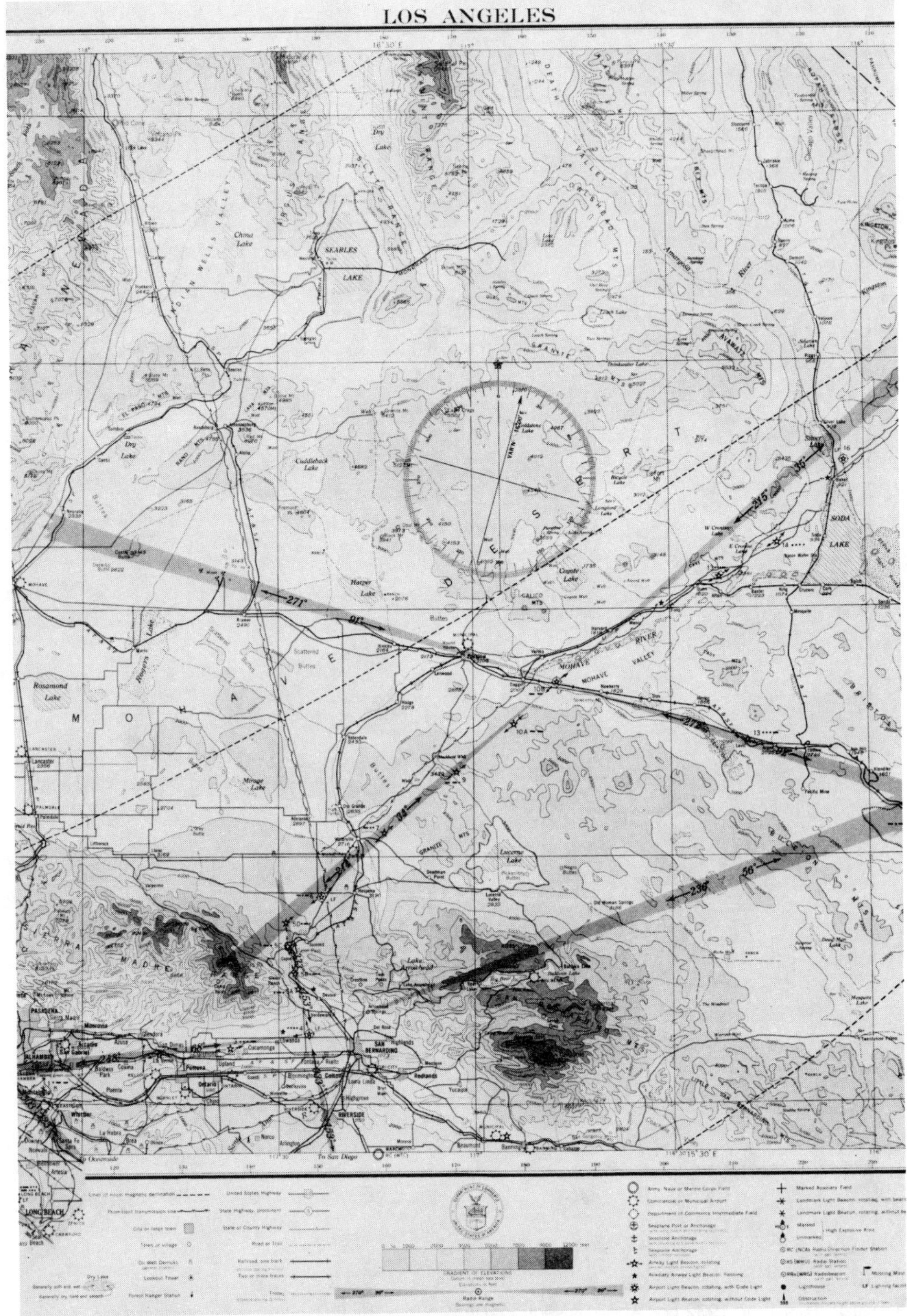

Fig. 1-2. Continued.

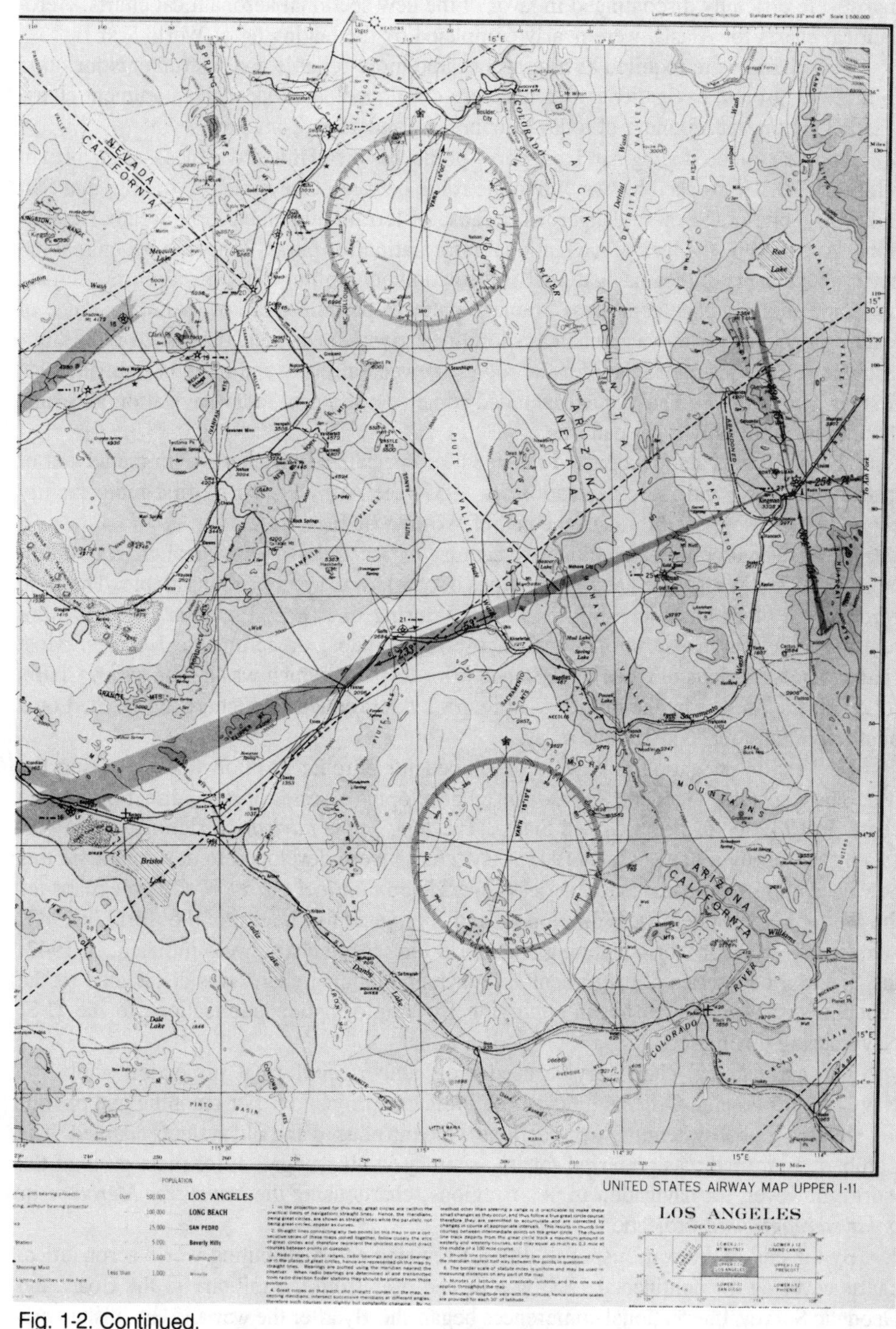

Fig. 1-2. Continued.

maps were officially discontinued in favor of the new sectional aeronautical charts. Aeronautical charts for Alaska were nearly completed by the beginning of World War II.

New charts were required as aircraft became more reliable and instruments for blind flying with ground navigational systems were developed. A pioneer of instrument charts was E.B. Jeppesen, founder of today's Jeppesen Sanderson Company.

Jeppesen joined Boeing Air Transport—predecessor of United Airlines—as an airmail pilot in 1930. He worked his way up to the route between Cheyenne, Wyoming, and Salt Lake City, one of the more dangerous because of terrain and changeable weather conditions. Appalled by the lack of navigational information for pilots, Jeppesen began making notes about every bit of navigational information, compiling data on airports, slopes, obstacles, and drainage patterns. (Drainage patterns pertain to the overall appearance of features associated with water, such as shorelines, rivers, lakes, and marshes, or any similar feature.) He developed data for the routes between airports and in 1934 published his first *Airway Manual*. The manual included routes via the new radio navigation aids plus the individual airport flight patterns.

The Air Corps also recognized the need for specialized air charts with the advent of instrument flying and radio navigation. In 1937, the Army issued its first radio facility charts, which were used through the end of World War II.

World War II compressed a quarter-century of peacetime aeronautical chart development into a few years. The need for all types of charts was urgent and insatiable. The term aeronautical chart became firmly established during this time. Previously, charts were referred to as air maps, aeronautical maps, flight maps, or aeronautical charts. Most charts were variations of those in existence prior to 1939, which was a time saver. Charts rolled off the presses by the millions. At its peak the production of charts in the St. Louis plant reportedly reached 10 tons a day.

New radio navigation aids were developed for enroute position finding during the war. The Coast and Geodetic Survey started developing a series of radio direction finding charts in 1939 to cover the United States. The *long range navigation* (loran) system was established on the East Coast in 1941. New charts were developed to accommodate this new navigation system and a new series of *world aeronautical chart* (WAC) scale maps for the Western Hemisphere were developed, and completed for the rest of the world in 1943. Additional series for world planning and world long range charts were initiated. In 1942, the first of a new series of instrument approach and landing charts was distributed by the geodetic survey. Jeppesen began supplying flight information publications to the U.S. Navy during this period.

The aeronautical chart service used the Lambert conformal projection while the Navy's Hydrographic Office employed the Mercator projection. For planning and operating within a global system, new projections were introduced and old systems adapted. The Lambert conformal was preferred for its accuracy in air navigation for most parts of the world; however, for navigation in polar regions, charts using the *transverse Mercator* or *polar stereographic* Projections were selected.

By the end of the war, responsibility for revising and distributing world aeronautical charts covering the continental U.S. was turned over from the military to the Coast and Geodetic Survey. International conferences began shortly after the war and established criteria and requirements to serve the needs of aircraft engaged in international flights.

From the middle to late forties, the Air Force published *Instrument Let Down* publications. These procedures were bound volumes consisting of four charts to each page produced by the Coast and Geodetic Survey. A new series of radio facility charts was introduced by the survey in 1947, superseded two years later by a series of 59 standard radio facility charts covering the entire United States.

Jeppesen introduced the *Standard Instrument Approach Procedures* in 1947. Prior to this time instrument approach procedures were designed by individual operators, for their own use, then approved by the Civil Aeronautics Authority (CAA), which was the predecessor to the Federal Aviation Administration (FAA). Jeppesen and the CAA developed a program where the CAA would provide standard approach procedures and authorize operators to use those procedures. The first *instrument landing system* (ILS) approach chart was developed in 1948, followed a year later with the first *very high frequency omni-directional radio range* (VOR) approach chart. Jeppesen introduced its 17-×-35-inch enroute chart in 1951.

Supplemental flight information documents were introduced in the early 1950s to keep pace with an increasing amount of navigational information.

Dunlap and Associates were given a contract by the U.S. Office of Naval Research in 1951 to study aeronautical charts and other graphic aids to navigation. Dunlap reported "two trends in the development of aeronautical charts. First, charts have become more complex because there has been a tendency to add new information to already existing charts. Second, there has been developed a wide variety of aeronautical charts so that the pilot must go to many sources to gather the information he needs. Both trends have been due to the increasing complexity of flight." Also during this period the *Airman's Guide*, *Directory of Airports*, and *Flight Information Manual* replaced *Airway Bulletins*.

The Coast and Geodetic Survey introduced a new family of aeronautical charts in the early 1950s to simplify high speed jet and transport navigation as well as lightplane visual flying. Series included planning, radio facility, approach and landing, as well as visual charts. These included *jet navigation charts* (JNC) for visual flying and the issuance of new experimental approach and landing charts for instrument operations the following year. With the new format that permitted two procedures to be printed on one side of the sheet it was hoped that the more than 1,100 instrument approach charts could be reduced to approximately 400. The smaller size sheets were also easier to handle in the aircraft. Specialized charts were also developed and tested: *operational navigation charts* (ONC), *global navigation charts* (GNC), and various long range navigation charts (loran, CONSOL, and CONSOLAN).

The radio facility charts proved unsatisfactory for jet aircraft flying at speeds faster than 500 knots. Consequently, in 1953, the Aeronautical Chart and Information Center introduced a new series of experimental radio facility charts that covered an area 750 nm by 250 nm and folded to 4½ × 9 inches for convenient handling.

The Aeronautical Chart and Information Center introduced the *Flight Information Publication—Planning* (FLIP) in 1958 to further eliminate nonessential material. This publication contained charts and textual data necessary for flight planning. FLIP, and other publications of this type, provide supplemental information that cannot be printed on the chart due to space constraints. The *Airman's Information Manual* (AIM) replaced the *Airman's Guide* in 1964. Since then the AIM has gone through various evolutionary stages, at times consisting of four documents, to its present form of *Basic Flight Information and ATC Procedures*, *Airport/Facility Directory*, and *Notices to Airmen*.

Jeppesen introduced high altitude enroute charts in 1959 to serve the emerging jet age. *Standard instrument departure* (SID) procedure charts were introduced in 1961, with *standard terminal arrival route* (STAR) procedure charts in 1967, followed a year later by *area navigation* (RNAV) enroute charts. RNAV approach charts were developed in 1971.

Area navigation prompted the National Oceanic and Atmospheric Administration (NOAA) to produce a series of high-altitude RNAV charts crisscrossing the United States with RNAV jet routes. These proved to be of limited value, however, and were subsequently discontinued. Jeppesen still offers RNAV enroute charts that, in effect, allow the pilot to design specific routes.

Sectional and world aeronautical charts evolved through several stages in the 1960s and '70s. Mostly because of economic considerations charts were printed on both sides reducing the total number of charts but unfortunately eliminating what many pilots considered useful information printed on the reverse side. Much of this information was transferred to various sections of the AIM.

Implementation of complex airspace configurations around major airports, the *terminal control area* (TCA) fostered development of the *terminal area chart* (TAC) for improved presentation of TCA dimensions and better resolution of ground references.

(Pilots had previously paid a nominal fee for charts; however, the government in the face of increasing deficits decided pilots should pick up more of the tab, and prices for charts and publications skyrocketed. At one point it was proposed that users pay the development as well as printing costs, which would have put the price of charts almost out of sight for many pilots. Fortunately, aviation organizations pressured the government into a compromise. Many still think prices are outrageous, but considering the information available, and the cost of charts in other countries, we're still getting a bargain.)

Jeppesen issued *profile descent* charts in 1976 and *loran RNAV approach* charts were introduced in 1990.

Today the U.S. Department of Commerce's National Oceanic and Atmospheric Administration's National Ocean Service (NOS) is responsible for preparing visual and instrument charts for the United States. Charts for most of the world are available from the Defense Mapping Agency's Combat Support Center. Commercially prepared charts and other navigational publications are also available, notably Jeppesen Sanderson.

Many individual states of the United States, and many international countries also produce aeronautical charts and publications.

2
Projections and limitations

ONLY A GLOBE CAN ACCURATELY PORTRAY LOCATIONS, DIRECTIONS, AND distances of the earth's surface. Although not actually a sphere, but a spheroid, the earth only approximates a true sphere due to the force of rotation that expands the earth at the equator and flattens it at the poles. This elliptical nature of the earth is a concern to cartographers; for most practical purposes of navigation, the earth can be considered a sphere. On the earth, meridians are straight and meet at the poles; parallels are straight and parallel (Fig. 2-1). Meridian spacing is widest at the equator and zero at the poles; parallels are equally spaced. Scale is true for every location. Because a globe is not possible for practical aeronautical charts, mathematicians and cartographers have devised a number of systems, known as projections, to describe features on the earth in the form of a plane or flat surface; several were mentioned in chapter 1. A map projection is a system used to portray the sphere of the earth, or a part thereof, on a plane or flat surface.

Locations on the earth are described by a system of latitude and longitude coordinates (Fig. 2-1). By convention, latitude is named first, then longitude. The reference point of latitude is the equator, with latitude measured in degrees north and south of the equator. Longitude, as discussed in chapter 1, is measured east and west of the prime, or Greenwich meridian.

Any point on the earth can be described using the system of latitude and longitude in degrees, minutes, and seconds. A degree being an arc 1/360 of a circle; therefore, a point with latitude 47 °N longitude 122 °W would be the intersection of the parallel 47 ° north of the equator and the meridian 122 ° west of the Greenwich meridian. Degrees can be further subdivided into minutes (′) which represent 1/60 of a degree and seconds (″) 1/60 of a minute. For example, the Seattle- Tacoma International Airport is located 4727N/12218W

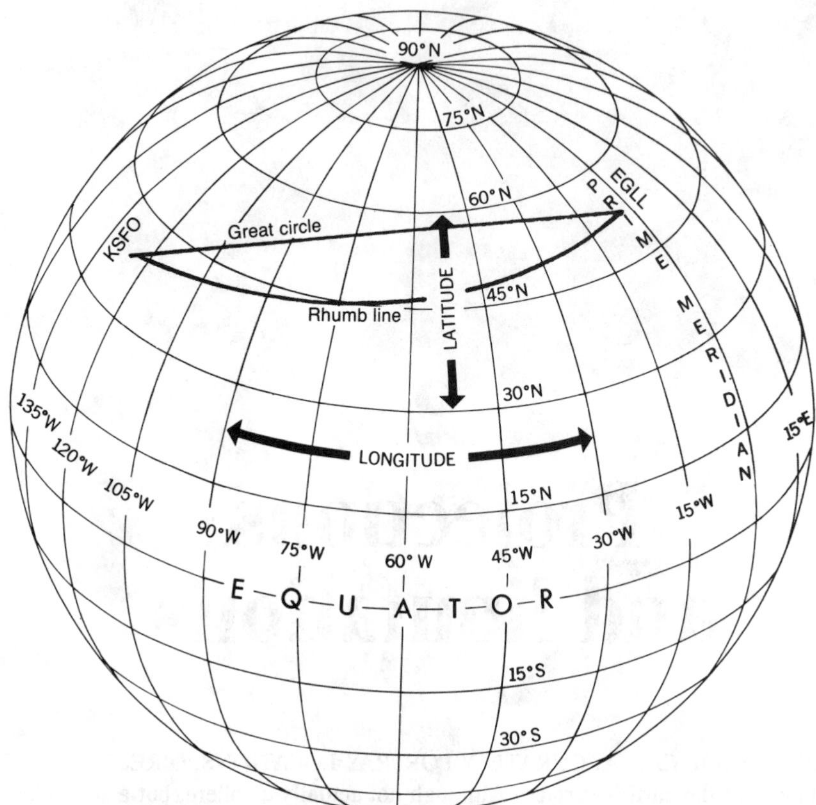

Fig. 2-1. Only on a globe are areas, distances, directions, and shapes true.

(47°27′ north/122°18′ west). For loran, or other users of coordinate navigation systems, the Seattle airport is described as N47*26.13′ W122*18.50′ (north 47°26.13 minutes; west 122°18.50 minutes).

Each degree of latitude equals 60 nautical miles (nm). Because meridians meet at the poles, a degree of longitude decreases in length with distance north or south of the equator; therefore, only for the special case of the equator does a degree of latitude equal 60 nautical miles.

PROJECTIONS

The goal of the map projection is to accurately portray true areas, shapes, distances, and directions. This includes the condition that lines of latitude are parallel and meridians of longitude pass through the earth's poles and intersect all parallels at right angles.

Areas. Any area on the earth's surface should be represented by the same area at the scale of the map. These projections are termed *equal-area* or *equivalent*.

Distances. The distance between two points on the earth should be correctly represented on the map. These projections are termed *equidistant*.

Directions. The direction or azimuth, from a point to other points on the earth should be correct on the map. These projections are termed *azimuthal* or *zenithal*.

Shapes. The shape of any feature should be correctly represented. The scale around any point must be uniform. These projections are termed *conformal*.

Because it is only possible to obtain all these properties on a globe, the cartographer must select the projection that preserves the most desired properties based on the chart's use. Figure 2-2 illustrates the four projections most often used on aeronautical charts: Mercator, transverse Mercator, Lambert conic conformal, and polar stereographic. Table 2-1, Chart Projections, compares the different projections used on aeronautical charts.

Mercator projection

The Mercator projection transfers the surface of the earth onto a cylinder tangent at the earth's equator. In the Mercator projection, meridians and parallels appear as lines crossing at right angles. Meridian *graticule spacing*, the network of parallels and meridians forming the map projection, is equal and the parallel spacing increases away from the equator.

Meridians are parallel on the Mercator projection, unlike meridians on the earth that meet at the poles. This results in increasingly exaggerated areas toward the poles. Scale, which is the relationship between distance on a chart and actual distance, changes with latitude.

Table 2-1. Chart projections

Projection	Lines of longitude (Meridians)	Lines of latitude (Parallels)	Graticule spacing	Scale	Uses
Globe and Earth	Straight and meet at poles	Straight and parallel	Meridian spacing maximum at equator, zero at poles; parallels equally spaced.	True	Impractical for navigation
Mercator	Straight and parallel	Straight and parallel	Meridian spacing equal; parallel spacing increases away from equator.	True only along equator; distortion increases away from equator.	Dead reckoning, celestial
Transverse Mercator	Curved and concave	Concave arcs	Meridian increase away from tangent meridian; parallels equally spaced.	True along line of tangency only.	Polar navigation
Lambert conic conformal	Straight converging at poles	Concave arcs	Parallels equally spaced.	True along standard parallels.	Pilotage, dead reckoning
Polar stereographic	Straight radiating from the pole	Concentric circles unequally spaced	Conformal	Increases away from pole.	Polar navigation

Mercator

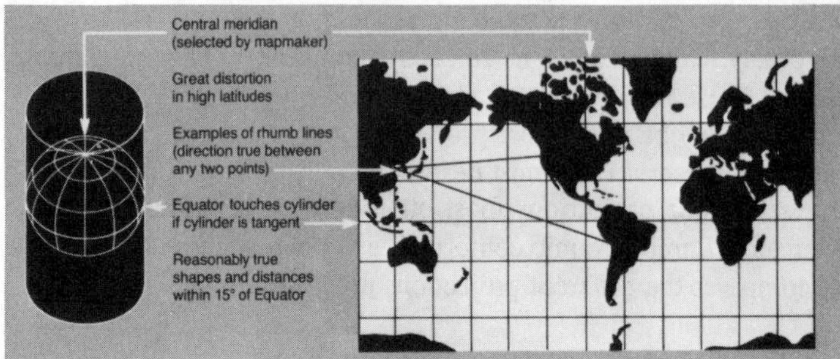

Transverse mercator

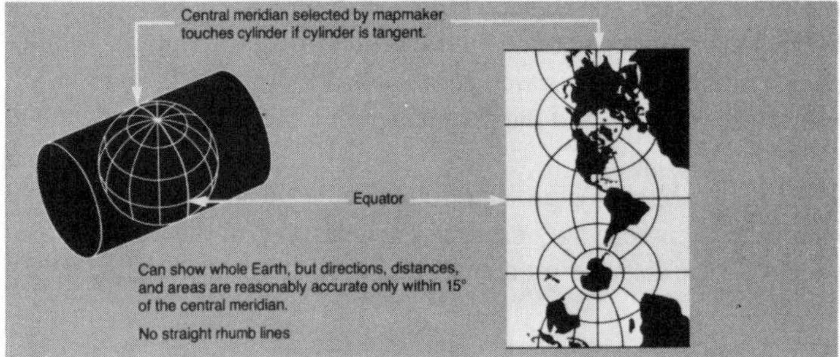

Lambert conic conformal

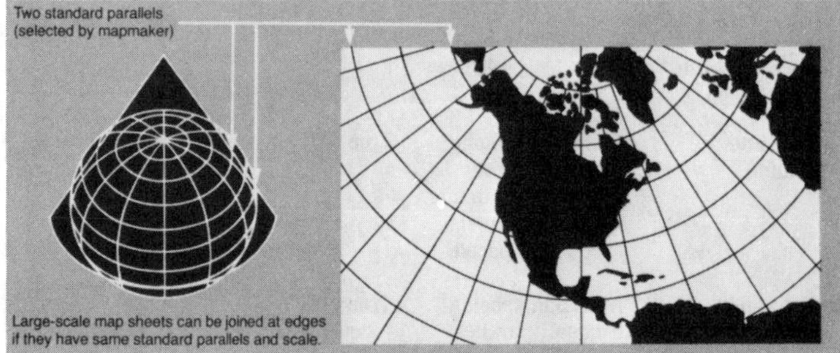

Polar stereographic

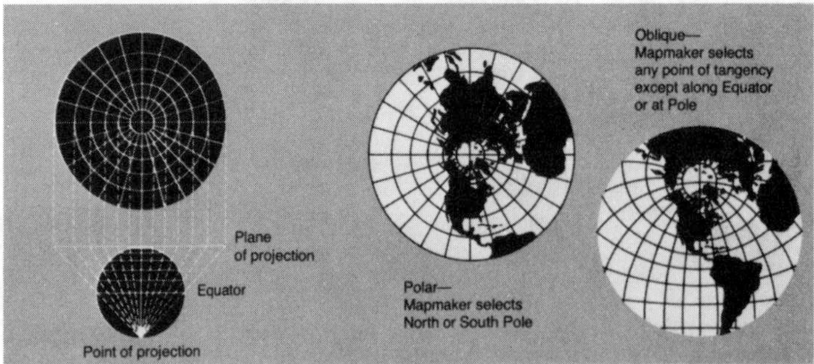

Fig. 2-2. The cartographer selects the projection that preserves the most desired properties based on the charts used.

The advantage of the Mercator is that a straight line on this projection crosses all meridians at the same angle. This allows the navigator to set a constant course from one point to another. A course crossing all meridians at a constant angle is known as a *rhumb line*. Figure 2-1 shows a rhumb line from San Francisco (KSFO) to London (EGLL).

A rhumb line is not normally the shortest distance between two places on the surface of the earth. The *great circle* distance is always the shortest distance between points on the earth. A great circle is an arc projected from the center of the earth through any two points on the surface (Fig. 2-1). A great circle, unlike the rhumb line, crosses meridians at different angles, except in two special cases: where the two points lie along the equator or the same meridian; in both cases the rhumb and great circle coincide.

For practical purposes at low latitudes, rhumb and great circle distances are nearly identical. As latitude and distance increase, differences become increasingly significant, as illustrated in Fig. 2-1; however, the projection is conformal in that angles and shapes within any small area are essentially true.

Transverse Mercator projection

The transverse Mercator projection rotates the cylinder so it becomes tangent to a meridian (Fig. 2-2). These projections are used in high north and south latitudes, and where the north-south direction is greater than the east-west direction. All properties of the regular Mercator are preserved, except the straight rhumb line. Parallels are no longer straight, becoming curved lines; meridians become complex curves. The projection is conformal. The line of true scale is no longer the equator, but the central meridian of the projection, where the cylinder is tangent.

Distances are true only along the central meridian selected by the cartographer, or else along two lines parallel to it, but all distances, directions, shapes, and areas are reasonably accurate within 15° of the central meridian. Distortion of distances, directions, and size of areas increases rapidly outside the 15° band.

Lambert conformal conic projection

A projection widely used for aeronautical charts is the Lambert conformal conic, with two standard parallels (Fig. 2-2). As the name implies, a cone is placed over the earth and intersects the earth's surface at two parallels of latitude. Scale is exact everywhere along the standard parallels, between the parallels scale decreases and beyond the parallels scale increases. Distortion of shapes and areas are minimal at the standard parallels, but distortion increases away from the standard parallels.

All meridians are straight lines that meet at a point beyond the map; parallels are concentric circles. Meridians and parallels intersect at right angles. The chart is considered conformal because scale is almost uniform around any point; scale error on any chart is so small that distances can be considered constant anywhere on the chart. A straight line from one point to another very closely approximates a great circle.

Polar stereographic projection

The standard Lambert is too inaccurate for navigation above a latitude of approximately 75° to 80°. The polar stereographic projection is sometimes used for polar regions. A plane tangent to the earth at the pole provides the projection (Fig. 2-2). Meridians are straight lines radiating from the pole and parallels are concentric circles. A rhumb line is curved and a great circle route is approximated by a straight line. Directions are true only from the center point of the projection. Scale increases away from the center point. The projection is conformal, with area and shape distortion increasing away from the pole.

HORIZONTAL DATUM

Cartographers need a defined reference point upon which to base the position of locations on a chart. This is known as the horizontal datum, or *horizontal constant datum* or *horizontal geodetic datum*. The horizontal datum used as a reference for position is defined by the latitude and longitude of this initial point. The horizontal datum for the United States was designated in 1927, located at Meades Ranch, Kansas, referred to as the North American Datum 1927 (NAD 27).

With the introduction of geodetic satellites for mapping the earth's surface and satellite navigation systems for innumerable applications, there have been recommendations to revise NAD 27. NOAA has redefined the North American Datum (NAD 83) and is readjusting the latitude and longitude of points within the system, resulting in differences of only approximately 1,000 feet between NAD 27 and NAD 83 positions.

LIMITATIONS

In addition to the limitations of chart projections, cartographers and chart users are faced with the problems of scale, simplification, and classification. Finally, chart users, especially pilots, are faced with the tremendous issue of current information. With visual charts only updated annually or semiannually, and instrument charts every 56 days, the pilot must understand the system used to provide the latest information between routine chart revisions.

Scale

Charts provide a reduced representation of the earth's surface. Scale defines the relationship between a distance on a chart and the corresponding distance on the earth. Scale is generally expressed as a ratio: the numerator, customarily 1, represents chart distance, and the denominator, a large number, represents horizontal ground distance. For example: 1:500,000 (sometimes written 1/500,000) states that any unit, whether inch, foot, yard, statute mile, nautical mile, or kilometer, on the chart, represents 500,000 units on the ground. That is, one inch on the chart equals 500,000 inches on the earth.

Chart makers provide scales for conversion of chart distance to statute or nautical miles or kilometers. Manufacturers of aeronautical plotters provide scales for standard aeronautical charts; however, the pilot must be familiar with the plotter used and chart scale for accurate calculations. While flying in the annual Hayward-Bakersfield-Las Vegas air race we use sectional charts (1:500,000), except in the Las Vegas area where a terminal area chart (1:250,000) is available. Sure enough in my haste I measured a leg using the wrong scale. This is disastrous in a race where the finishing order is a matter of seconds and tenths of a gallon of fuel.

The smaller the scale of a chart, the less detail it can portray. For example, a chart with a scale of 1:1,000,000 cannot provide the detail of a chart with a scale of 1:250,000. Charts with a smaller scale increase the size of the area covered (assuming a constant size), but reduce the detail that can be shown.

Simplification and classification

Because scale reduces the size of the earth, information must be generalized. Making the best use of available space is a major problem in chart development. The detail of the real world cannot be shown on the chart. The crowding of lines and symbols beyond a specific limit renders the chart unreadable, yet the amount of information that might be useful or desirable is almost unlimited. The smaller the chart scale, the more critical and difficult the problem; therefore, the cartographer is forced to simplify and classify information.

Simplification is the omission of detail that would clutter the map and prevent the pilot from obtaining needed information. The necessity for detail is subjective and not all will agree on what should, or should not, be included. The inclusion of too much detail runs the risk of confusing the reader by obscuring more important information. For example, the chart producer might have to decide whether to include a prominent landmark, the limit of controlled airspace, or a symbol indicating a parachute jump area. The problem of simplification has led directly to the use of aeronautical publications, such as the *Airport/ Facility Directory*.

The *Airport/Facility Directory* is divided into seven booklets that cover the United States, including Puerto Rico and the Virgin Islands. Alaska is covered by the *Alaska Supplement*, areas of the Pacific by the *Pacific Chart Supplement*. These directories are a pilot's manual containing data on airports, seaplane bases, heliports, navigational aids, communications, special notices, and operational procedures. They provide information that cannot be readily depicted on charts. For example, airport hours of operation, types of fuel available, runway widths, lighting information, and other data, as well as a means to update charts between editions. These directories, as well as charts and other related publications, may be obtained directly from NOAA. Free catalogs are available.

Classification is necessary in order to reduce the amount of information into a usable form. The cartographer must classify towns, rivers, and highways of different appearance on the ground into a common symbol for the chart. The pilot must then be able to interpret this information.

On a flight from Phillipsburg, Penn., to Huntington, W.Va., we were forced to fly below a 1,500-foot overcast because of radio trouble. Due to poor planning on my part, we only had a world aeronautical chart (WAC), instead of the larger scale sectional chart. I misidentified the Kanawha River as the Ohio River. To verify the position I checked the highways, railroads, and power lines adjacent to the river. Nothing matched. After a few moments of utter confusion I checked the time from last known position and determined we could not have made it all the way to the Ohio. Based on this estimate of distance, I reevaluated our position as over the Kanawha; now everything matched. On the WAC the Kanawha was represented by a thin blue line, the Ohio by a wide blue line. From the air both rivers appeared identical.

To maximize the amount of information on a chart the cartographer uses symbols. Symbol shape, size, color, and pattern are used to convey specific information. The pilot must be able to interpret these symbols. Lack of chart symbol knowledge can lead to misinterpretation, confusion, and wandering into airspace where one has no business.

Currency

With the ever changing environment, charts are almost outdated as soon as they are printed and become available. A pilot's first task when using any chart or publication is to ensure its currency.

Visual charts are revised and reissued semiannually or annually. Changes to visual charts are supplemented by the "Aeronautical Chart Bulletin," in the *Airport/Facility Directory*, revised every 56 days.

NOS IFR charts contained in the *Terminal Procedures Publication* (TPP) are published every 56 days. A 28-day midcycle change notice volume contains revised procedures that occur between the 56 day publication cycle. These changes are in the form of new charts. The subsequent publication of the TPP incorporates change notice volume revisions and any new changes since change notice issuance.

Visual and instrument charts are further supplemented by the FAA's *Notice to Airmen* system (NOTAMs). The notices publication is published every 14 days and supplements the "Aeronautical Chart Bulletin" of the *Airport/Facility Directory*, and TPP and change notice volumes. A detailed discussion of the NOTAM publication is in chapter 9.

Aeronautical information not received in time for publication is distributed through the FAA's telecommunications systems. These include unanticipated or temporary changes, or hazards when their duration is for a short period or until published. A NOTAM is classified into one of three groups:

- NOTAM (D)
- NOTAM (L)
- FDC NOTAM

NOTAM (D)s consist of information that requires wide distribution and pertains to enroute navigational aids, civil public use landing areas listed in the *Airport/Facility Directory*, and aeronautical data that relates to IFR operations.

NOTAM (L)s include information that requires local dissemination, but does not qualify as a NOTAM (D), such as bird activity, moored balloons, airport beacons, taxiway lights, and the like.

FDC NOTAMs consist of information that is regulatory in nature pertaining to charts, procedures, and airspace. This includes such items as temporary flight restrictions, and revisions to visual and instrument charts.

NOTAMs from each category are routinely provided as part of a standard flight service station (FSS) weather briefing; however, the FSS only provides FDC NOTAMs for locations within 400 miles of the facility, the pilot must request FDC NOTAMs in areas beyond 400 miles. NOTAMs, except (L) are also available through direct user access terminals (DUAT) and other commercial weather vendors. When FDC NOTAMs are associated with a specific facility identifier, they are included as part of the DUAT briefing; however, most enroute chart changes are not associated with a specific facility identifier. DUAT users are faced on every briefing with a disclaimer: "FDC NOTAMs that are not associated with an affected facility identifier will now be presented unless you specifically choose to decline to receive such information." It's almost like looking for the proverbial "needle in the haystack," but to be safe, these NOTAMs must be reviewed. Once published, unlike the FSS where the pilot has the option to request the data, DUAT users must remember that: "Published FDC NOTAM Data are not available, and must be obtained from other publications/charts/etc." Pilots using NOS charts must be aware of these limitations and plan their flight briefings accordingly. This might mean a call to the FSS specifically for any pertinent FDC NOTAMs. Loran NOTAMs are now available on DUATs, but only available from the FSS on request.

Once a new chart becomes effective, the NOTAMs, including those carried in the *Notice to Airmen* publication and "Aeronautical Chart Bulletin" of the *Airport/Facility Directory*, are canceled. Pilots are presumed to be using current charts.

A major advantage of Jeppesen's manuals, and other suppliers of chart services, is a more timely revision schedule than is available through government products. Jeppesen uses individual sheets that fit into a seven-ring, looseleaf binder. Jeppesen subscribers, basically, have access to the *Notice to Airmen* publication and much of the information contained in FDC NOTAMs through frequent revisions, and the enroute and terminal chart NOTAMs sections of their *Airway Manual*. Immediate and short term changes to the National Airspace System (NAS) must still be obtained, but this information is normally provided as part of an FSS standard briefing or DUAT briefing.

It is not within the scope of this book to provide a detailed explanation of the NOTAM system. For the reader who would like additional information on this subject, including decoding, translating, and interpreting NOTAMs, refer to TAB's Practical Flying Series book, *The Pilot's Guide to Weather Reports, Forecast & Flight Planning*.

A British friend with whom I flew in England in the middle 1960s was astonished by the frequent revisions of U.S. aeronautical charts. The British civil charts we had to fly with were infrequently updated and only contained about a fifth of the information of U.S. charts. For example, all NAVAIDs were shown by a single symbol without any indication of the type or frequency. And, oh yes, during this period the English charts cost about $5 in American currency.

NOS has made every effort to ensure that each piece of information shown on NOAA's charts and publication is accurate. Information is verified to the maximum extent possible. According to NOS, "You, the pilot, are perhaps our most valuable source of information. You are encouraged to notify NOAA, National Ocean Service, of any discrepancies you observe while using our charts and related publications. Postage-paid chart correction cards are available at FAA Flight Service Stations for this purpose (or you may write directly to NOAA, at the address below). Should delineation of data be required, mark and clearly explain the discrepancy on a current chart (a replacement copy will be returned to you promptly)."

National Ocean Service
NOAA, N/CG31
6010 Executive Blvd.
Rockville, MD 20852
800-626-3677

NOS emphasizes that the, "Use of obsolete charts or publications for navigation may be dangerous. Aeronautical information changes rapidly, and it is vitally important that pilots check the effective dates on each aeronautical chart and publication to be used. Obsolete charts and publications should be discarded and replaced by current editions." One pilot called an FSS and requested a briefing from Bishop to Santa Cruz, Calif. The briefer explained that the airport was closed. The pilot responded, "Oh, I must be using an old chart." Indeed, the airport had been closed for over two years. There are no valid reasons for using obsolete charts.

The use of current charts and publications, and obtaining a complete preflight briefing cannot be overemphasized. It's like using the restroom before a flight, we know we should, but sometimes it's a little inconvenient. Failure, however, often leads to a very uncomfortable flight. Only by understanding the system can pilots ensure they meet their obligation as pilot in command.

3

Visual chart terminology and symbols

TERMS AND SYMBOLS DISCUSSED IN THIS CHAPTER NOT ONLY APPLY TO standard visual charts—world aeronautical charts, sectional charts, and terminal area charts—but most carry over to planning charts, instrument charts, and charts available through the Defense Mapping Agency. While reading this chapter the reader should keep in mind chart limitations and any problems facing the cartographer, specifically scale, simplification, and classification as discussed in chapter 2.

TOPOGRAPHY AND OBSTRUCTIONS

The elevation and configuration of the earth's surface are of prime importance to visual navigation. Cartographers devote a great deal of attention to portray relief and obstructions in a clear, concise manner.

Topography is the configuration of the surface of the earth; subdivided into *hypsography*, *hydrography*, and *culture*. Hypsography is the science or art of describing elevations of land surfaces with reference to a datum, usually sea level; it is that part of topography dealing with relief or elevation of terrain. Hydrography is the science that deals with the measurements and description of the physical features of the oceans, seas, lakes, rivers, and their adjoining coastal areas, with particular reference to their use for navigational purposes. It is that part of topography pertaining to water and drainage features, where drainage features are those associated with shorelines, rivers, lakes, marshes, and the like, and the overall appearance made by these features on a map or chart. Culture is features of the terrain that have been constructed by man: roads, buildings, canals, and boundary lines.

Relief is inequalities of elevation and the configuration of land features on the surface of the earth. Relief might be represented on charts by contours, colored tints representing elevations or gradients (*hypsometric* tints), shading, spot elevations, or short disconnected lines drawn in the direction of the slopes (*hachures*).

Obstructions are manmade vertical features that might affect navigable airspace. NOAA maintains a file of more than 50,000 obstacles in the U.S., Canada, Mexico, and the Caribbean. Normally, only obstacles more than 200 feet above ground level (agl) are charted, unless the obstacle is considered significant, for example near an airport or much higher than surrounding terrain.

Terrain

Three methods are used on aeronautical charts to display relief: *contour lines*, *shaded relief*, and *color tints*. Contour lines, as the name implies, connect points of equal elevation above mean sea level (MSL) on the earth's surface. Contours graphically depict terrain and are the principal means used to show the shape and elevation of the surface. Contours are depicted by continuous lines—except where elevations are approximate, then with broken lines—labeled in feet MSL. On sectional charts basic contours are spaced at 500 foot intervals, although intermediate contours might be shown at 250-foot intervals in moderately level or gently rolling areas. Occasionally, auxiliary contours portray smaller relief features at 50-, 100-, 125-, or 150-foot intervals.

Figure 3-1 shows how contours, shaded relief, and color tints depict terrain. Contours show the direction of the slope, gradient, and elevation. For example, in Fig. 3-1 valley floors have little or no gradient, while the mountains have steep gradients. The contours are labeled with their elevation. Shaded relief depicts how terrain might appear from the air. The cartographer shades the areas that would appear in shadow if illuminated from the northwest. Shaded relief enhances and supplements contours by drawing attention to canyons and mountain ridges (Fig. 3-1). Color tints depict bands of elevation. These colors range from light green for the lowest elevations to dark brown for higher elevations. Color tints in Fig. 3-1 range from light brown in the valley to dark brown—you'll have to take my word for it—over the mountain ranges, supplementing the contours and enhancing recognition of rapidly rising terrain.

In addition to contours, shading, and tints, significant elevations are depicted as *spot* elevations, *critical* elevations, and *maximum* elevation figures. Spot elevations represent a point on the chart where elevation is noted. They usually indicate the highest point on a ridge or mountain range. A solid dot depicts the exact location when known. An "x" denotes approximate elevations; where elevation is known, but location approximate, only the elevation appears, without the dot or "x" symbol. Critical elevation is the highest elevation in any group of related and more-or-less similar relief formations. Critical elevations are depicted by larger elevation numerals and dots than are used for spot elevations. Figure 3-1 illustrates the difference between spot and critical elevations.

Maximum elevation figures (MEF) represent the highest elevation, including terrain and other vertical obstacles—natural and manmade—bounded by the ticked lines of the latitude/longitude grid on the chart. Depicted to the nearest 100 foot value, the last two digits of the number are omitted. The center of the grid in Fig. 3-1 shows 112. The MEF for this grid is 11,200 feet MSL. This figure is determined from the highest elevation or obstacle, corrected upward for any possible vertical error (including the addition of 200 feet for any natural or manmade obstacle not portrayed), then rounded upward to the next

Fig. 3-1. Chart relief is represented by contour lines, shaded relief, and color tints.

higher hundred foot level; therefore, almost all MEFs will be higher than any elevation or obstacle portrayed within the grid on the chart. Pilots should note that these figures cannot take into account altimeter errors and should be considered as any other terrain elevation figure for flight planning purposes.

Latitude and longitude are labeled in degrees. Lines of latitude and longitude are subdivided by lines representing 10 minutes, and half lines representing one minute of arc. Because longitude represents the same distance anywhere on the earth, unlike latitude where distances are shorter toward the poles, one minute of longitude anywhere on the earth equals one nautical mile (nm); therefore, lines of longitude can be used for quick estimates of distance.

Other topographical relief features considered suitable for navigation are contained in Fig. 3-2. They include lava flows, sand and gravel areas, rock strata and quarries, mines, craters, and other relief information usable for visual checkpoints.

Lava flows, and sand ridges and dunes are quite pronounced when seen from the air and make excellent checkpoints, especially if they are isolated by other terrain features. Unfortunately, most of these features only appear in the Western United States. Strip mines and large quarries also make excellent checkpoints because of their visibility.

Large craters, where they appear, make excellent checkpoints. I was taking a Cessna 182 from California to Wichita where three other pilots and I would pick up three new Cessna 150s. The Cessna 182 had no navigational radios. East of Prescott, Arizona, we

RELIEF	
UNSURVEYED AREAS Label appropriately as required	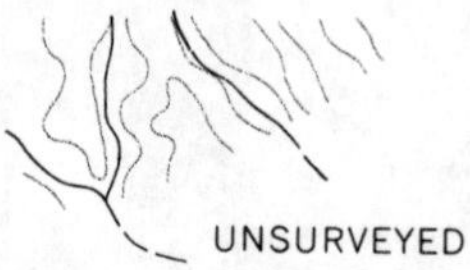 UNSURVEYED
DISTORTED SURFACE AREAS	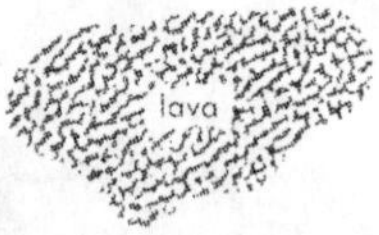 lava
LAVA FLOWS	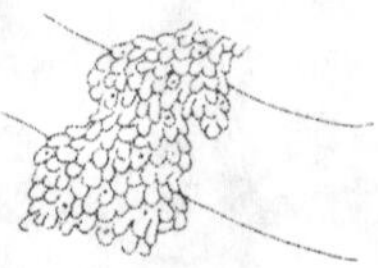
SAND OR GRAVEL AREAS	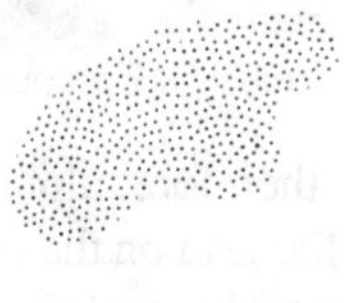
SAND RIDGES To Scale	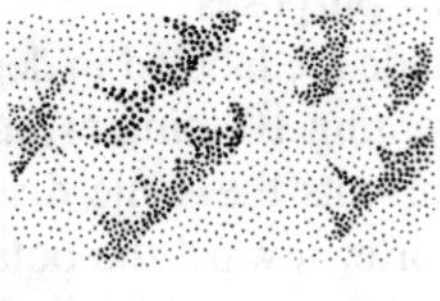
SAND DUNES To Scale	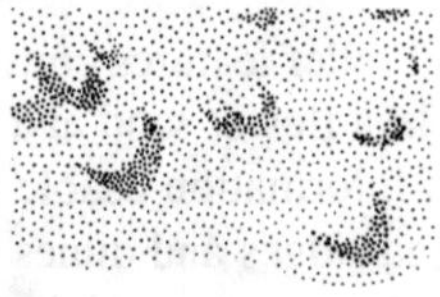

Fig. 3-2. Supplemental relief features aid in visual flying.

RELIEF	
SHADED RELIEF	
ROCK STRATA OUTCROP	rock strata
QUARRIES TO SCALE	quarry
STRIPMINES, MINE DUMPS AND TAILINGS To Scale	strip mine mine dump
CRATERS	crater crater
ESCARPMENTS, BLUFFS, CLIFFS, DEPRESSIONS, ETC.	
LEVEES AND ESKERS	levee

Fig. 3-2. Continued.

climbed over an undercast and proceeded eastward. My dead reckoning navigation plan was to find a hole or return to Prescott. Forty-five minutes into the leg we found a hole and descended. As luck would have it we descended over Arizona's Meteor Crater, which was just about the most prominent landmark in that area. From there we were able to proceed on course.

HYDROGRAPHY

Hydrography pertains to water and drainage features. Hydrographic features on aeronautical charts are represented in blue: stream, river, or aqueduct depicted by a single blue line; lakes and reservoirs by a blue tint. Small dots or hatching indicate where streams and lakes fan out (or are not perennial), or where reservoirs are under construction.

Figure 3-3 shows how shorelines, lakes, streams, reservoirs, and aqueducts are depicted. Shorelines usually make excellent checkpoints, except where they are relatively straight without features. Pilots need to pay attention to shoreline orientation. For example, most people assume that California's coastline is north-south; however, in certain areas, such as around Santa Barbara, the coast is actually east-west. This has led to much confusion for student pilots and others unfamiliar with the area. Lakes usually make good checkpoints, especially when their shape is unique or they are dammed, as discussed in chapter 4.

Caution needs to be exercised with all lakes, perennial and nonperennial. A perennial lake contains water year round; a nonperennial lake is intermittently dry, usually during the dry season. There can be confusion during periods of drought when perennial lakes will be dry. Other discrepancies result from human decisions to drain, expand, or abandon reservoirs. Streams should, if at all possible, only be used to support other checkpoints. That is, there should be other landmarks that establish position, supported by the position of the stream. Perennial and nonperennial streams should be treated with the same cautions as perennial and nonperennial lakes. Reservoirs are similar to lakes, and can be treated in the same way; however, reservoirs are usually perennial.

Figure 3-4 describes other hydrographical features contained on aeronautical charts. Among them are symbols for swamps, marshes, and bogs.

The terrain for a flight from Lufkin, Texas, to New Orleans' Lakefront Airport was flat, but contained widespread areas of swamp. I plan direct flights unless there is a significant reason to do otherwise. Spending most of my life in the West, I didn't know what a swamp was. Fortunately, a concerned specialist at Lufkin FSS (decommissioned) explained the significance of this hydrographical feature. A swamp is nothing more than a lake with trees growing out of it. He was certainly right and I appreciated the information. Pilots flying over unfamiliar terrain, as in this case, would be well advised to seek the advice of local pilots or the local FSS—as local as possible during these times.

Tundra describes a rolling, treeless, often marshy plain, usually associated with arctic regions. Hummocks and ridges describe a wooded tract of land that rises above an adjacent marsh or swamp. Mangroves are any of a number of evergreen shrubs and trees growing in marshy and coastal tropical areas; a nipa is a palm tree indigenous to these areas. Bogs are areas of moist, soggy ground, usually over deposits of peat. Flumes, penstocks, and similar features depict water channels used to carry water as a source of power, such as a waterwheel. Pilots flying in northwestern Montana and especially Alaska can expect

(Continued on page 36.)

HYDROGRAPHY	
SHORELINES **Definite**	
Fluctuating	
Unsurveyed Indefinite	
Man-made	
LAKES Label as required **Perennial** When too numerous to show individual lakes, show representative pattern and descriptive note.	numerous small lakes 756 618
Non-Perennial (dry, intermittent, etc.) Illustration includes small perennial lake	

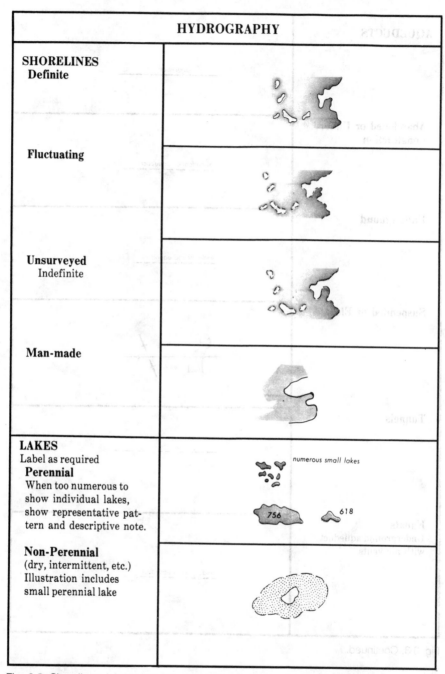

Fig. 3-3. Shorelines, lakes, streams, reservoirs, and aqueducts have landmark value, with specific limitations.

AQUEDUCTS	aqueduct
Abandoned or Under Construction	abandoned aqueduct
Underground	underground aqueduct
Suspended or Elevated	
Tunnels	
Kanats Underground aqueduct with air vents	underground aqueduct

Fig. 3-3. Continued.

HYDROGRAPHY	
RESERVOIRS **Natural Shorelines**	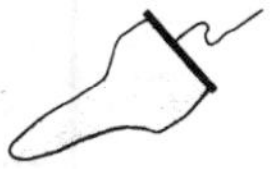
Man-made Shorelines Label when necessary for clarity	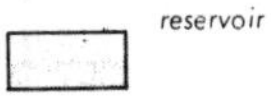 reservoir
Too small to show to scale	 reservoir
Under Construction	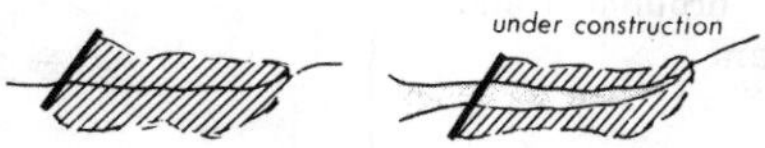 under construction
STREAMS **Perennial**	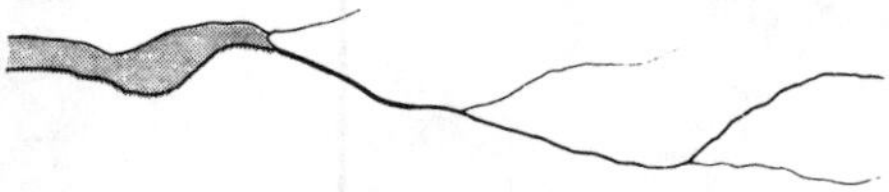
Non-Perennial	

Fig. 3-3. Continued.

STREAMS (Continued) **Fanned Out** Alluvial fan	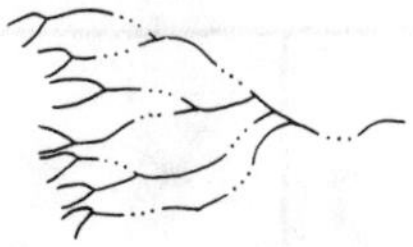
Braided	
Disappearing	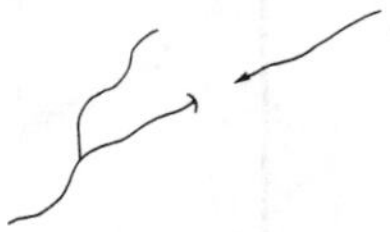
Seasonally Fluctuating with undefined limits	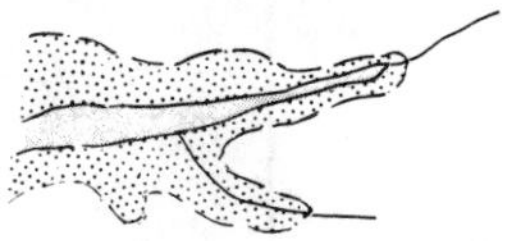
with maximum bank limits, prominent and constant	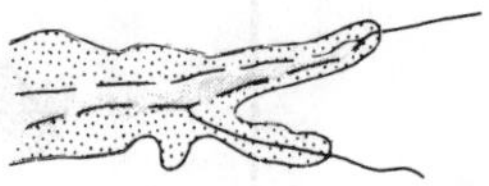
Sand Deposits In and Along Riverbeds	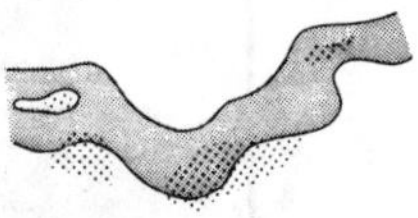
WET SAND AREAS Within and adjacent to desert areas	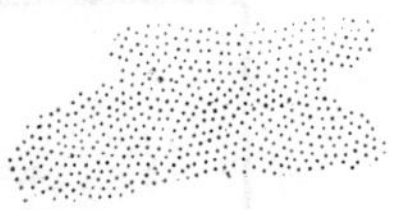

Fig. 3-3. Continued.

HYDROGRAPHY	
SALT EVAPORATORS AND SALT PANS MAN EXPLOITED	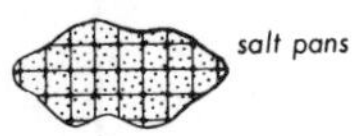 salt pans
SWAMPS, MARSHES AND BOGS	
HUMMOCKS AND RIDGES	
MANGROVE AND NIPA	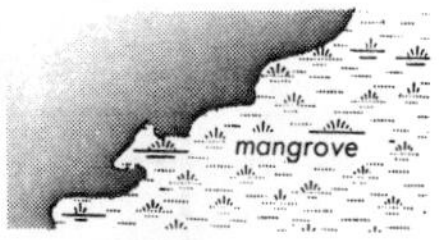 mangrove
PEAT BOGS	 peat bog
TUNDRA	tundra
CRANBERRY BOGS	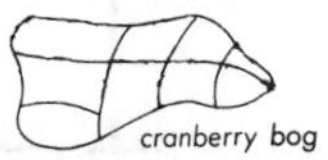 cranberry bog

Fig. 3-4. Some hydrographical features warn of danger, such as swamps, others provide supplemental landmark value.

FLUMES, PENSTOCKS AND SIMILAR FEATURES	flume
Elevated	flume
Underground	underground flume
FALLS Double-Line	falls
Single-Line	falls
RAPIDS Double-Line	rapids
Single-Line	rapids

Fig. 3-4. Continued.

HYDROGRAPHY	
RICE PADDIES Extensive areas indicated by label only.	
LAND SUBJECT TO INUNDATION	
SPRINGS, WELLS AND WATERHOLES	•
GLACIERS	
Glacial Moraines	
ICE CLIFFS	
SNOWFIELDS, ICE FIELDS AND ICE CAPS	9000 8000 7000

Fig. 3-4. Continued.

CANALS	ERIE
To Scale	
Abandoned or Under Construction	abandoned
To Scale	abandoned
SMALL CANALS AND DRAINAGE/IRRIGATION DITCHES **Perennial**	
Non-Perennial	
Abandoned or Ancient	abandoned
Numerous Representative pattern and/or descriptive note.	
Numerous	numerous canals and ditches

Fig. 3-4. Continued.

(Continued from page 28.)
to see glaciers and glacial moraines (debris carried by the glacier), ice cliffs, snow and ice fields, and ice caps. Other than canals, the other features in Fig. 3-4 might be difficult to verify and should normally only be used to support other checkpoints.

Figure 3-5 contains the remaining hydrographical features contained on aeronautical charts. Ice peaks, polar ice, and pack ice are features restricted to polar and arctic regions. Boulders, wrecks, reefs, and underwater features are displayed because they have certain landmark value. Some of these features might be small and difficult to identify.

One other topographical feature should be mentioned. Mountain passes are depicted by black curved lines outlining the pass. The name of the pass and elevation are shown. Sorona Pass in Fig. 3-1, lower left, is shown with an elevation of 9,628 MSL.

CULTURE

Land features constructed by man include roads and highways, railroads, buildings, canals, dams, boundary lines, and the like. Many landmarks that can be easily recognized from the air, such as stadiums, racetracks, pumping stations, and refineries, are identified by brief descriptions adjacent to a small black square or circle marking exact location. Depictions might be exaggerated for improved legibility.

Figure 3-6 contains a description of railroads, roads, bridges, and tunnels, contained on aeronautical charts. Differences between sectionals and WACs are noted. Single track railroads have one crosshatch, double and multiple railroads have a double crosshatch. Railroads often make excellent checkpoints. A word of caution. Numerous railroads emanate like spokes from many large cities. Pilots navigating exclusively by the "iron compass" have become hopelessly confused when they inadvertently took the wrong track.

Never navigate solely by one landmark. Major highways (category 1) also make excellent checkpoints, but they do suffer from the same problems as the railroad. Secondary roads (category 2, and especially secondary category 2) are often difficult to positively identify, especially when flying over sparse areas of desert or plains. Bridges, viaducts, and causeways are often very good checkpoints.

Figure 3-7 shows populated areas (large cities from Fig. 3-6), boundaries, water features, and miscellaneous cultural features. Large and medium cities are shown by their outline as it appears on the ground. This helps significantly with identification. Towns and villages are only represented by a small circle. Especially where several towns or villages are in the same general area, this symbology makes them hard to positively identify. Political boundaries are shown using standard map symbols. Cultural coastal features are depicted because of their landmark value. Small mines and quarries are shown by a small crossed picks symbol.

Pilots should pay particular attention to the symbol for catenaries (Fig. 3-7). A catenary, depicted on aeronautical charts, is a cable, power line, cable car, or similar structure suspended between peaks, a peak and valley below, or across a canyon or pass. A catenary is normally 200 feet or higher above terrain, which poses a very serious hazard to low flying aircraft, perhaps marked with orange balls or lights. Catenaries are not shown on WAC charts.

Cultural features are not revised as often as aeronautical information; therefore, especially in areas of rapid metropolitan development, cultural features might not be correctly depicted.

(Continued on page 48.)

HYDROGRAPHY	
ICE PEAKS	
FORESHORE FLATS Tidal flats exposed at low tide.	
ROCKS - ISOLATED **Bare or Awash**	*
WRECKS **Exposed**	
REEFS - ROCKY OR CORAL	coral
MISCELLANEOUS UNDERWATER FEATURES NOT OTHERWISE SYMBOLIZED	shoals
FISH PONDS AND HATCHERIES	fish hatchery

Fig. 3-5. Most ice features are only applicable to arctic regions.

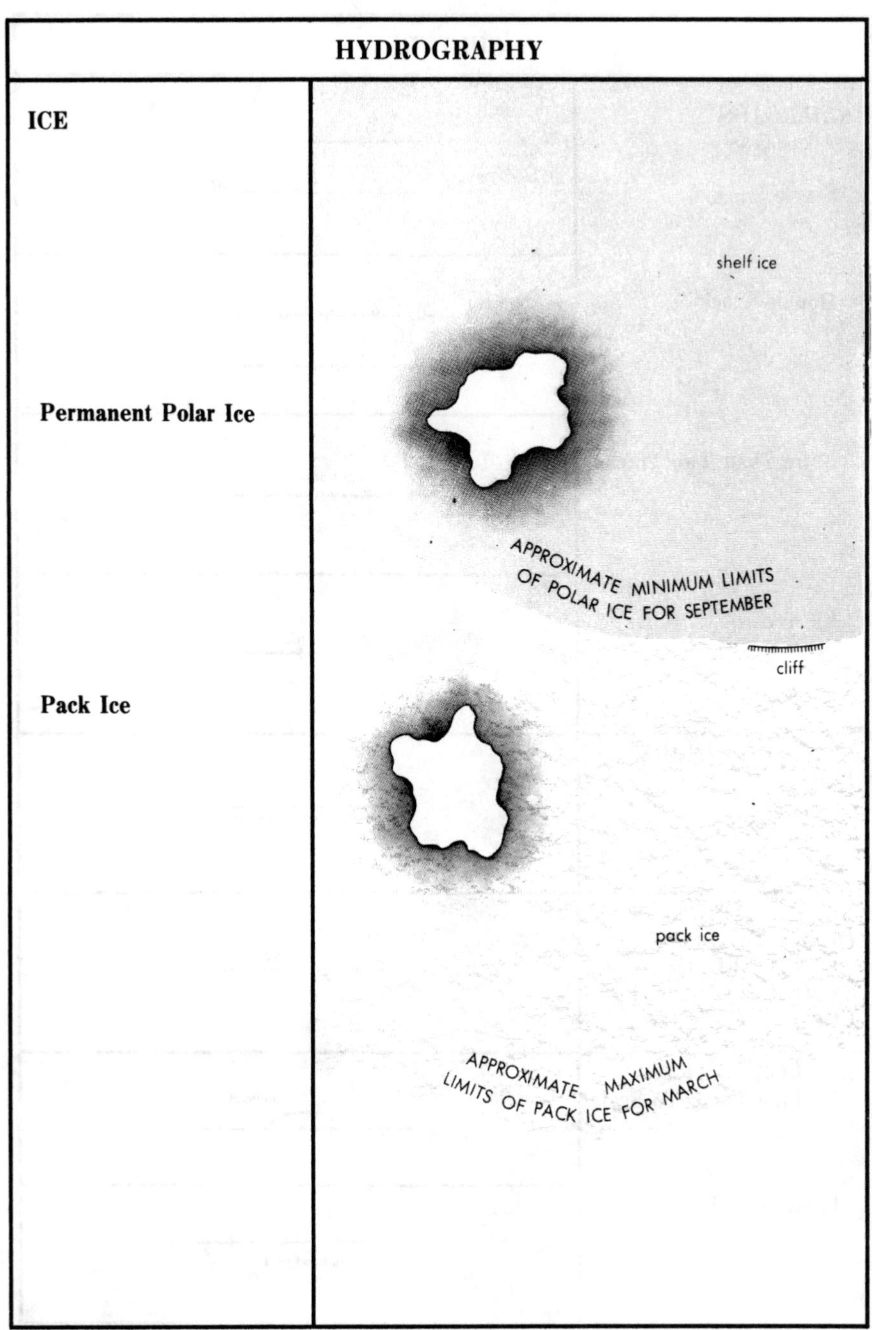

Fig. 3-5. Continued.

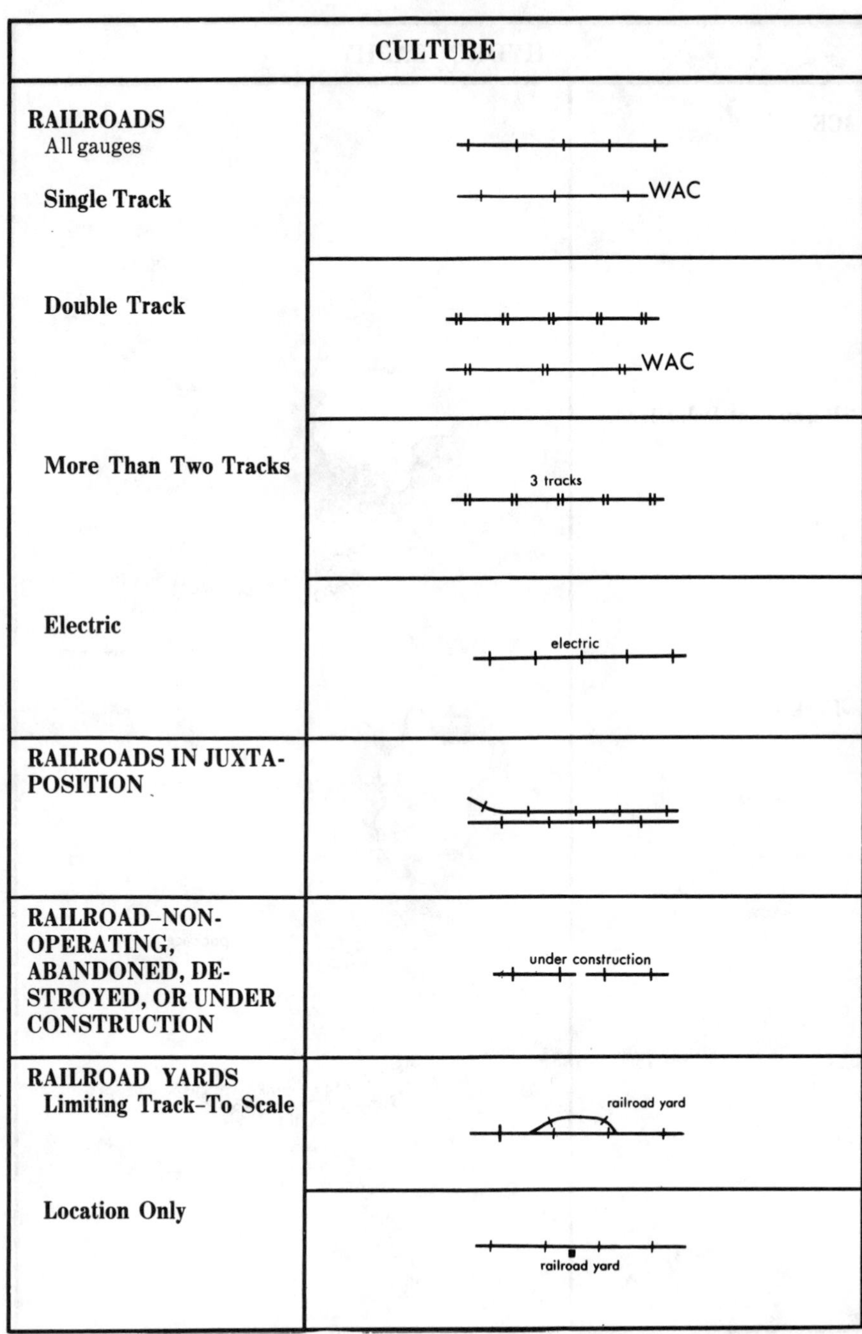

CULTURE
RAILROADS
All gauges
Single Track
WAC
Double Track
WAC
More Than Two Tracks
3 tracks
Electric
electric
RAILROADS IN JUXTA-POSITION
RAILROAD-NON-OPERATING, ABANDONED, DE-STROYED, OR UNDER CONSTRUCTION
under construction
RAILROAD YARDS
Limiting Track-To Scale
railroad yard
Location Only
railroad yard

Fig. 3-6. Major railroads, roads, bridges, and cities make good visual landmarks.

ROAD NAMES	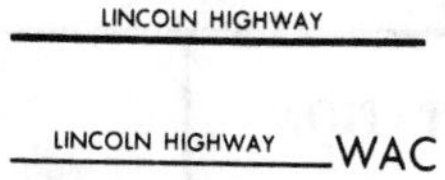 LINCOLN HIGHWAY LINCOLN HIGHWAY WAC
ROADS - UNDER CONSTRUCTION	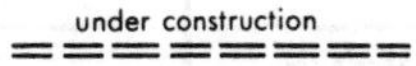 under construction
BRIDGES AND VIADUCTS **Railroad**	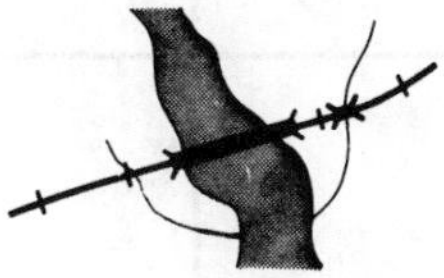
Road	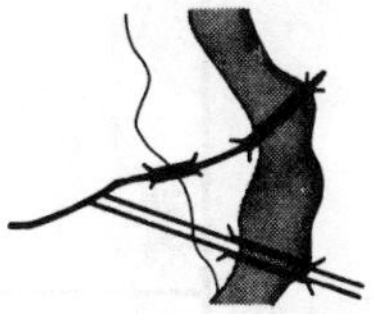
OVERPASSES AND UNDERPASSES	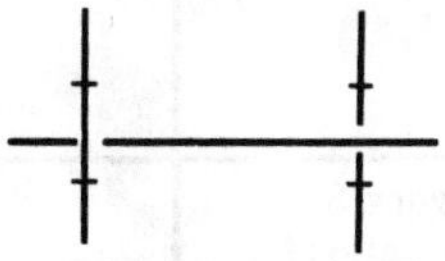
CAUSEWAYS	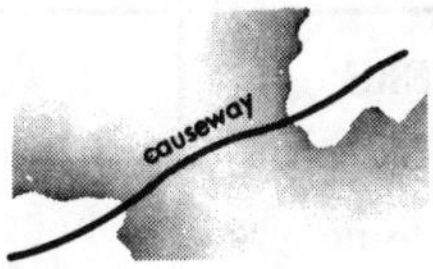 causeway

Fig. 3-6. Continued.

CULTURE	
RAILROAD STATIONS	 station station
RAILROAD SIDINGS AND SHORT SPURS	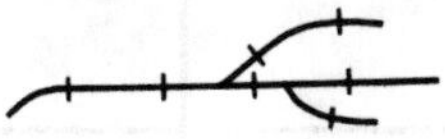
ROADS **Dual Lane** **Category 1**	 WAC
Primary **Category 2**	 WAC
Secondary **Category 2**	
TRAILS Category 3 Provides symbolization for dismantled railroad when combined with label "dismantled railroad."	
ROAD MARKERS **U.S. route no.** **Interstate route no.** **Air Marked Identification Label**	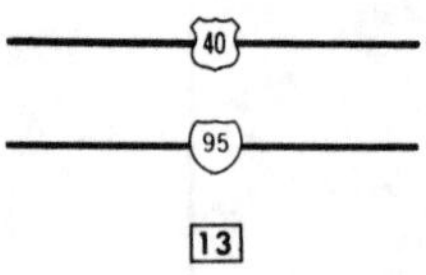 40 95 13

Fig. 3-6. Continued.

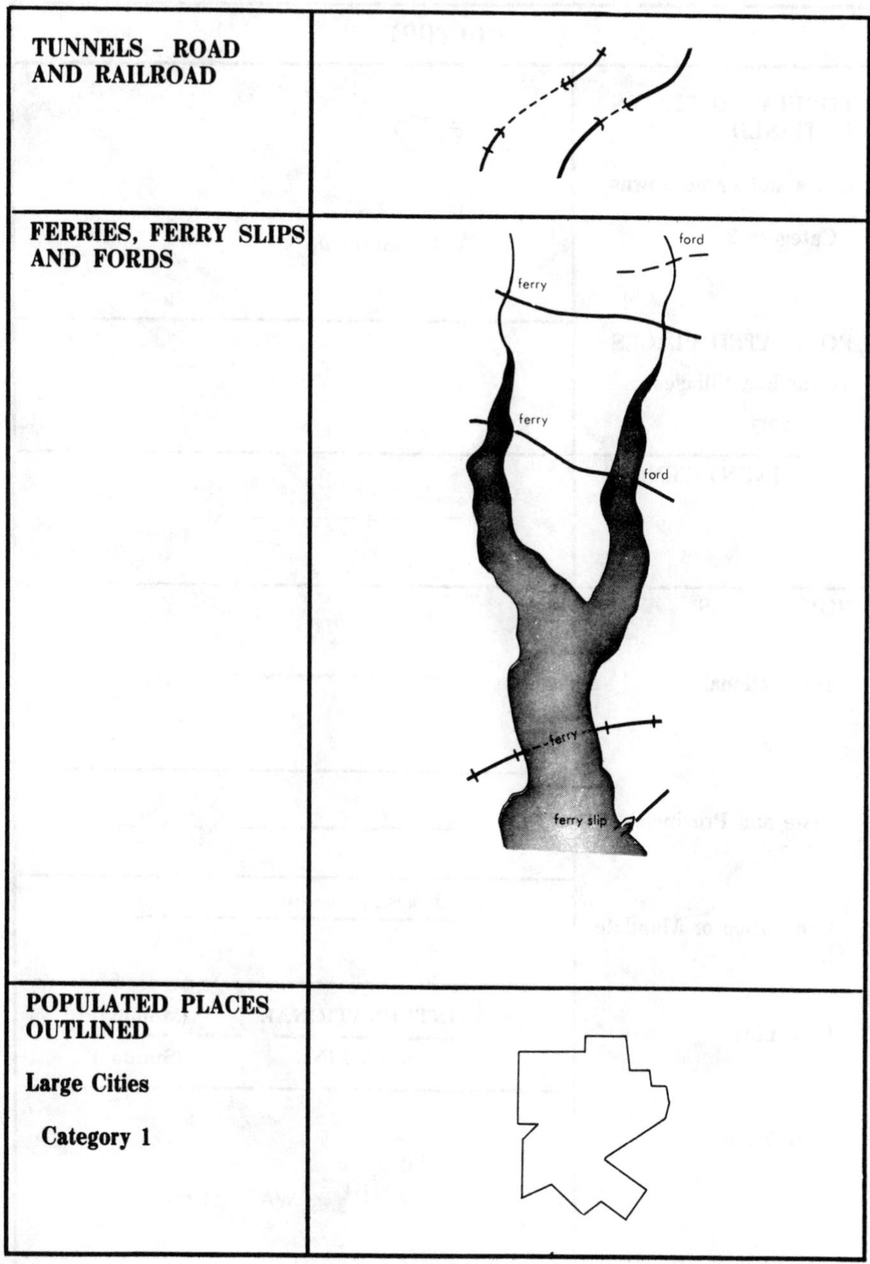

TUNNELS - ROAD AND RAILROAD	
FERRIES, FERRY SLIPS AND FORDS	
POPULATED PLACES OUTLINED **Large Cities** **Category 1**	

Fig. 3-6. Continued.

CULTURE	
POPULATED PLACES OUTLINED **Cities and Large Towns** **Category 2**	WAC not shown
POPULATED PLACES **Towns and Villages** **Category 3**	
PROMINENT FENCES	
BOUNDARIES **International**	
State and Provincial	
Convention or Mandate Line	US–RUSSIA CONVENTION LINE OF 1867 
Date Line	INTERNATIONAL (Monday) DATE LINE (Sunday)
Time Zone	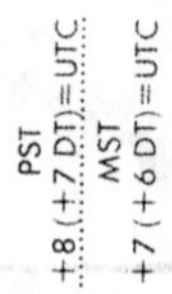 PST +8 (+7 DT)=UTC MST +7 (+6 DT)=UTC WAC not shown

Fig. 3-7. Towns and villages, secondary roads, and small cultural features should only be used as supplemental landmarks.

WEIRS AND JETTIES	jetties
SEAWALLS	seawall
BREAKWATERS	breakwater breakwater
PIERS, WHARFS, QUAYS, ETC.	piers piers
MISCELLANEOUS CULTURE FEATURES	■ stadium
FORTS CEMETERIES	■ fort ■ cemetery
OUTDOOR THEATER	

Fig. 3-7. Continued.

CULTURE	
MINES AND QUARRIES **Shaft Mines and Quarries**	⚒
POWER TRANS-MISSION, TELEPHONE & TELEGRAPH LINES	WAC
Catenaries	WAC not shown
PIPELINES	pipeline
Underground	underground pipeline
DAMS	
DAM CARRYING ROAD	
PASSABLE LOCKS	locks
SMALL LOCKS	

Fig. 3-7. Continued.

WELLS **Other Than Water**	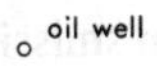 oil well
RACE TRACKS	
LOOKOUT TOWERS Air marked identification	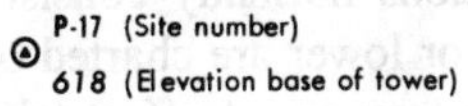 P-17 (Site number) *618* (Elevation base of tower)
LANDMARK AREAS	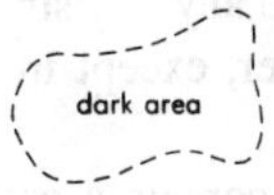 dark area
TANKS	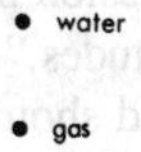 water gas
COAST GUARD STATION	 CG
AERIAL CABLEWAYS, CONVEYORS, ETC.	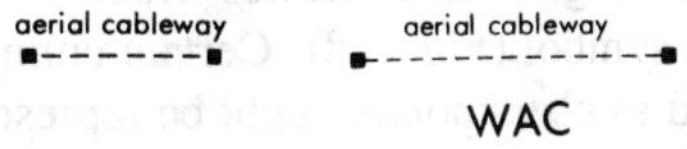 aerial cableway aerial cableway WAC

Fig. 3-7. Continued.

Intentionally blank.

(Continued from page 37.)

Power transmission lines (high tension lines) are depicted for their landmark and safety value. Often, transmission lines can be used to verify the identification of other landmarks. Although not normally qualifying as an obstruction, their depiction alerts pilots flying at low altitudes to this sometimes almost invisible hazard. Transmission lines are shown as small black towers connected by a single line (Fig. 3-1).

OBSTRUCTIONS

Charted obstructions normally consist of features extending higher than 200 feet agl. Objects 200 feet or lower are charted only if considered hazardous, for instance close to an airport where they might affect takeoffs and landings. Federal Aviation Regulations (FAR) require that airplane pilots, even over sparsely populated areas, cannot "operate closer than 500 feet to any . . . structure," except for takeoff or landing; therefore, objects 200 feet high or lower, except in the case of an emergency, should have no operational impact.

Sectional charts contain a caution note: "This chart is primarily designed for VFR navigational purposes and does not purport to indicate the presence of all telephone, telegraph and power transmission lines, terrain or obstacles which may be encountered below reasonable and safe altitudes." The fact that objects of 200 feet can exist without the requirement to be charted should be a sobering thought when considering scud running. Also keep in mind that many obstructions have guy wires that extend some distance outward from the structure. How about helicopter pilots? Yes, helicopters can operate at less than these minimums "if the operation is conducted without hazard to persons or property on the surface." Helicopter, balloon, and ultralight pilots need to understand these charting limitations and plan their flights accordingly.

Obstacles lower than 999 feet agl are shown by the standard obstruction symbol as illustrated in Fig. 3-8. Obstacles 1,000 feet or higher agl are shown by the elongated obstruction symbol (Fig. 3-8). Certain cultural features that can be clearly seen from the air and used as checkpoints might be represented with pictorial symbols (black) with elevation data in blue, such as the prominent Golden Gate Bridge.

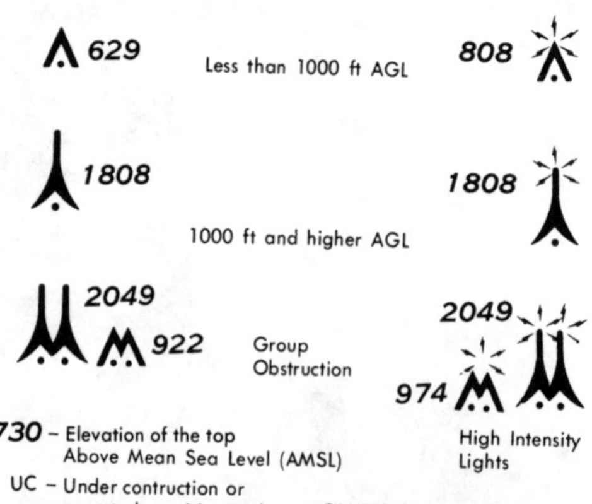

Fig. 3-8. Normally, only obstructions higher than 200 feet are charted.

Height of the obstacle agl and elevation of the top of the obstacle MSL are shown when known or when they can be reliably determined. The height agl is shown in parentheses below the MSL elevation of the obstacle [which is a logical arrangement]:

2468
(1200)

The height of the obstacle is 1,200 feet above ground level (1,200). The top of the obstacle above mean sea level is 2,468 feet. The height agl might be omitted in extremely congested areas to avoid confusion. Within high density groups of obstacles, only the highest obstacle in the area will be shown using the group obstacle symbol (Fig. 3-8). Obstacles under construction are indicated by the letters "UC" immediately adjacent to the symbol. When available, the eventual agl height of the obstruction will appear in parentheses. Obstacles with strobe lighting systems are shown as indicated in Fig. 3-8.

AERONAUTICAL

Aeronautical information on visual charts consists of airports, radio aids to navigation (NAVAIDs), airspace, and navigational information. Airports are identified for size (runway lengths), use (civil, military, private), and services (availability of fuel and repairs, which are typically detailed in a directory rather than on a chart). The type and frequency of NAVAIDS are depicted. Controlled airspace within the scope of the chart is shown (below 18,000 feet). Navigational information such as magnetic variation, airway intersections, and lighting aids are depicted.

Airports

Visual charts depict civil, military, and some private, landplane, helicopter, and seaplane airports (Fig. 3-9). Hard-surfaced runways of 1,500 to 8,000 feet are enclosed within a circle depicting runway orientation. All recognizable runways, including some that might be closed, are shown for visual identification. Hard-surfaced runways greater than 8,000 feet do not conveniently fit in a circle; the circle is omitted, but runway orientation is preserved. Airports with other than hard-surfaced runways, such as dirt, sod, gravel, and the like, are depicted as open circles. Airports served by a control tower (CT), and having an airport traffic area, are shown in blue, all others in magenta—a purplish red color. Symbols indicate the availability of fuel and repair facilities. Pilots should keep in mind that types of fuel and repair facilities are contained in the *Airport/Facility Directory*, with changes or nonavailability of services mentioned in NOTAMs.

Restricted, private, and abandoned airports are shown for emergency or landmark purposes only. Pilots wishing to use restricted or private landing facilities must obtain permission from that airport authority. A check of your insurance policy might also be in order. Some policies restrict landings to public airports, except in emergencies. Airports are labeled unverified when available for public use, but warranting more than ordinary precaution due to lack of current information on field conditions, or available information indicates peculiar operating limitations. Selected ultralight flight parks appear only on sectional charts as an "F" within the airport circle. *(Continued on page 52.)*

AIRPORTS		
LANDPLANE-MILITARY Refueling and repair facilities for normal traffic. All recognizable runways, including some which may be closed, are shown for visual identification. Airports having Airport Traffic Area (CT) are shown in blue, all others in magenta.	PAPAGO AAF *1270* *L 30 NAS MOFFETT CT – **118.3** *40* L 92	WAC
SEAPLANE-MILITARY Refueling and repair facilities for normal traffic.	NAS ALAMEDA *00* *L 100	WAC
LANDPLANE-CIVIL Refueling and repair facilities for normal traffic.	SCOTT VALLEY *2728* *L 37 *122.8* FSS SISKIYOU CO *2648* L 75 *123.0* SAN FRANCISCO INTL CT – **120.5** ATIS 115.8 113.7 *12* L 106 *123.0*	WAC
SEAPLANE-CIVIL Refueling and repair facilities for normal traffic.	HARTUNG LODGE *00* L 150	WAC
LANDPLANE CIVIL AND MILITARY Refueling and repair facilities for normal traffic.	SIOUX CITY *1097* L 90 *123.0* SANTA MONICA CT – **118.3** ATIS 110.8 *175* L 82 *123.0*	WAC

Fig. 3-9. Airport symbols provide information on size, use, and services.

AIRPORTS	
SEAPLANE CIVIL AND MILITARY Refueling and repair facilities for normal traffic.	PORT ARBOUR *05* – 150 WAC
LANDPLANE-EMERGENCY No facilities or complete information is not available. Add appropriate notes as required: "closed, approximate position, existence unconfirmed".	ELMA **20** L 21 PUBLIC USE – limited attendance or no service available OMH (Pvt) **200** – 17 RESTRICTED OR PRIVATE – use only in emergency, or by specific authorization AIRPORT UNVERIFIED – a landing area available for public use but warranting more than ordinary precaution due to: (1) lack of current information on field conditions, and/or (2) available information indicates peculiar operating limitations ABANDONED – depicted for landmark value or to prevent confusion with an adjacent useable landing area. (Normally at least 3000′ paved) WAC
SEAPLANE-EMERGENCY No facilities or complete information is not available.	COMMAND *00* – 100 WAC
HELIPORT (Selected)	PENTAGON (ARMY) *995* WAC
ULTRALIGHT FLIGHT PARK (Selected)	WAC not shown

Fig. 3-9. Continued.

(Continued from page 49.)

Figure 3-10 decodes standard airport information. As of late 1991, the letter R preceding the airport name indicates the availability of airport surveillance radar, and the airport location identifier follows the airport name: R Oakland (OAK). If the airport name is in a box, as in Fig. 3-10, the airport has a nonstandard airport traffic area. Federal Aviation Regulation (FAR) Part 93 "Special Air Traffic Rules and Airport Traffic Patterns" contains details on the few airports where this regulation applies. Pilots can also obtain specifics on special traffic areas from the "Regulatory Notices" portion of the A/FD.

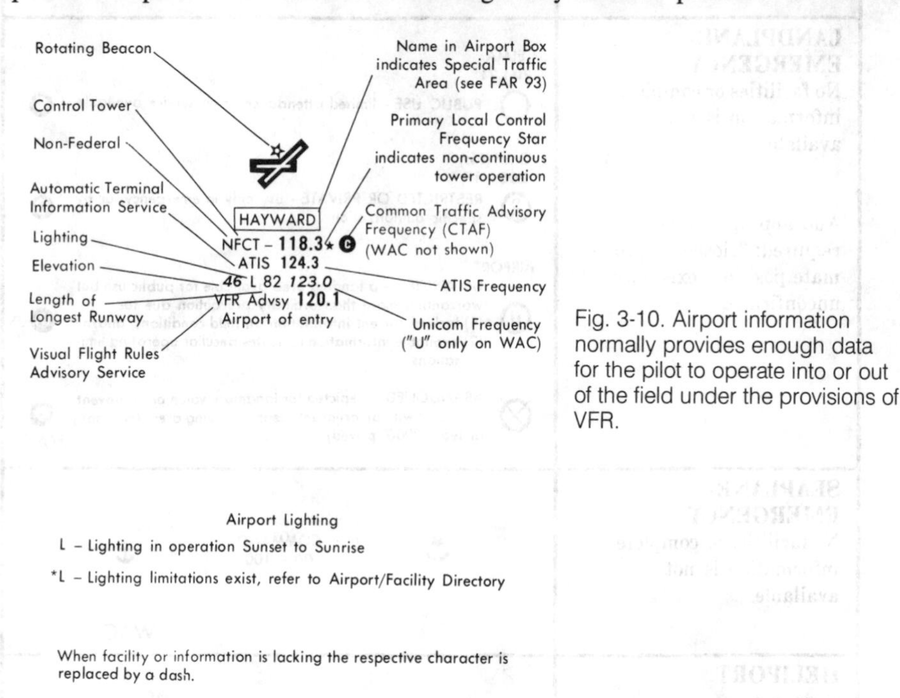

Fig. 3-10. Airport information normally provides enough data for the pilot to operate into or out of the field under the provisions of VFR.

FSS indicates a flight service station on the field, RFSS a remote flight service station. These facilities might provide airport advisory service at selected airports. The decision whether AFSSs will provide airport advisory services is still in question. Airports with nonfederal control towers are identified by the contraction NFCT. Only the primary local control frequency appears. A star following the local control frequency indicates a part-time tower. Supplemental and additional frequencies, such as approach, secondary local control, and ground frequencies, and tower hours of operation are contained on the end panels or margin of the chart, and in the *Airport/Facility Directory*. Automatic terminal information service (ATIS), where available, is always shown. Supplemental frequencies, such as an aeronautical advisory station (UNICOM) or VFR advisory service might be listed. The letter "C" within a circle indicates the common traffic advisory frequency (CTAF): not shown on WACs. This frequency is usually the tower frequency at airports with part-time towers.

Airport elevation is always in feet above mean sea level and never abbreviated, as in Fig. 3-10, 46 feet MSL. The length of the longest runway is shown in hundreds of feet; in the example the longest runway is 8,200 feet. Airport lighting, indicated by the letter "L," operates sunset to sunrise, unless preceded by an asterisk, which indicates limitations exist. All lighting codes refer to runway lights. The lighted runway might not be the longest or lighted full length. Pilots must refer to the *Airport/Facility Directory* for specific limitations, such as pilot controlled lighting. Other remarks are added as required, such as airport of entry. When information is not available, the respective character is replaced by a dash.

Radio aids to navigation

Figure 3-11 shows standard symbols for the very high frequency omnidirectional radio range (VOR), a VOR collocated with distance measuring equipment (DME), and a VOR collocated with a tactical air navigation (TACAN) facility. A TACAN provides azimuth information similar to a VOR, but on an ultra high frequency (UHF) band used by military aircraft, and distance information from the DME. When the NAVAID is located on an airport (Fig. 3-11), the type of facility (in this case VOR only) appears above the NAVAID box, otherwise, the appropriate symbol indicates the type of facility (VOR, VOR/DME, or collocated VOR and TACAN).

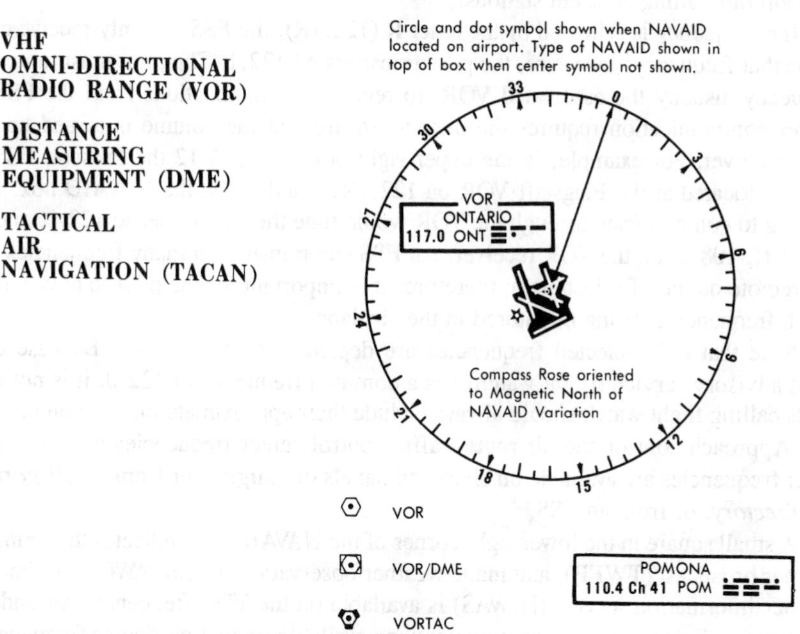

Fig. 3-11. NAVAID symbols provide type of facility and frequency.

Note in Fig. 3-11 that the NAVAID box provides frequency and identification information for the Ontario VOR. The VOR frequency is 117.0 MHz, identification ONT, followed by a representation of its aural Morse code signal. The lower right corner of Fig. 3-11 shows a VORTAC NAVAID box. This is the Pomona VORTAC. The VOR frequency is 110.4 MHz, DME and TACAN channel (Ch) 41. (Because VOR frequencies and DME channels are paired, when a pilot chooses the VOR frequency the paired DME channel is automatically selected.) The identification of the NAVAID is POM, followed by the representation of the aural Morse code signal.

Locations were abbreviated with two letters in the early days of aviation, for instance NK was Newark. As the number of NAVAIDs and airports increased, three letter identifiers came into use. All VOR, VOR/DME, VORTAC, and many low frequency radio beacons have three-letter identifiers.

Figure 3-12 contains other standard NAVAID and flight service station communication symbols. Low and medium frequency NAVAIDs are shown in magenta; VOR, VOR/DME, and VORTACs in blue. Low frequency/medium frequency radio ranges still exist in Canada and Mexico; the charting symbol is depicted in Fig. 3-12. Nondirectional and marine beacons and broadcast station symbols are also shown.

Heavy line boxes indicate standard flight service station (FSS) communication frequencies (121.5 and 122.2 MHz, which are *simplex*—the pilot transmits and receives on the same channel). Other FSS frequencies are printed above the box. For example, 123.6 for airport advisory service and FSS discrete frequencies. Routine communications should be accomplished on the station's discrete frequency. These frequencies are unique to individual facilities and locations. Their use will usually avoid frequency congestion with aircraft calling adjacent stations.

If a frequency is followed by the letter R (122.1R), the FSS has only receive capability on that frequency; therefore, the pilot transmits on 122.1. The pilot must tune another frequency, usually the associated VOR, to receive communications from the FSS. This *duplex* communication requires the pilot to ensure that the volume is turned up on the VOR receiver. For example, in the upper right box of Fig. 3-12 the Prescott FSS has a receiver located at the Flagstaff VOR on 122.1R, noted above the NAVAID box. A pilot wishing to communicate through the VOR would tune the transmitter to 122.1, and select Flagstaff, 108.2, on the VOR receiver. An FSS can transmit on many frequencies (VORs and remote outlets, for instance); therefore, it is important for the pilot to advise the FSS which frequency is being monitored in the airplane.

Note that only selected frequencies are depicted on these charts. Because enroute flight advisory service (flight watch) has a common frequency of 122.0, it is not shown. Pilots calling flight watch should always include their approximate location on initial contact. Approach control and air route traffic control center frequencies are also omitted. Other frequencies are available on chart end panels or margins, or from the *Airport/Facility Directory*, or from an FSS.

A small square in the lower right corner of the NAVAID box indicates that transcribed weather broadcast (TWEB), automatic weather observation system (AWOS), or hazardous weather information service (HIWAS) is available on the VOR frequency. An underlined frequency indicates no voice communications available on that particular frequency.

Airspace

For our purposes, airspace on visual charts can be divided into two basic categories: controlled airspace and special use airspace. Airspace types affect VFR pilots to a much greater degree than IFR pilots; therefore, most types of airspace must be displayed on visual charts. Controlled airspace specifies visibility and cloud separation, radio communi-

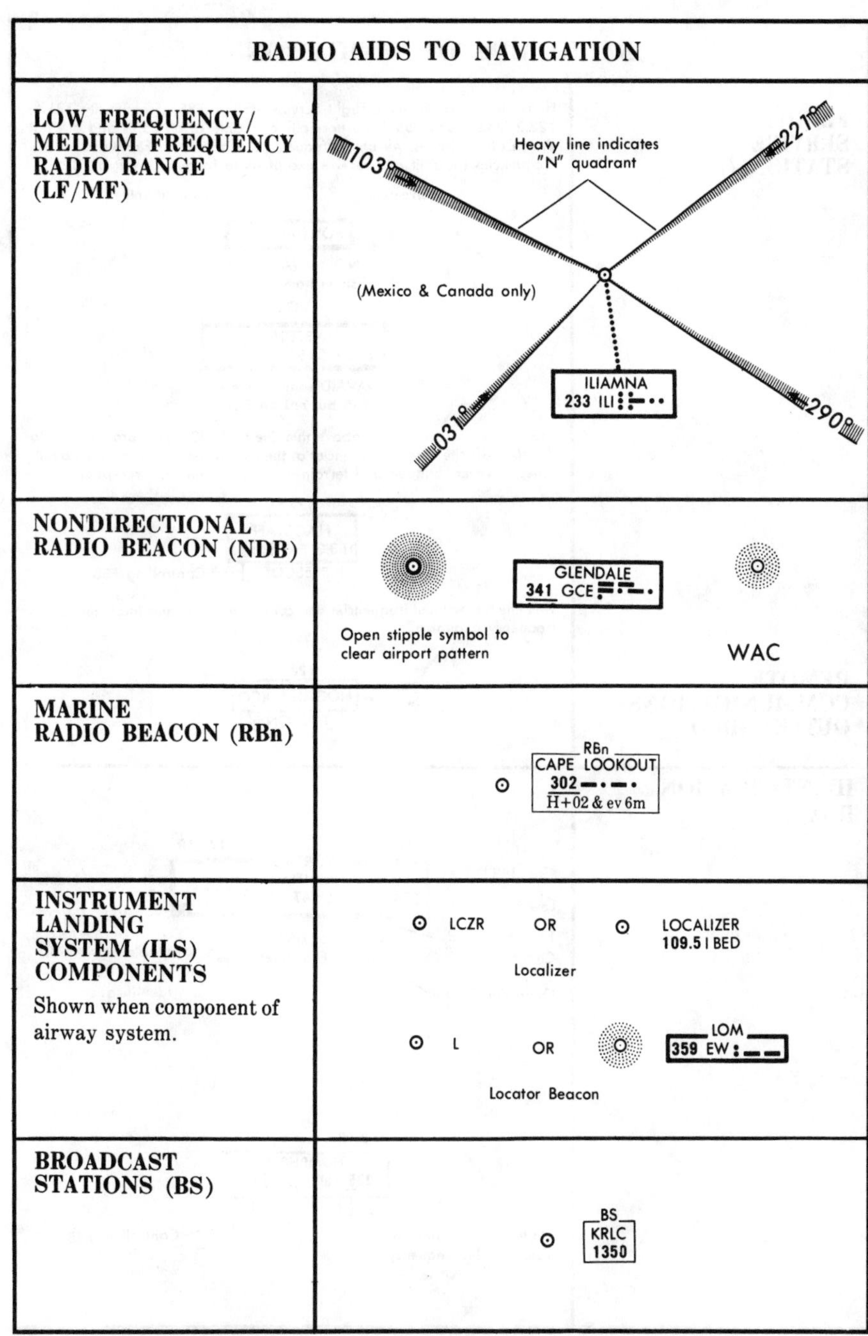

Fig. 3-12. As well as NAVAID information, visual charts contain FSS communication frequencies.

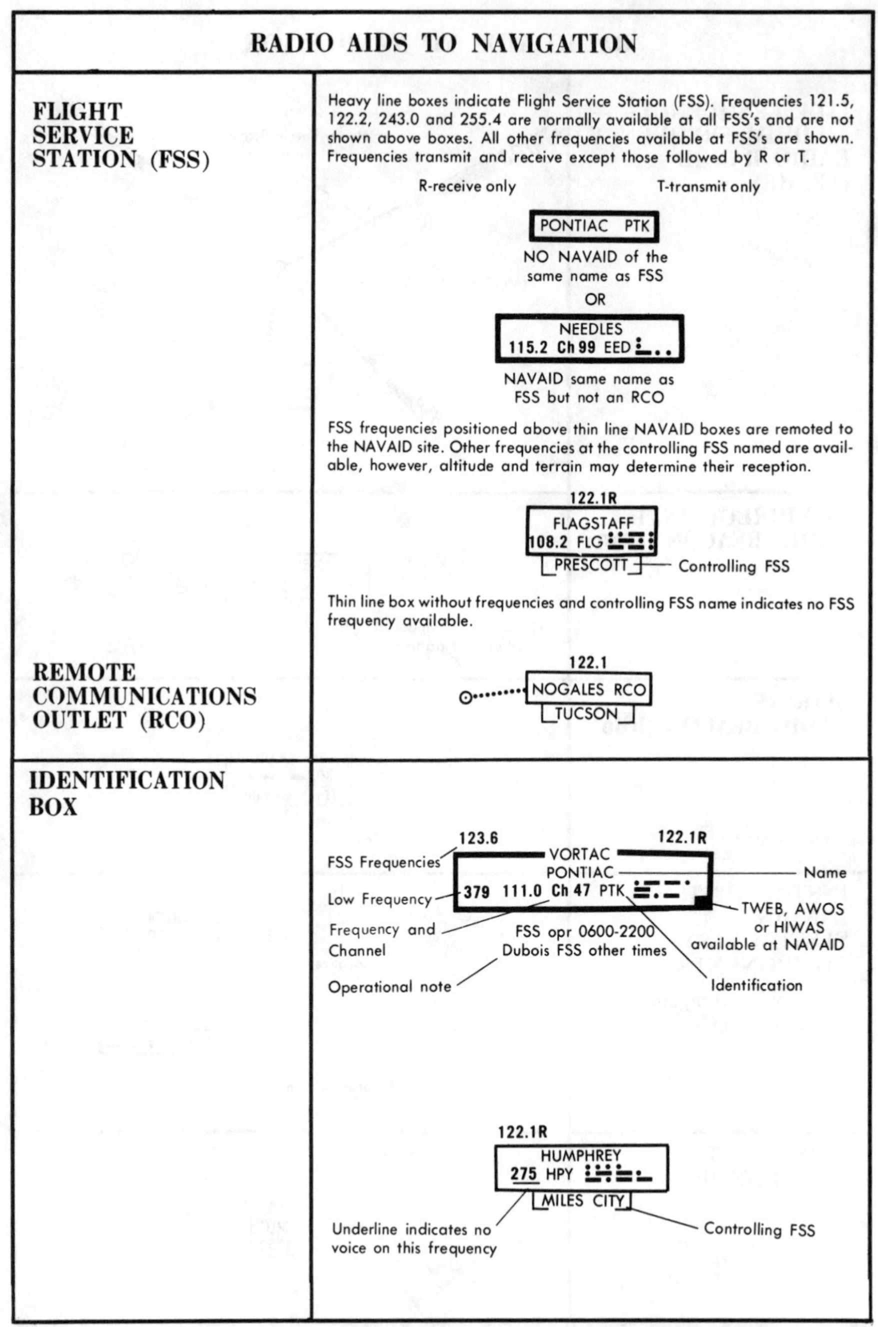

RADIO AIDS TO NAVIGATION

FLIGHT SERVICE STATION (FSS)	Heavy line boxes indicate Flight Service Station (FSS). Frequencies 121.5, 122.2, 243.0 and 255.4 are normally available at all FSS's and are not shown above boxes. All other frequencies available at FSS's are shown. Frequencies transmit and receive except those followed by R or T. R-receive only T-transmit only PONTIAC PTK NO NAVAID of the same name as FSS OR NEEDLES 115.2 Ch 99 EED NAVAID same name as FSS but not an RCO FSS frequencies positioned above thin line NAVAID boxes are remoted to the NAVAID site. Other frequencies at the controlling FSS named are available, however, altitude and terrain may determine their reception. 122.1R FLAGSTAFF 108.2 FLG PRESCOTT — Controlling FSS Thin line box without frequencies and controlling FSS name indicates no FSS frequency available.
REMOTE COMMUNICATIONS OUTLET (RCO)	122.1 NOGALES RCO TUCSON
IDENTIFICATION BOX	123.6 122.1R FSS Frequencies — VORTAC PONTIAC — Name Low Frequency — 379 111.0 Ch 47 PTK TWEB, AWOS or HIWAS available at NAVAID Frequency and Channel FSS opr 0600-2200 Dubois FSS other times Operational note Identification 122.1R HUMPHREY 275 HPY MILES CITY Underline indicates no voice on this frequency Controlling FSS

Fig. 3-9. Continued.

cations, and air traffic control (ATC) clearance requirements for the VFR pilot. Controlled airspace designated on visual charts is control zones, transition areas, control areas, airport radar service areas, terminal radar service areas, and terminal control areas. Controlled airspace not depicted is airport traffic areas, the continental control area, and the positive control area. Special use airspace is designated as prohibited, restricted, warning, alert, and military operations areas and military training routes. Figures 3-13 and 3-14 depict controlled airspace shown on visual charts.

Control zones establish controlled airspace from the surface upward to the base of the continental control area. The control zone is based on a primary airport, but might include other airports. A zone is normally circular with a radius of five miles, and might include extensions when necessary for departures and arrivals. Their primary purpose is to protect IFR aircraft during the takeoff and landing phase when weather conditions do not allow see-and-avoid separation. Control zones are depicted by a dashed blue line on visual charts. Fixed-wing special VFR flight is prohibited within certain control zones. These control zones are shown with a "T" control zone symbol as illustrated in Fig. 3-13.

The controlled airspace portion of Fig. 3-13 illustrates the depiction of transition areas and control areas on sectional and terminal area charts. Transition areas are designed to protect IFR operations during that portion of flight between the terminal and enroute phase. Transition areas extend upward from 700 feet agl (magenta vignette), 1,200 agl (blue vignette), or higher (floor specified, for instance 4,000 MSL). Transition areas extend to the base of the overlying controlled airspace, usually the continental control area. Control areas consist of that airspace designated as airways and control area extensions. Control area floors, blue vignette, are depicted in the same manner as transition areas and normally terminate at the base of the overlying controlled airspace. Like control zones, their primary purpose is to protect IFR aircraft when weather conditions do not allow see and avoid separation.

Figure 3-14 contains visual chart depictions of terminal control areas (blue), and airport radar service areas and terminal radar service areas (magenta). Terminal control areas (TCA) extend upward from the surface or higher to specified altitudes. TCAs require certain pilot qualifications, navigational equipment, and an ATC clearance.

Airport radar service area (ARSA) standard design consists of an inner five nautical mile circle and an outer 10 nm circle. The inner circle extends from the surface to 4,000 feet agl, the outer circle from 1,200 feet agl to the altitude of the inner circle. ARSA dimensions and altitudes might vary from this standard due to terrain and other factors. ARSAs require certain aircraft equipment and two-way radio communications.

Terminal radar service areas (TRSA) designate airspace where traffic advisories, vectoring, sequencing, and separation of VFR aircraft are provided. TRSAs are designated Stage I, Stage II, or Stage III, which specify radar services that are available. The type of TRSA (Stage I, II, or III) can be found in the *Airport/Facility Directory*.

The continental control area consists of that airspace above the 48 contiguous states, the District of Columbia, and Alaska—except the peninsula west of 160 degree longitude—at and above 14,500 feet MSL, with some minor exclusions. (A detailed explanation of controlled airspace is contained in the *Airman's Information Manual*.) Because the

(Continued on page 62.)

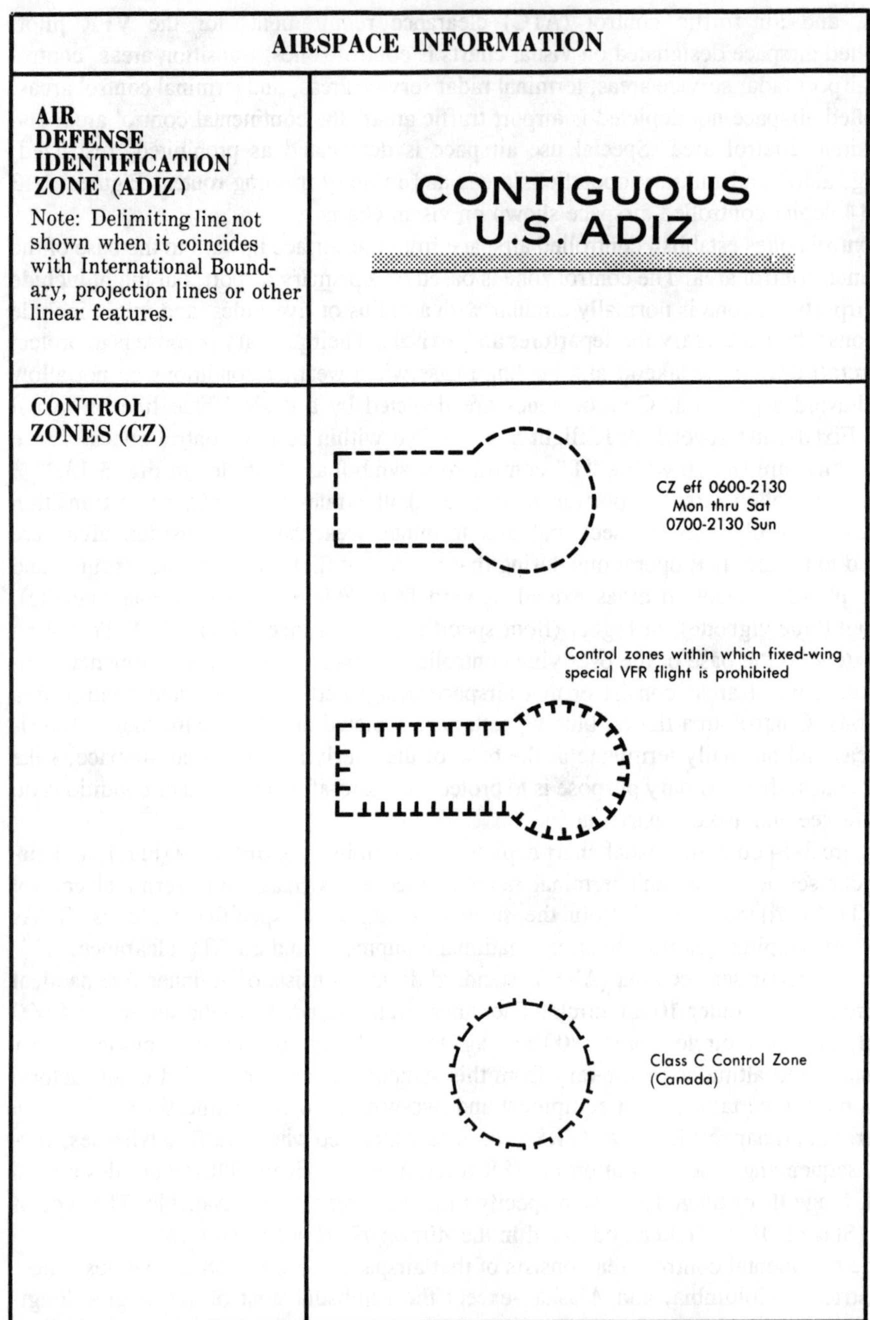

AIRSPACE INFORMATION
AIR DEFENSE IDENTIFICATION ZONE (ADIZ)
Note: Delimiting line not shown when it coincides with International Boundary, projection lines or other linear features.
CONTIGUOUS U.S. ADIZ
CONTROL ZONES (CZ)
CZ eff 0600-2130
Mon thru Sat
0700-2130 Sun
Control zones within which fixed-wing special VFR flight is prohibited
Class C Control Zone
(Canada)

Fig. 3-13. Visual charts must provide enough airspace information to allow the VFR pilot to operate safely within an ever increasingly complex system.

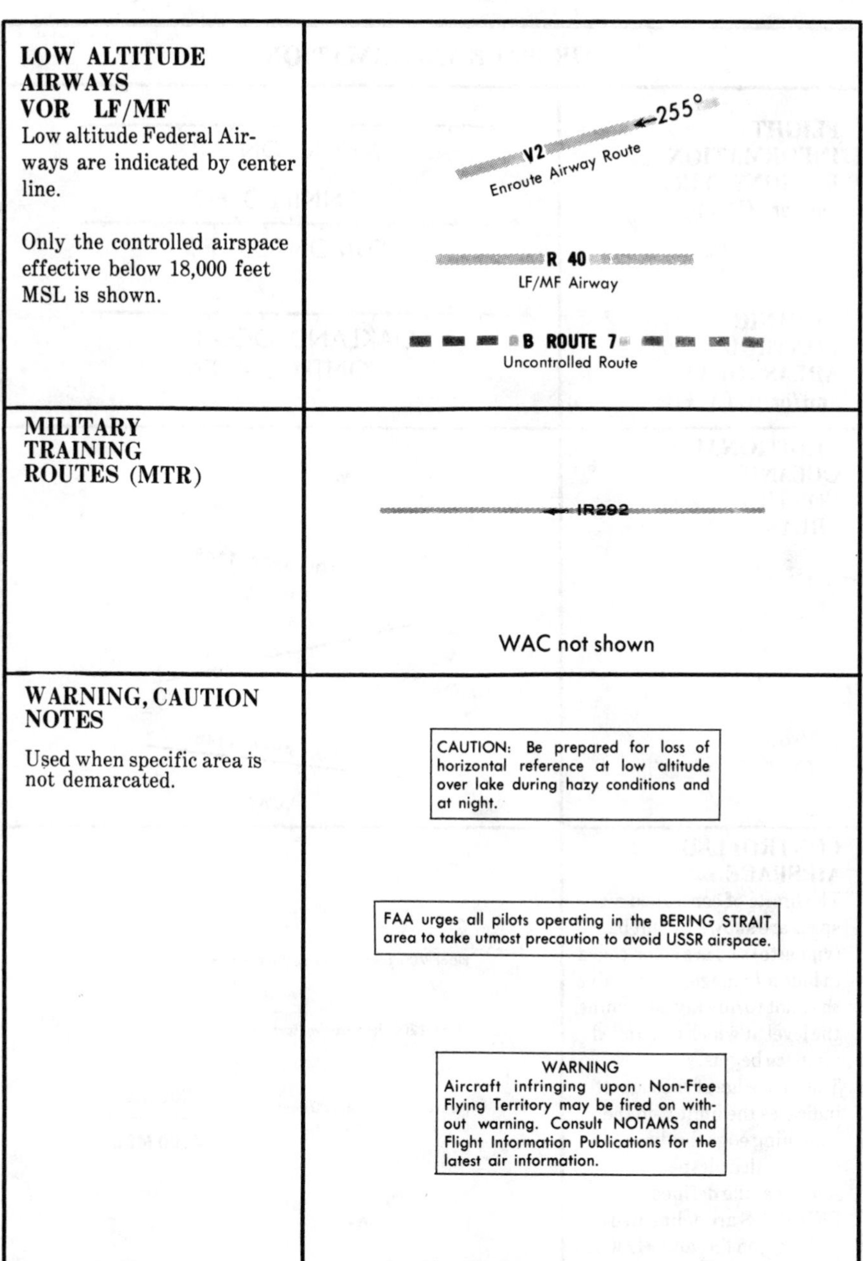

LOW ALTITUDE AIRWAYS VOR LF/MF Low altitude Federal Airways are indicated by center line. Only the controlled airspace effective below 18,000 feet MSL is shown.	←255° V2 Enroute Airway Route R 40 LF/MF Airway B ROUTE 7 Uncontrolled Route
MILITARY TRAINING ROUTES (MTR)	← IR292 WAC not shown
WARNING, CAUTION NOTES Used when specific area is not demarcated.	CAUTION: Be prepared for loss of horizontal reference at low altitude over lake during hazy conditions and at night. FAA urges all pilots operating in the BERING STRAIT area to take utmost precaution to avoid USSR airspace. WARNING Aircraft infringing upon Non-Free Flying Territory may be fired on without warning. Consult NOTAMS and Flight Information Publications for the latest air information.

Fig. 3-13. Continued.

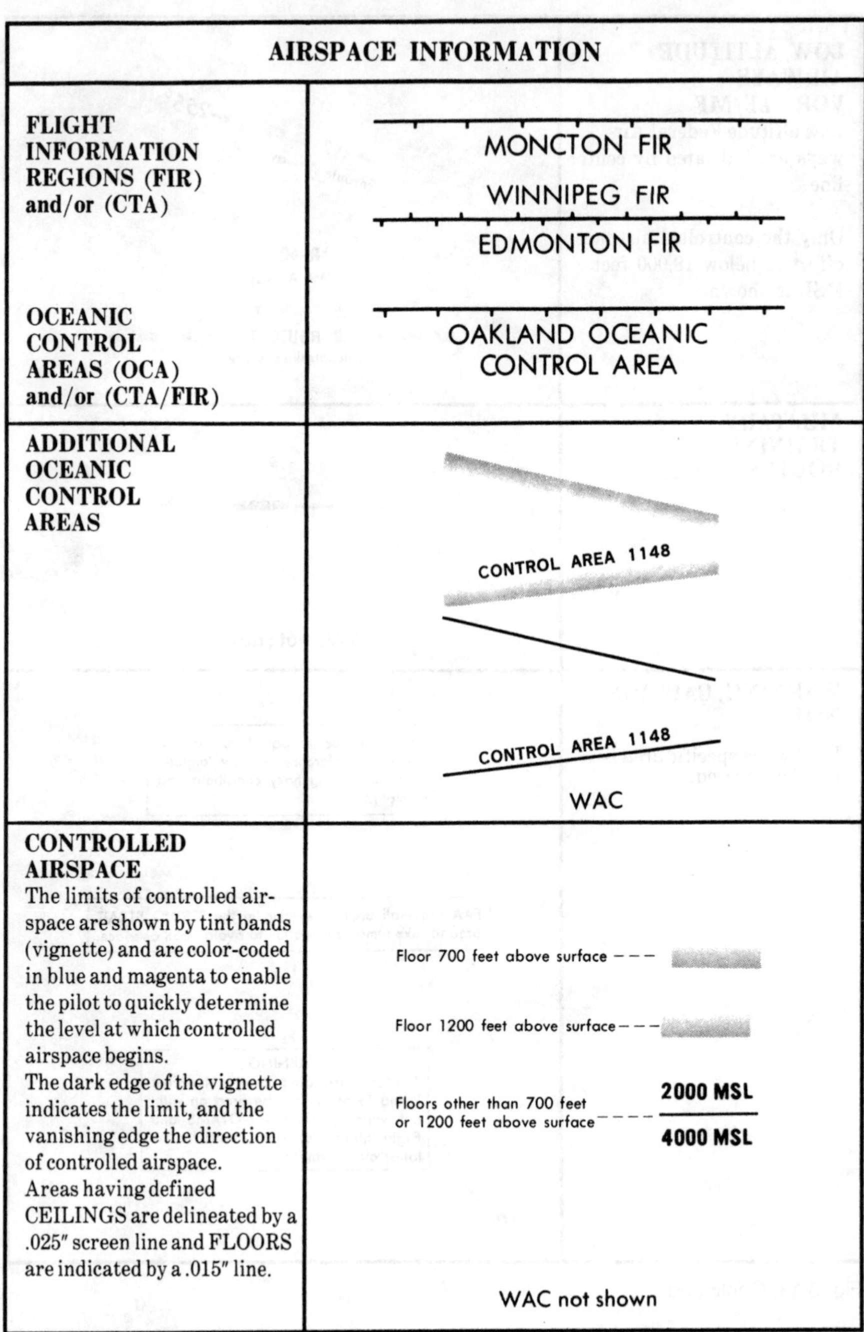

AIRSPACE INFORMATION	
FLIGHT INFORMATION REGIONS (FIR) and/or (CTA)	MONCTON FIR WINNIPEG FIR EDMONTON FIR
OCEANIC CONTROL AREAS (OCA) and/or (CTA/FIR)	OAKLAND OCEANIC CONTROL AREA
ADDITIONAL OCEANIC CONTROL AREAS	CONTROL AREA 1148 CONTROL AREA 1148 WAC
CONTROLLED AIRSPACE The limits of controlled air-space are shown by tint bands (vignette) and are color-coded in blue and magenta to enable the pilot to quickly determine the level at which controlled airspace begins. The dark edge of the vignette indicates the limit, and the vanishing edge the direction of controlled airspace. Areas having defined CEILINGS are delineated by a .025″ screen line and FLOORS are indicated by a .015″ line.	Floor 700 feet above surface Floor 1200 feet above surface Floors other than 700 feet or 1200 feet above surface 2000 MSL / 4000 MSL WAC not shown

Fig. 3-13. Continued.

SPECIAL USE AIRSPACE Only the airspace effective below 18,000 feet MSL is shown. The type of area shall be spelled out in large areas if space permits.	PROHIBITED, RESTRICTED OR WARNING AREA P-56 OR R-6401 OR W-518 ALERT AREA ALERT AREA A-631 CONCENTRATED STUDENT HELICOPTER TRAINING MILITARY OPERATIONS AREA (MOA) VANCE 2 MOA
TERMINAL AREA CHART COVERAGE	Sectional only

Fig. 3-13. Continued.

(Continued from page 57.)
continental control area is virtually continuous at and above 14,500 MSL, it is not depicted on visual charts. Also, the positive control area is not depicted on charts because for practical purposes it exists over the contiguous U.S. from 18,000 feet to and including flight level (FL) 600 (60,000 feet); aircraft flying in this area must be operating in accordance with an IFR clearance.

One other type of airspace, important to the VFR pilot is not directly depicted on visual charts; the airport traffic area exists around all airports with an operating control tower within a five statute mile (sm) radius of the airport, from the surface up to but not including 3,000 feet agl, with scattered exceptions according to FAR 93. Basically, these areas require two-way radio communication with the tower.

Visual charts depict special use airspace (SUA) below 18,000 feet MSL. SUA consists of airspace where activities must be confined because they pose a hazard to aircraft operations. Prohibited, restricted, warning, alert, military operations areas and military training routes are shown in Fig. 3-13, with special military activity routes in Fig. 3-14. Aircraft operations are prohibited within prohibited areas. These areas are established for security or other reasons associated with the national welfare. Aircraft operations are prohibited within Restricted Areas when the area is active. Restricted Areas are established for unusual, often invisible, hazardous activities, such as artillery firing, aerial gunnery practice, or guided missile firing. Warning areas are established for the same hazards as restricted areas, but over international waters. Alert areas inform nonparticipating pilots of areas that might contain a high volume of training or unusual activity. Pilots should exercise extra caution within these areas.

Military operation areas (MOA) alert pilots to military training activities. In addition to possible high concentration of aircraft, military pilots might conduct acrobatic flight and operate at speeds in excess of 250 knots below 10,000 feet. High speed, low level military operations are conducted along military training routes (MTR). MTRs where operations are conducted IFR are designated as IR; VFR operations are designated VR; IR and VR routes operated at or below 1,500 feet agl will be identified by four-digit numbers (IR 1007, VR 1009). Operations that are conducted above 1,500 feet agl are identified by three digit numbers (IR 205, VR 257). Special military activity routes alert pilots to areas where cruise missile tests are conducted.

Figure 3-14 contains symbols that alert pilots to parachute jumping, glider operations, and ultralight activity. An additional symbol has been added for hang gliding activity. The symbol resembles a hang glider in flight. Where these symbols appear, pilots cannot expect to be alerted to the activity through NOTAMs. Details on the activity are normally found in the *Airport/Facility Directory*.

Navigational information

Navigational information consists of isogonic lines and values, local magnetic disturbance notes, aeronautical lights, airway intersection depictions, and VFR checkpoints.

Isogonic lines (lines of equal magnetic declination for a given time) provide the pilot with the difference between true north and magnetic north in degrees. These lines and values are updated every five years. Local magnetic notes alert pilots to areas where the magnetic compass might be unreliable, often due to large deposits of iron ore (Fig. 3-15).

(Continued on page 68.)

AIRSPACE INFORMATION	
PARACHUTE JUMPING AREA	WAC not shown
GLIDER OPERATING AREA	WAC not shown
ULTRALIGHT ACTIVITY	WAC not shown
TERMINAL CONTROL AREA (TCA) Appropriate notes as required may be shown.	LAS VEGAS TCA (Outer limit only shown on WAC)
	20 NM -- Distance from facility (TAC) 70 ---- Ceiling of TCA in hundreds of feet MSL 50 ---- Floor of TCA in hundreds of feet MSL 124.3 ---ATC Sector Frequency (Los Angeles TCA only) WAC not shown CONTACT LAS VEGAS APPROACH CONTROL ON 121.1 OR 257.8 (TAC only)
MODE C AREA (See FAR 91.24/AIM) Appropriate notes as required may be shown.	MODE C 30 NM Distance from facility All mileages are nautical (NM)
TERMINAL RADAR SERVICE AREA (TRSA) Appropriate notes as required may be shown.	BILLINGS TRSA WAC not shown
	80 ----- Ceiling of TRSA in hundreds of feet MSL 40 ----- Floor of TRSA in hundreds of feet MSL SEE TWR FREQ TAB WAC not shown

Fig. 3-14. As well as airspace information, visual charts include potential hazards, such as parachute jumping areas, glider, ultralight, and hang glider activity, and locations of high speed military operations.

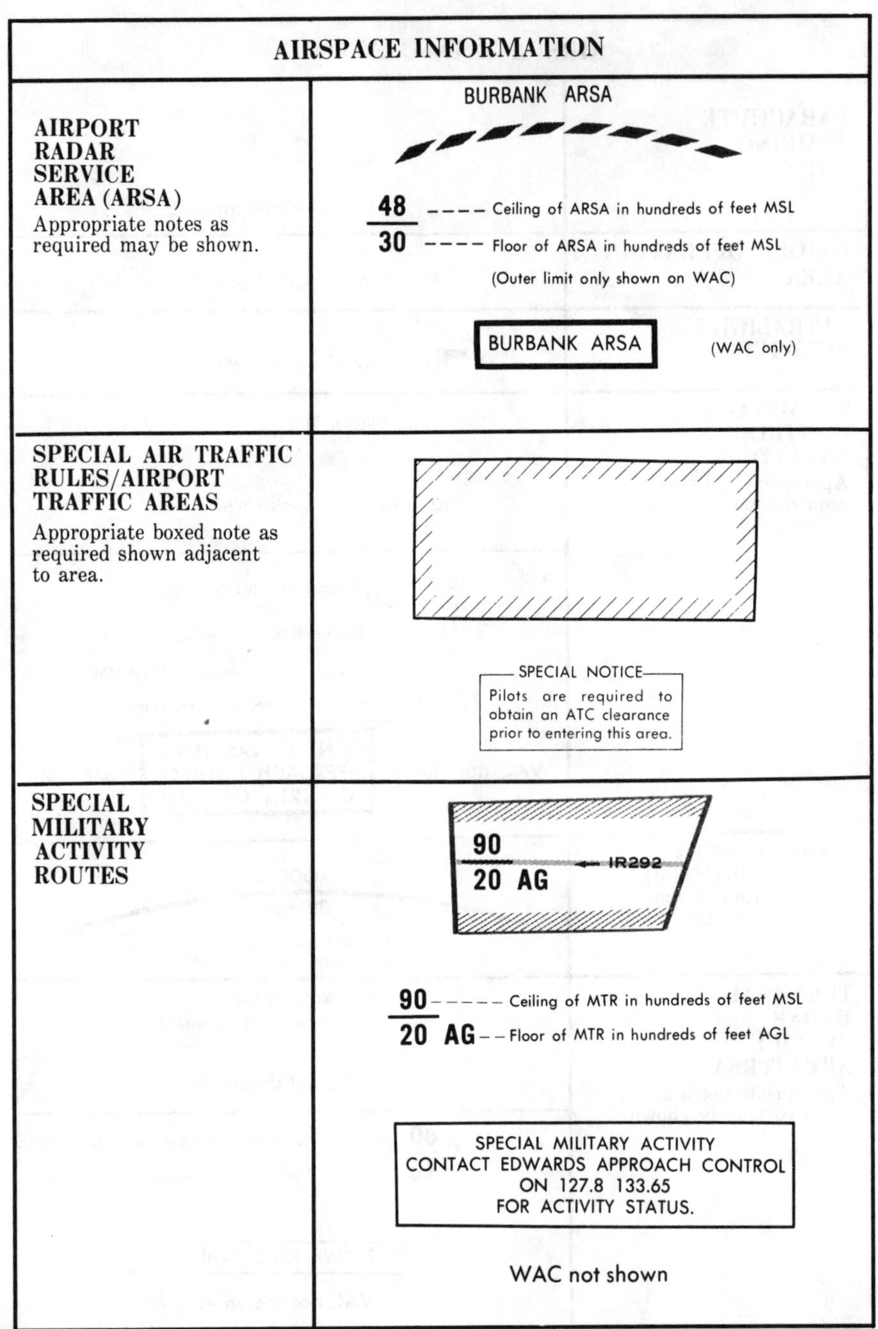

Fig. 3-14. Continued.

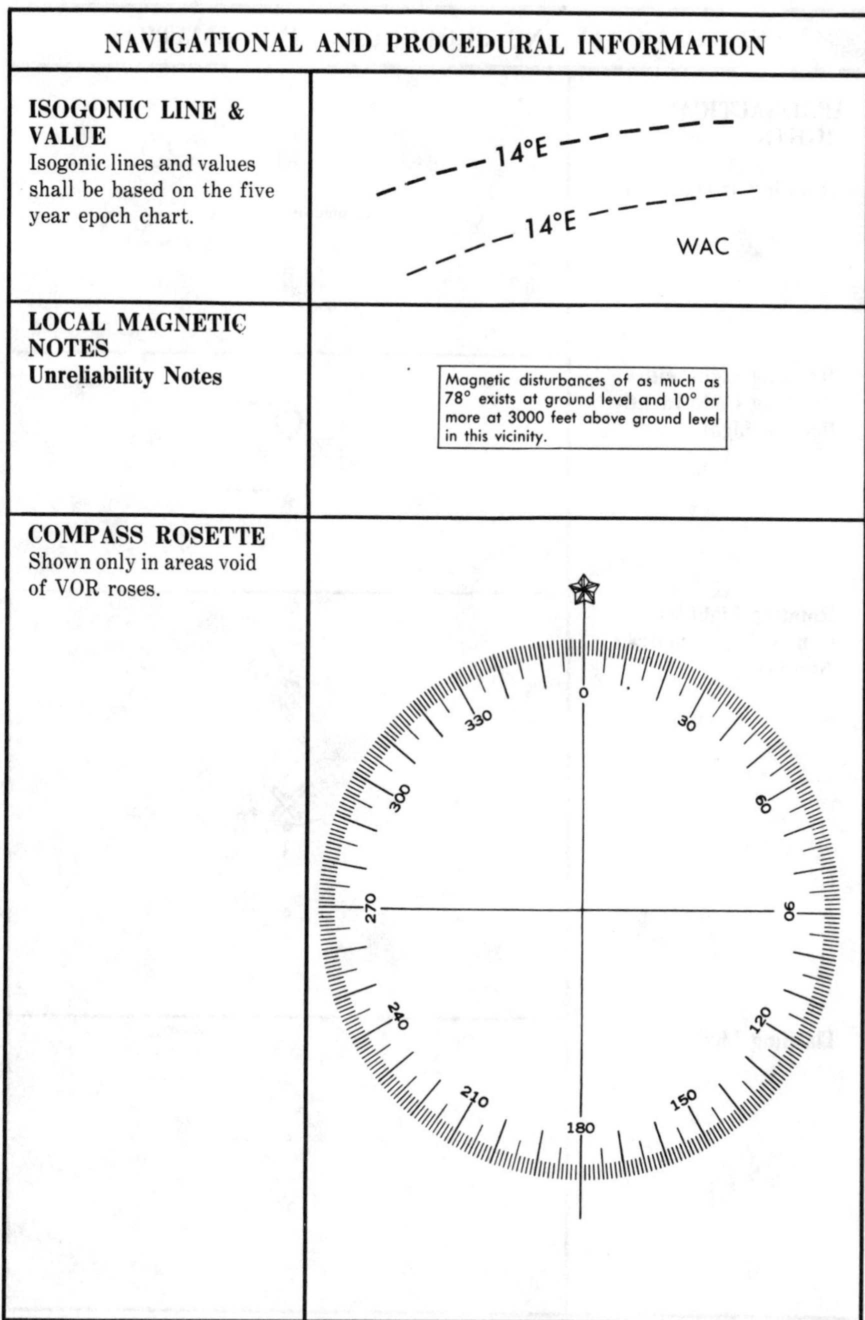

NAVIGATIONAL AND PROCEDURAL INFORMATION
ISOGONIC LINE & VALUE
Isogonic lines and values shall be based on the five year epoch chart.
14°E
14°E
WAC
LOCAL MAGNETIC NOTES
Unreliability Notes
Magnetic disturbances of as much as 78° exists at ground level and 10° or more at 3000 feet above ground level in this vicinity.
COMPASS ROSETTE
Shown only in areas void of VOR roses.
0
30
60
90
120
150
180
210
240
270
300
330

Fig. 3-15. Navigational and procedural information includes lines of magnetic variation and aeronautical lights.

NAVIGATIONAL AND PROCEDURAL INFORMATION	
AERONAUTICAL LIGHTS **Rotating or Oscillating**	Located at aerodrome In isolated location
Rotating Light with Flashing Code Identification Light	
Rotating Light with Course Lights and Site Number	5 18 4B
Flashing Light	Rotating Beacon Fl

Fig. 3-15. Continued.

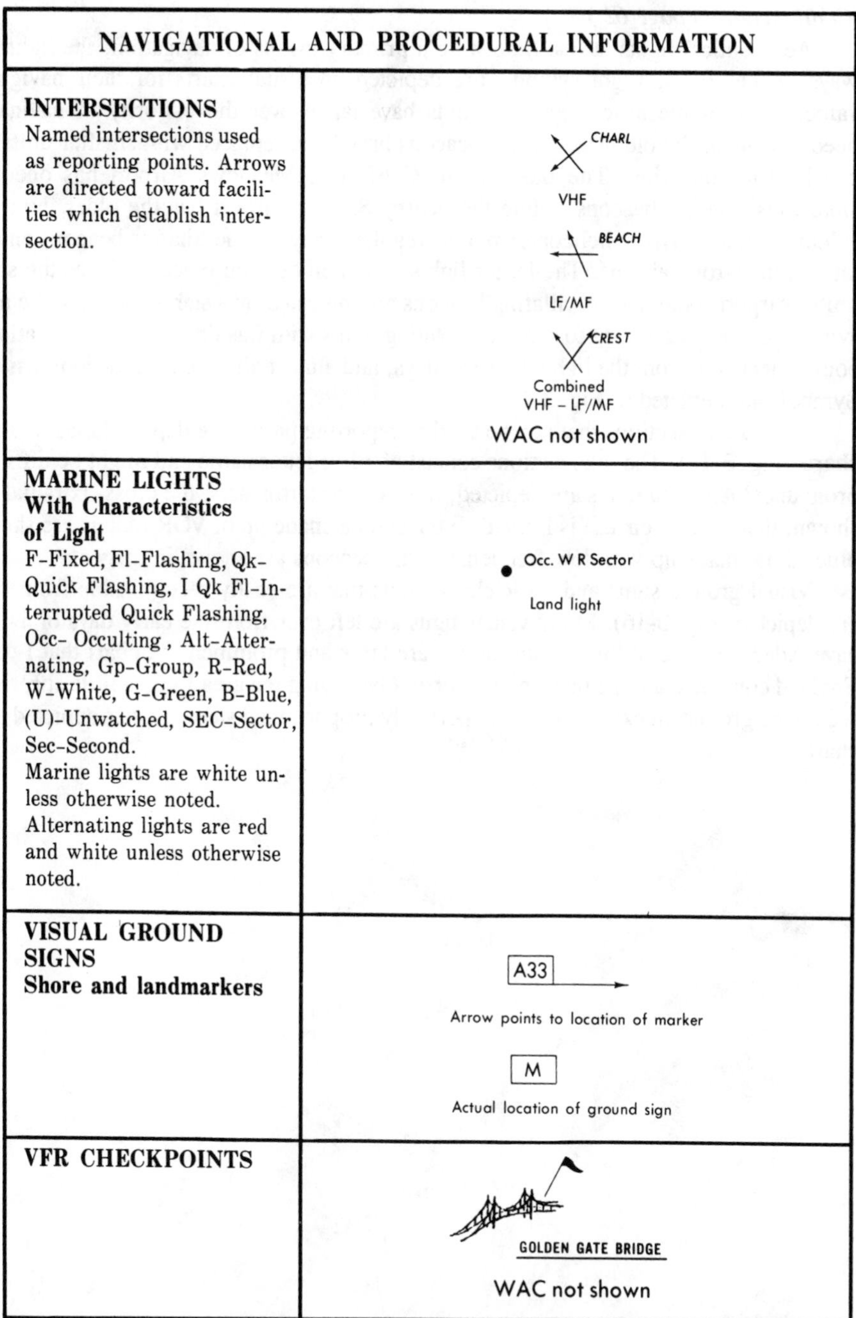

NAVIGATIONAL AND PROCEDURAL INFORMATION	
INTERSECTIONS Named intersections used as reporting points. Arrows are directed toward facilities which establish intersection.	CHARL VHF BEACH LF/MF CREST Combined VHF – LF/MF WAC not shown
MARINE LIGHTS With Characteristics of Light F-Fixed, Fl-Flashing, Qk-Quick Flashing, I Qk Fl-Interrupted Quick Flashing, Occ- Occulting , Alt-Alternating, Gp-Group, R-Red, W-White, G-Green, B-Blue, (U)-Unwatched, SEC-Sector, Sec-Second. Marine lights are white unless otherwise noted. Alternating lights are red and white unless otherwise noted.	Occ. W R Sector Land light
VISUAL GROUND SIGNS **Shore and landmarkers**	A33 Arrow points to location of marker M Actual location of ground sign
VFR CHECKPOINTS	GOLDEN GATE BRIDGE WAC not shown

Fig. 3-16. Visual charts contain airway intersections, visual ground signs, and VFR checkpoints to aid in visual navigation.

(Continued from page 62.)

Aeronautical lights at one time were a primary means of navigation, the lighted airways of the 1930s. Light symbols are depicted on visual charts for their navigational value; however, because electronics aids have taken over the majority of navigational needs, many of the old large airport beacons have been replaced with smaller units. This can lead to confusion. The Bakersfield, California, Meadows Airport has one of the small, less intense beacons, while the nearby Shafter Airport has the older, large unit. Pilots, including Army helicopter pilots, regularly key on the Shafter beacon, and even land at the wrong airport. The larger lights can often be seen twice as far as the smaller units. Airport rotating or oscillating beacons are indicated by a star adjacent to the airport symbol and operate sunset to sunrise. Rotating lights with flashing code identification and course lights are from the lighted airway days, and almost all have been decommissioned. Symbols are depicted in Fig. 3-15.

Named intersections that can be used as reporting points are depicted on some visual charts (Fig. 3-16). The intersections consist of a five letter name and might be difficult to pronounce. Airway radials are depicted; however, unfortunately, the cross fix radial is not shown; therefore, their use is limited. Intersections made up of VOR radials are shown in blue, those made up with low frequency radio beacons are shown in magenta.

Visual ground signs and VFR checkpoints that are easily recognizable from the air are depicted (Fig. 3-16). Many visual signs are left over from the early days of aviation, have faded, and are of little value, others are large and prominent. A chart that I used in England contained a large picture of a horse. Overflying the area I was a little surprised to see on the ground an extremely large, perfectly proportioned steed, just as depicted on the chart.

4
Standard visual charts

THE NATIONAL OCEAN SERVICE (NOS) PUBLISHES A NUMBER OF AERONAUTICAL charts specifically designed to assist the pilot with visual or VFR navigation: planning charts, sectional charts, terminal area charts, and world aeronautical charts. Interestingly, these charts are often more complex than those used for instrument or IFR navigation. Chapter 3 discusses the dozen or so types of airspace plus a description of terminology and symbols common to visual charts; therefore, in this chapter only those terms and symbols unique to individual charts are presented.

PLANNING CHARTS

Planning charts, as the name implies, are designed for the initial portion of flight preparation, as opposed to operational charts—sectionals, WACs—used for preflight planning and navigation. These charts are most useful for planning long trips, usually with several or more legs. For example, I have planned and executed two VFR crossings of the U.S., one from California to New York and the other from California to Florida in a Cessna 150. Each trip was nothing more than a series of short cross-country flights. Here is where planning charts are of most value. The two most common planning charts are the VFR wall planning chart and the flight case planning chart. Area of coverage is illustrated in Fig. 4-1.

VFR wall planning chart

This large 56-×-82-inch chart, which can be folded to $8^{1}/_{2}$ × 14 inches, is often found in planning rooms of flight schools and fixed based operators (FBO). Produced in two halves, east and west, it can be assembled to form a VFR planning chart on one side or a low altitude IFR planning chart on the other. Coverage includes the continental U.S.

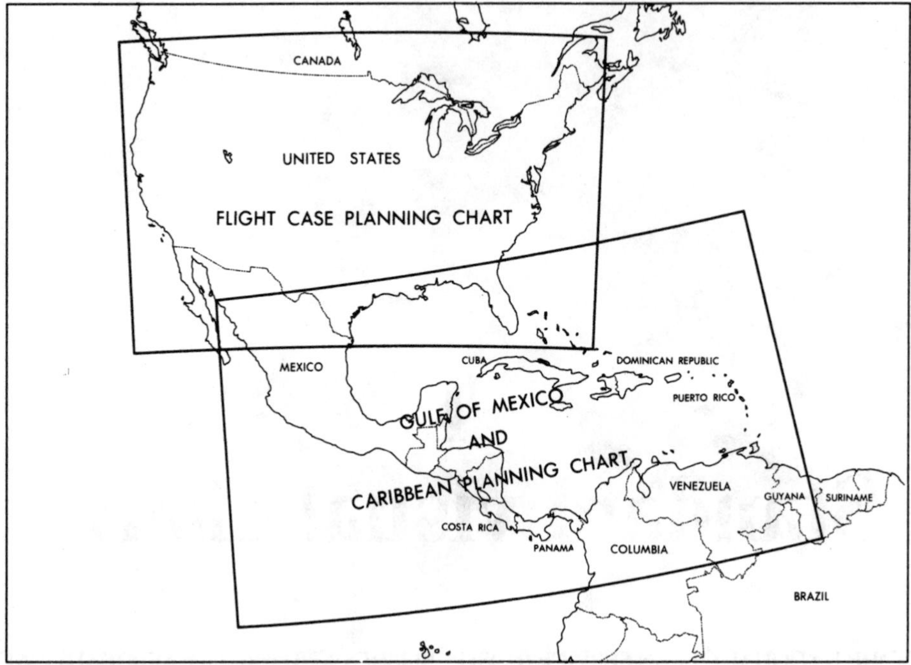

Fig. 4-1. Planning charts are designed for the initial portion of flight preparation.

and the Gulf of Mexico. Its small scale 1:2,333,232 (1 inch equals 32 nautical miles) can only provide an overview. It is revised every 56 days. Figure 4-2 contains airspace and navigational symbols unique to this chart. Features include:

- Low altitude airways
- Mileages
- Radio aids to navigation
- Airports
- Special use airspace
- State boundaries
- Large bodies of water
- Time zones
- Military training routes
- Military refueling tracks
- Shaded relief and terrain elevations
- Index of sectional charts

Flight case planning chart

Smaller in scale than the wall planning chart, the flight case planning chart provides a more portable product for preflight and in-flight planning below 18,000 feet. The chart is

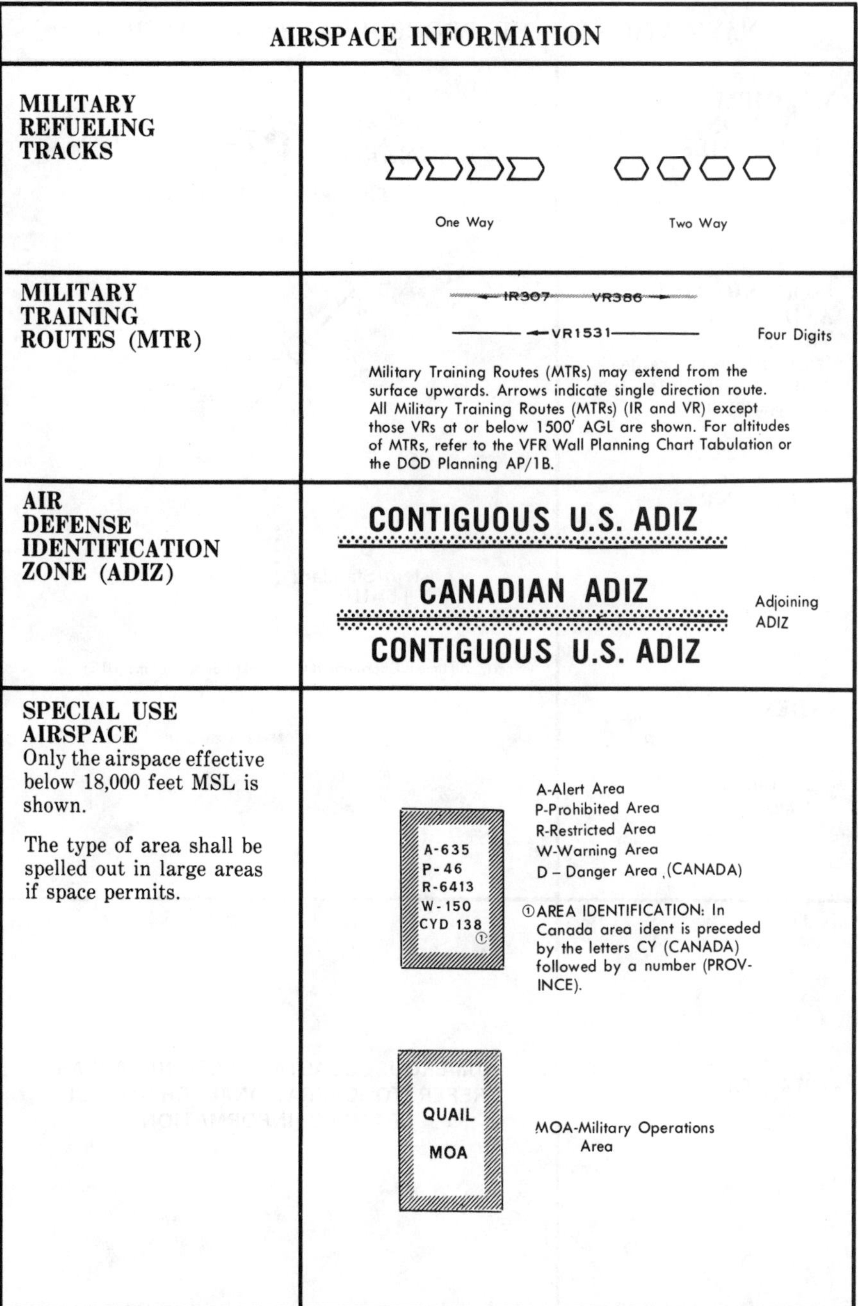

AIRSPACE INFORMATION
MILITARY REFUELING TRACKS
One Way
Two Way
MILITARY TRAINING ROUTES (MTR)
IR307
VR386
VR1531
Four Digits
Military Training Routes (MTRs) may extend from the surface upwards. Arrows indicate single direction route. All Military Training Routes (MTRs) (IR and VR) except those VRs at or below 1500′ AGL are shown. For altitudes of MTRs, refer to the VFR Wall Planning Chart Tabulation or the DOD Planning AP/1B.
AIR DEFENSE IDENTIFICATION ZONE (ADIZ)
CONTIGUOUS U.S. ADIZ
CANADIAN ADIZ
CONTIGUOUS U.S. ADIZ
Adjoining ADIZ
SPECIAL USE AIRSPACE
Only the airspace effective below 18,000 feet MSL is shown.
The type of area shall be spelled out in large areas if space permits.
A-635
P-46
R-6413
W-150
CYD 138
A-Alert Area
P-Prohibited Area
R-Restricted Area
W-Warning Area
D – Danger Area (CANADA)
①AREA IDENTIFICATION: In Canada area ident is preceded by the letters CY (CANADA) followed by a number (PROVINCE).
QUAIL
MOA
MOA-Military Operations Area

Fig. 4-2. These symbols are unique to the VFR wall planning chart.

NAVIGATIONAL AND PROCEDURAL INFORMATION	
MAXIMUM ELEVATION FIGURE (MEF)	(Thousands of Feet) 6 7 (Hundreds of Feet) (Highest within each Sectional Chart)
ISOGONIC LINE AND VALUE Isogonic lines and values shall be based on the five year epoch chart.	14°E
TIME ZONE	Eastern Standard +5=UTC · Atlantic Standard +4=UTC All time is Coordinated Universal (Standard) Time (UTC)
INDEX **Sectional Charts**	WASHINGTON
NOTE **Warning**	NUMEROUS COASTAL WARNING AREAS REFER TO OPERATIONAL CHART FOR DETAILED INFORMATION

Fig. 4-2. Continued.

30 × 50 inches, which can be folded to the standard 5-×-10-inch size and has a scale of 1:4,374,803 (1 inch equals 60 nautical miles). This chart is revised every 24 weeks. Airports depicted have a minimum 5,000 foot hard-surface runway; in noncongested areas, airports with a minimum 3,000 foot hard-surface runway are depicted. Figure 4-3 contains navigational symbols unique to this chart. Features include:

- Shaded relief and critical elevations
- Selected flight service stations
- Parachute jumping areas
- Special use airspace table
- Mileage table between 174 major airports
- City/airport location index
- Indexes for sectional, Canadian, Pacific, and South American charts
- Time zones

Gulf of Mexico and Caribbean planning chart

The Gulf of Mexico and Caribbean planning chart has a still smaller scale than either the wall planning chart or flight case planning chart. This chart is designed for preflight planning of flights through and around the Gulf of Mexico and Caribbean. Intended to be used in conjunction with world aeronautical charts, it is printed on the back of the Puerto Rico-Virgin Islands VFR terminal area chart. The area of coverage is shown in Fig. 4-1. The chart is 20 × 34 inches, which can be folded to the standard 5 × 10, and has a scale of 1:6,192,178 (1 inch equals 85 nautical miles). This chart is revised annually. Features include:

- Airports of entry
- Special use airspace below 18,000 feet
- Significant bodies of water
- International boundaries
- Large islands and island groups
- Capital cities and cities where an airport is located
- Selected other major cities
- Air mileage between airports of entry
- Index of world aeronautical charts
- Directory of airports, including facilities, servicing, and fuel
- Department of Defense requirements for civilian use of military airports
- Checklist for ditching
- Runway visual range (RVR) conversion table from feet to meters
- Emergency procedures

Charted VFR flyway planning charts

Charted VFR flyway planning charts are designed to assist pilots planning flights through or around high density areas such as terminal control areas and airport radar ser-

NAVIGATIONAL AND PROCEDURAL INFORMATION	
MAXIMUM ELEVATION FIGURE (MEF)	(Thousands of Feet) — 14^3 — (Hundreds of Feet) (Highest within each Sectional Chart)
TIME ZONE	Eastern Standard +5=UTC ⋮ Atlantic Standard +4=UTC All time is Coordinated Universal (Standard) Time (UTC)
INDEXES **Sectional Charts**	TWIN CITIES
Canada Charts	GASPE — Designed for visual flights of short duration primarily for pilotage.
Tactical Pilotage Charts	H-23A — Published by Defense Mapping Agency Aerospace Center. Designed for detailed pre-flight planning. Emphasis is on ground features significant in visual and radar low-level high speed navigation.
TERMINAL CONTROL AREA (TCA)	City name in GREEN indicates location is listed on Mileage Table located on back of chart. Washington Green underline indicates Terminal Control Area (TCA) Chart is available

Fig. 4-3.

vice areas. These two color charts are printed on the back of selected terminal area charts (TACs) with coverage corresponding to the TACs. The following TACs contain VFR flyway planning charts:

- Atlanta
- Baltimore-Washington
- Dallas-Fort Worth
- Los Angeles
- Miami
- San Diego

Charted VFR flyway planning charts, as the name implies, are not to be used for navigation, or a substitute for the TAC or sectional chart. Features include:

- Airports
- NAVAIDs
- Special use airspace
- Terminal control areas and airport radar service areas
- Control zones
- VFR flyways (suggested headings and altitudes)
- Procedural notes
- Military training routes
- Selected obstacles
- VFR checkpoints
- Hydrographic features
- Cultural features
- Terrain relief designated as VFR checkpoints
- Critical spot elevations

USING PLANNING CHARTS

Recall that a long cross-county flight is nothing more than a series of individual legs. Take for example a flight from Van Nuys, California, to Jamestown, New York, and return. The first consideration is the aircraft, the second consideration is the pilot and passengers. It doesn't do much good to plan four hour legs with two hour bladders. In a Cessna 150, 250-mile legs are comfortable; in a Bonanza, depending upon fuel load, 350- to 400-mile legs would be comfortable; in a Hughes 269 helicopter, with its speed and limited fuel, only 100 mile legs are practical. It all depends on the aircraft, fuel, and pilot/passenger endurance.

With average legs in mind we proceed to the planning chart, realize aircraft performance, and consider terrain, airports and services, and controlled and special use airspace. I don't like flying over high, rough terrain, or unpopulated areas of deserts or swamps, if at all possible. Enroute destinations are selected for the services available. We can plan the flight to major airports, if we have the proper electronic equipment, or to uncontrolled fields, should we wish. We have to avoid prohibited and restricted areas and we might wish to avoid MOAs.

The flight from California to New York was preplanned with routes, altitudes, alternates, and services considered. For example, ARSA, TRSA, control towers, and FSS frequencies were logged. Of 24 individual legs, only one, from Huntington, W.V., to Cincinnati, Ohio, was not on the original itinerary. The unscheduled stop was required to repair the radio.

From the planning chart we determine which sectional charts or WACs are needed. From these charts we can either plan to avoid terminal control areas or obtain required terminal area charts. From these charts we develop a plan to negotiate or avoid the TCAs and their heavy concentrations of traffic.

SECTIONAL CHARTS

Sectional charts are designed for visual navigation of slow to medium speed aircraft. These multicolored charts provide the most accurate means of pilotage—navigating the aircraft by means of ground reference—because of their scale. The chart is 20 × 60 inches, which can be folded to the standard 5 × 10 inches and has a scale of 1:500,000 (1 inch equals 7 nautical miles). They are revised semiannually, except for some Alaskan charts that are revised annually. Sectional charts are named for a major city within the area of coverage. Figure 4-4 contains sectional chart coverage for the contiguous U.S., and Fig. 4-5 shows coverage for Alaska. Features include:

- Visual aids to navigation
- Radio aids to navigation
- Airports
- Controlled airspace
- Restricted areas
- Obstructions
- Topography
- Shaded Relief
- Latitude and longitude lines
- Airways and fixes
- Other low level related data

The Hawaii sectional is the only chart of this series that is not oriented to true north. This is necessary to portray all the islands of the group on a standard size sheet, which also contains the Mariana and Samoa Island groups.

Before using any chart, or aeronautical publication, a pilot's first task is to determine currency. That means reviewing the cover pages for effective and obsolescent dates. Figure 4-6 is the cover page and data panel from the Las Vegas Sectional Aeronautical Chart. The chart is a Lambert conformal conic projection with standard parallels 33°20′ and 38°40′, based on the North American Datum of 1927. Topographic data has been corrected to January 1991. The significance of this data is discussed in chapter 2.

This is the 45th edition of the chart, which became effective April 4, 1991, including airspace amendments that would include changes in the lateral limits of controlled and special use airspace, and other aeronautical data received by February 7, 1991, such as

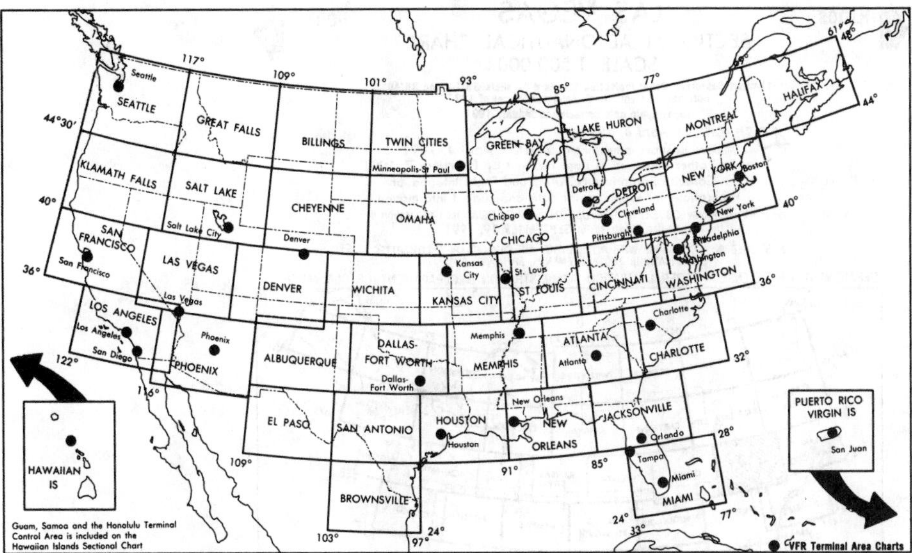

Fig. 4-4. Sectional charts are named for a major city within their area of coverage.

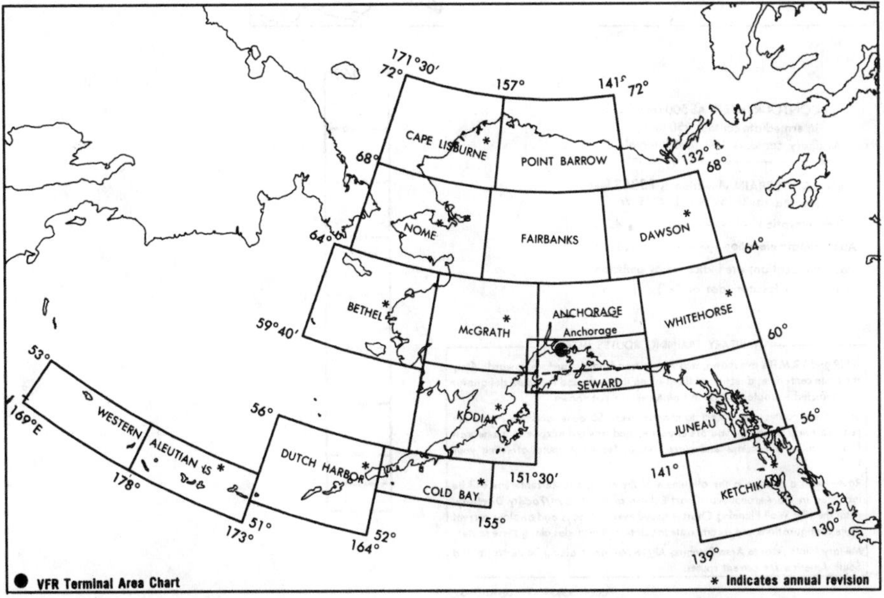

Fig. 4-5. Sectional charts are revised semiannually, except for some Alaskan charts that are revised annually.

radio navigation aids and frequencies, and radio communication frequencies and locations. Note that it cautions pilots to consult appropriate NOTAMs and flight information publications for supplemental data and current information. The obsolescent statement is

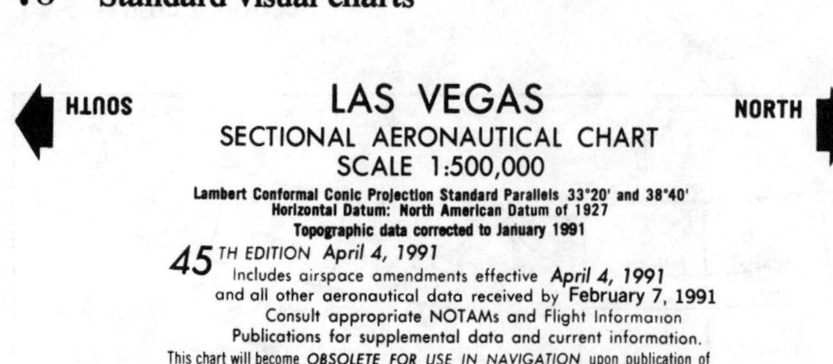

SOUTH
LAS VEGAS
NORTH
SECTIONAL AERONAUTICAL CHART
SCALE 1:500,000
Lambert Conformal Conic Projection Standard Parallels 33°20' and 38°40'
Horizontal Datum: North American Datum of 1927
Topographic data corrected to January 1991
45TH EDITION April 4, 1991
Includes airspace amendments effective April 4, 1991
and all other aeronautical data received by February 7, 1991
Consult appropriate NOTAMs and Flight Information
Publications for supplemental data and current information.
This chart will become OBSOLETE FOR USE IN NAVIGATION upon publication of
the next edition scheduled for SEPTEMBER 19, 1991
PUBLISHED IN ACCORDANCE WITH INTERAGENCY AIR CARTOGRAPHIC COMMITTEE
SPECIFICATIONS AND AGREEMENTS, APPROVED BY:
DEPARTMENT OF DEFENSE • FEDERAL AVIATION ADMINISTRATION • DEPARTMENT OF COMMERCE

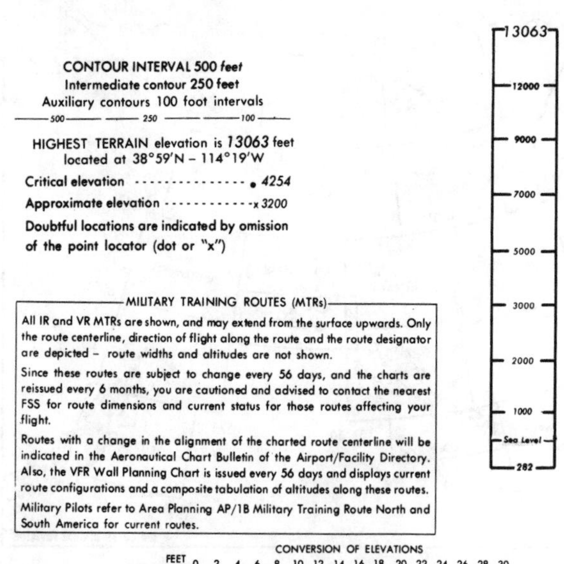

CONTOUR INTERVAL 500 feet
Intermediate contour 250 feet
Auxiliary contours 100 foot intervals
500
250
100
HIGHEST TERRAIN elevation is 13063 feet
located at 38°59'N – 114°19'W
Critical elevation • 4254
Approximate elevation x 3200
Doubtful locations are indicated by omission
of the point locator (dot or "x")
MILITARY TRAINING ROUTES (MTRs)
All IR and VR MTRs are shown, and may extend from the surface upwards. Only the route centerline, direction of flight along the route and the route designator are depicted – route widths and altitudes are not shown.
Since these routes are subject to change every 56 days, and the charts are reissued every 6 months, you are cautioned and advised to contact the nearest FSS for route dimensions and current status for those routes affecting your flight.
Routes with a change in the alignment of the charted route centerline will be indicated in the Aeronautical Chart Bulletin of the Airport/Facility Directory. Also, the VFR Wall Planning Chart is issued every 56 days and displays current route configurations and a composite tabulation of altitudes along these routes.
Military Pilots refer to Area Planning AP/1B Military Training Route North and South America for current routes.
13063
12000
9000
7000
5000
3000
2000
1000
Sea Level
282

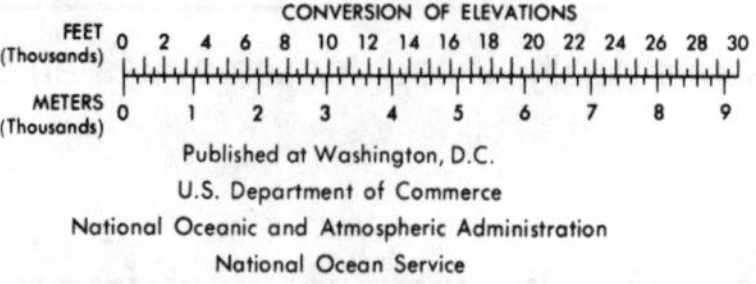

Fig. 4-6. Chart cover pages display effective dates and data panels with supplemental information.

CONTROL TOWER FREQUENCIES ON LAS VEGAS SECTIONAL CHART

Airports which have control towers are indicated on this chart by the letters CT followed by the primary VHF local control frequency. Selected transmitting frequencies for each control tower are tabulated in the adjoining spaces, the low or medium transmitting frequency is listed first followed by a VHF local control frequency, and the primary VHF and UHF military frequencies, when these frequencies are available. An asterisk (*) follows the part-time tower frequency remoted to the collocated full-time FSS for use as Airport Advisory Service (AAS) during hours tower is closed. Hours shown are local time. Ground control frequencies listed are the primary ground control frequencies.

Automatic Terminal Information Service (ATIS) frequencies, shown on the face of the chart are normal primary arrival frequencies. ATIS operational hours may differ from control tower operational hours.

ASR and/or PAR indicates Radar Instrument Approach available.

"Mon-Fri" indicates Monday thru Friday.

CONTROL TOWER	OPERATES	TWR FREQ	GND CON	ATIS	ASR/PAR
CHINA LAKE NWC/ ARMITAGE FIELD	0630-2230	126.2 340.2	360.2	265.2	
GRAND CANYON NATIONAL PARK	1 MAY-30 SEP 0700-1900 1 OCT-30 APR 0800-1800	119.0	121.9	124.3	
INDIAN SPRINGS AF AUX	0600-1700 MON-FRI EXC HOL	118.3 358.3	118.3 275.8		
McCARRAN INTL	CONTINUOUS	119.9 257.8	121.9 257.8	ARR 132.4 DEP 125.6	ASR
NELLIS AFB	CONTINUOUS	126.2 324.3	121.8 275.8	270.1	ASR/PAR
NORTH LAS VEGAS	0600-2000	125.7	121.7	118.05	

TCA, ARSA, TRSA, AND SELECTED RADAR APPROACH CONTROL FREQUENCIES

LAS VEGAS TCA	119.4 335.5 (360°-080°) 133.95 295.0 (280°-360°) 125.9 387.0 (W. of INTERSTATE 15; S. of 280°) 118.4 387.0 (E. of INTERSTATE 15; S. of 080°)
NELLIS AFB	124.95 279.7 NORTH

SPECIAL USE AIRSPACE ON LAS VEGAS SECTIONAL CHART

Unless otherwise noted altitudes are MSL and in feet; time is local.
Contact nearest FSS for information.
†Other time by NOTAM contact FSS

The word "TO" an altitude means "To and including."
"MON-FRI" indicates "Monday thru Friday"
FL – Flight Level
NO A/G – No air to ground communications

U.S. P-PROHIBITED, R-RESTRICTED, A-ALERT, W-WARNING, MOA-MILITARY OPERATIONS AREA

NUMBER	LOCATION	ALTITUDE	TIME OF USE	CONTROLLING AGENCY**
R-2502 N	FORT IRWIN, CA	UNLIMITED	CONTINUOUS	ZLA CNTR
R-2505	CHINA LAKE, CA	UNLIMITED	CONTINUOUS	ZLA CNTR
R-2506	CHINA LAKE SOUTH, CA	TO 6000	SR-SS MON-FRI	ZLA CNTR
R-2524	TRONA, CA	UNLIMITED	CONTINUOUS	ZLA CNTR
R-4806 E	LAS VEGAS, NV	100 AGL TO UNLIMITED	0500-2000 MON-SAT†	ZLA CNTR
R-4806 W	LAS VEGAS, NV	UNLIMITED	CONTINUOUS	ZLA CNTR
R-4807 A	TONOPAH, NV	UNLIMITED	CONTINUOUS	ZLA CNTR
R-4807 B	TONOPAH, NV	UNLIMITED	CONTINUOUS	ZLA CNTR
WARNING: R-4806 W AND R-4807 A AND R4807 B CONTAIN MANY UNEXPLODED BOMBS AND ROCKETS, AND OTHER ORDNANCE THAT MAY EXPLODE IF DISTURBED. NO AIRFIELD OR ACTIVE RUNWAY EXISTS WITHIN THESE AREAS; HOWEVER, SIMULATED AIRFIELDS UTILIZED AS BOMBING TARGETS DO EXIST. AIRCRAFT LANDINGS IN R-4806 W AND R-4807 A AND R4807 B ARE AT THE PILOTS OWN RISK.				
R-4808 N	LAS VEGAS, NV	UNLIMITED	CONTINUOUS	NO A/G
R-4808 S	LAS VEGAS, NV	UNLIMITED	CONTINUOUS	ZLA CNTR
R-4809	TONOPAH, NV	UNLIMITED	CONTINUOUS	NO A/G
R-4816 N	DIXIE VALLEY, NV	1500 AGL TO BUT NOT INCL FL 180	0715-2330	ZOA CNTR
R-4816 S	DIXIE VALLEY, NV	500 AGL TO BUT NOT INCL FL 180	0715-2330	ZOA CNTR
R-6402 A	DUGWAY PROVING GROUND, DUGWAY, UT	TO FL 580	CONTINUOUS	ZLC CNTR
R-6405	WENDOVER, UT	100 AGL TO FL 580	CONTINUOUS	ZLC CNTR
R-6407	HILL AFB, UT	TO FL 580	CONTINUOUS	ZLC CNTR
A-481	SHEEP RANGE, NV	7000 TO 17,000	SR-2200†	NO A/G

**ZLA-Los Angeles, ZLC-Salt Lake City, ZOA-Oakland

MOA NAME	ALTITUDE OF USE*	TIME OF USE†	CONTROLLING AGENCY**
AUSTIN 1	200 AGL	INTERMITTENT BY NOTAM	ZLC CNTR
AUSTIN 2	200 AGL	INTERMITTENT BY NOTAM	ZLC CNTR
COMPLEX 1	200 AGL	0600-2200 MON-FRI SR-1200 SAT	ZLA CNTR OR EDWARDS RAPCON
COMPLEX 2	200 AGL	0600-2200 MON-FRI	ZLA CNTR OR EDWARDS RAPCON
COMPLEX 3	200 AGL	0600-2200 MON-FRI	ZLA CNTR OR EDWARDS RAPCON
COMPLEX 4 & 4 ALPHA	200 AGL	0600-2200 MON-FRI	ZLA CNTR OR EDWARDS RAPCON
DESERT EAST	100 AGL	DAYLIGHT HRS MON-SAT	ZLA CNTR
DESERT REVEILLE HIGH	11,000	INTERMITTENT SR-SS MON-SAT	ZLC CNTR
DESERT REVEILLE LOW	100 AGL TO BUT NOT INCL 11,000	INTERMITTENT SR-SS MON-SAT	ZLC CNTR
DESERT SALLY	100 AGL	DAYLIGHT HRS MON-SAT	ZLA CNTR
GABBS CENTRAL	100 AGL	0715-2330	ZOA CNTR
GABBS N, S	100 AGL	0715-2330	ZOA CNTR
GANDY	100 AGL	0500-2000 MON-SAT	ZLC CNTR
SEVIER A	100 AGL TO 14,500	0500-2000 MON-SAT	ZLC CNTR
SEVIER B	100 AGL TO 9500	0500-2000 MON-SAT	ZLC CNTR
SEVIER C	14,500	BY NOTAM 6 HRS IN ADVANCE	ZLC CNTR
SEVIER D	9500	BY NOTAM 6 HRS IN ADVANCE	ZLC CNTR
SILVER	200 AGL TO 7000	0600-2200 MON-FRI	ZLA CNTR
SUNNY	12,000	0600-1900 MON-FRI	ZAB CNTR

*Altitudes indicate floor of MOA. All MOA's extend to but do not include FL 180 unless otherwise indicated in tabulation or on chart.
†Other time by NOTAM contact FSS.
**ZAB-Albuquerque, ZLA-Los Angeles, ZLC-Salt Lake City, ZOA-Oakland

Fig. 4-6. Continued.

straightforward: "This chart will become OBSOLETE FOR USE IN NAVIGATION upon publication of the next edition scheduled for SEPTEMBER 19, 1991."

The margins of sectional charts contain data panels with supplemental control tower information, selected radar approach control frequencies, and SUA information. Refer to Fig. 4-6. The China Lake NWC (Naval Weapons Center) control tower operates between 6:30 a.m. and 10:30 p.m. on frequencies 126.2 for civilian aircraft and 340.2 for military aircraft. Ground control (GND CON), automatic terminal information service (ATIS), and the availability of radar approaches are also noted. This information is followed by selected TCA, ARSA, TRSA, and radar approach control facilities.

Data regarding special use airspace includes altitudes, effective times, and controlling agency. Some restricted areas are routinely released to VFR and IFR operations when not in use. Pilots can obtain this information from the local flight service station or controlling agency, usually an approach control or the respective air route traffic control center. (I queried a military airport tower about the status of a restricted area along my route. The controller advised that it was "cold." Later, in the middle of the restricted area, I contacted an approach controller. This controller advised to remain clear of all restricted areas. It turned out the area was not in use, but would be in a few hours.) Some restricted areas are reported via the NOTAM (D) category and will be part of a standard FSS or DUAT briefing.

Military operation areas (MOA) are confusing; they are not restricted areas; their purpose is to alert pilots to the existence of military operations. Pilots should exercise additional vigilance while transiting active MOAs. Flight service station specialists often receive requests for published MOA activity. Pilots should first refer to the chart for this information. Only if the MOA is active will departure and enroute FSSs be able to provide its status. Because MOA NOTAMs are local, they will only be available from an FSS within approximately 100 miles of the area; therefore, a pilot will have to check with FSSs enroute for MOA activity beyond this distance. Like restricted areas, MOA status can also be obtained from the respective controlling agency.

With preliminary planning completed it's time to open the sectionals or WACs and prepare a navigation log with routes, true courses, distances, and magnetic variation, along with communication and navigation frequencies. This can be done on anyone of many commercially available forms, such as the preflight planner navigation log illustrated in Fig. 4-7.

Planning a long trip several days, even several weeks, in advance is not a difficult chore. What about the weather? The general weather patterns of the U.S. are well documented. For example, we know about the winter storms of the Midwest and East, the convective weather of the Midwest in the spring and summer, the heat of the southwestern deserts in summer, and coastal low clouds of the Pacific states in late spring, summer, and fall. If needed, a call to the area's FSS will often provide the general weather conditions for a certain area and time of year. Please, don't expect specifics.

How can we use this general weather for flight planning? With a VFR-only Cessna 150, flying out of the Los Angeles Basin, I always plan to depart on the first leg in the afternoon, after the fog clears. Another solution is to move the airplane inland out of the affected coastal areas, which also permits flight over the desert in the late afternoon or early evening when the turbulence has diminished. Departures were then planned early the next morning to avoid desert convective activity and turbulence. Time of year is also

PREFLIGHT PLANNER™

NAVIGATION LOG

Estimated CAS ____ Knots   Fuel Consumption ____ GPH/PPH

ROUTE	TC	CRUISE ALT	CRUISE Temp	WIND DIR	WIND SPD	TAS	TH	+− VAR	MH	+− DEV	CH	GS	Dist	Time	Fuel	REMARKS
			C									K	nm	:		
			C									K	nm	:		
			C									K	nm	:		
			C									K	nm	:		
			C									K	nm	:		
			C									K	nm	:		
TOTAL													nm	:		

WT & BALANCE

ITEM	WT	× ARM =	MOM
1. Aircraft			
2. Pilot & Front Seat			
3. Rear Seat/Cargo			
4.			
5. Fuel ____ gal*/lb			
6. Fuel ____ gal*/lb			
7. Oil (7.5 lbs/gal)			
8. Baggage			
9.			
10. RAMP			
11. (−) Fuel start-runup			
12. TAKE OFF			
13. (−) Fuel to DESTN			
14. LANDING			

*6 lbs/gal

mom/wt = ____ CG

NAVCOM

LOCATION	CPT* ATIS	DEP** APCH	TWR	GND	VOR NDB	FSS	REMARKS
OAKLAND	128.5		118.3	121.9		122.5	
RENO	124.35	120.8	118.7	121.9		122.5	
							* CPT-CLEARANCE PRETAXI (CLNC DELIVERY) ** DEP-DEPARTURE CONTROL

DEP ATIS:
CODE: __

DESTN ATIS:
CODE: __

COMMON FREQUENCIES/CODES

121.5 EMERGENCY
122.2 COMMON FSS
122.0 FLIGHT WATCH
122.75 AIR TO AIR

1200 VFR
7600 LOST COM
7700 EMERGENCY

CRUISING ALTITUDES

EASTBOUND MAGNETIC COURSE
IFR - ODD THOUSANDS
VFR - ODD THOUSANDS + 500*

WESTBOUND MAGNETIC COURSE
IFR - EVEN THOUSANDS
VFR - EVEN THOUSANDS + 500*
*ABOVE 3,000 AGL

SPECIAL EQUIPMENT CODES

X - NONE
T - TRANSPONDER
U - TRANSPONDER/ALTITUDE ENCODING
D - DME ONLY
B - TRANSPONDER & DME
A - TRANSPONDER/ALTITUDE ENCODING & DME
W - RNAV ONLY
C - TRANSPONDER & RNAV
R - TRANSPONDER/ALTITUDE ENCODING & RNAV

TIME CONVERSION UTC (Z)

EST +5 = UTC   EDT +4 = UTC
CST +6 = UTC   CDT +5 = UTC
MST +7 = UTC   MDT +6 = UTC
PST +8 = UTC   PDT +7 = UTC

A - ALPHA, B - BRAVO, C - CHARLIE, D - DELTA, E - ECHO, F - FOXTROT, G - GOLF, H - HOTEL, I - INDIA, J - JULIETT, K - KILO, L - LIMA, M - MIKE, N - NOVEMBER, O - OSCAR, P - PAPA, Q - QUEBEC, R - ROMEO, S - SIERRA, T - TANGO, U - UNIFORM, V - VICTOR, W - WHISKEY, X - XRAY, Y - YANKEE, Z - ZULU

U.S. DEPARTMENT OF TRANSPORTATION FEDERAL AVIATION ADMINISTRATION
FLIGHT PLAN
(FAA USE ONLY) ☐ PILOT BRIEFING ☐ VNR ☐ STOPOVER — TIME STARTED — SPECIALIST INITIALS

1 TYPE (VFR, IFR, DVFR) | 2 AIRCRAFT IDENTIFICATION | 3 AIRCRAFT TYPE/SPECIAL EQUIPMENT | 4 TRUE AIRSPEED KTS | 5 DEPARTURE POINT | 6 DEPARTURE TIME PROPOSED (Z) ACTUAL (Z) | 7 CRUISING ALTITUDE
8 ROUTE OF FLIGHT
9 DESTINATION (Name of airport and city) | 10 EST. TIME ENROUTE HOURS MINUTES | 11 REMARKS
12 FUEL ON BOARD HOURS MINUTES | 13 ALTERNATE AIRPORT(S) | 14 PILOT'S NAME, ADDRESS & TELEPHONE NUMBER & AIRCRAFT HOME BASE | 15 NUMBER ABOARD
17 DESTINATION CONTACT/TELEPHONE (OPTIONAL)
16 COLOR OF AIRCRAFT | CIVIL AIRCRAFT PILOTS. FAR Part 91 requires you file an IFR flight plan to operate under instrument flight rules in controlled airspace. Failure to file could result in a civil penalty not to exceed $1,000 for each violation (Section 901 of the Federal Aviation Act of 1958, as amended). Filing of a VFR flight plan is recommended as a good operating practice. See also Part 99 for requirements concerning DVFR flight plans.

FAA Form 7233-1 (8-82)   CLOSE VFR FLIGHT PLAN WITH ______ FSS ON ARRIVAL

Fig. 4-7. A navigation log with navcom section is ideal for noting frequencies to be used enroute.

important. If you plan winter operations, VFR only, be prepared for delays. May, June, September, and October seem to have the best flying weather.

What about winds aloft? If we don't try to stretch our trip legs, 10 to 20 knots of wind either way shouldn't present a problem. For example, we planned a leg from Phoenix to Albuquerque; however, because of the distance we could not tolerate any headwind component; therefore, we planned an alternate, using Gallup. If we were not on time at a specific point, about halfway, we would divert to our planned alternate. Fortunately, the winds were with us and we proceeded to our planned destination.

Despite the best plans of mice and men, things go wrong. On a leg from Kalamazoo to Detroit, flight service advised of a thunderstorm over Jackson, Michigan. Further checking indicated that to the south, towards Toledo, was clear. A slight diversion and pilotage navigation took us safely to the new destination.

Study the charts in advance to determine best routes, comfortable legs, adequate services at destinations, and possible alternates. With everything planned, if a problem occurs, a pilot is in a much better position to evaluate the situation and develop a sound alternative.

TERMINAL AREA CHARTS

Terminal area charts (TAC), also known as TCA or terminal control area charts, replaced the local chart series beginning in the early 1970s. These multicolored charts depict terminal control areas and provide much more detail than is available with sectional charts because of their larger scale. They are designed for pilots operating from airports within or near a TCA or transiting the vicinity. Charts are 20 × 25 inches, which can be folded to the standard 5 × 10 inches, and have a scale of 1:250,000 (1 inch equals 3 nautical miles). TAC charts are revised semiannually. Charts are named for the TCA they depict, and locations are shown in Fig. 4-4 and Fig. 4-5. Note that the Honolulu TAC is on the Hawaii Sectional Chart. Features include:

- Visual aids to navigation
- Radio aids to navigation
- Airports
- Controlled airspace
- Restricted areas
- Obstructions
- Topography
- Shaded relief
- Latitude and longitude lines
- Airways and fixes
- Other low level related data

Improved scale allows for a great deal of topographical detail. Along with depicted NAVAIDs, TACs should allow a pilot to safely navigate in the vicinity of, and remain clear of, the TCA. Even with this detail, in marginal weather conditions new pilots might not have the experience to navigate in these areas. Remembering that FAR minimums are just that, minimums, not necessarily equating to safe, each pilot must set standards, based on experience and training. This might mean avoiding terminal airspace altogether, only flying in clear weather, or obtaining additional training from a qualified instructor. Every new pilot planning to fly into a TCA, or other congested airspace, should make at least one trip with an instructor or an experienced pilot.

HELICOPTER ROUTE CHARTS

Helicopter route charts are designed primarily to depict helicopter routes in and around major metropolitan areas. These charts are available for the following locations:

- Boston
- Chicago
- Los Angeles
- New York
- Washington

Scale is the same as a TAC's and dimensions are similar, except for the New York chart, which includes a larger scale inset of Lower Manhattan, and the Hudson and East Rivers, and the Boston chart with its downtown Boston inset. Charts contain specific route descriptions as illustrated in Fig. 4-8. The inset in Fig. 4-8 contains symbols unique to this chart. In addition to specific helicopter routes, features include:

- Pictorial symbols of prominent landmarks
- Public, private, and hospital heliports
- NAVAID and communications frequencies
- Selected obstructions
- Roads
- Spot elevations
- Commercial broadcast stations
- TCA, ARSA, and control zone boundaries

Large metropolitan areas without published helicopter route charts often have local procedures that accomplish the same purpose. Letters of agreement between air traffic control facilities in these areas designate helicopter checkpoints, routes, and route names. Pilots planning operations in these areas should contact local pilots or ATC facilities for details on these procedures. This might require the pilot or operator to become a signatory to the letter of agreement, stating that he or she understands and will comply with its provisions.

WORLD AERONAUTICAL CHART
GULF COAST VFR AERONAUTICAL CHART

World aeronautical charts (WAC) are designed for visual navigation by moderate speed aircraft and aircraft operating at higher altitudes, up to 17,500 feet MSL. Because of their smaller scale these charts cannot show the detail of sectionals and TACs, for example, the limits of controlled airspace. WACs are normally not recommended for student or new

(Continued on page 86.)

HELICOPTER ROUTE CHART
BOSTON
SCALE 1:125,000

Lambert Conformal Conic Projection Standard Parallels 42°05′ and 42°45′
Topographic data corrected to December 1988

1ST EDITION April 6, 1989
Includes airspace amendments effective April 6, 1989
and all other aeronautical data received by February 9, 1989
Consult appropriate NOTAMs and Flight Information
Publications for supplemental data and current information.
This chart will become *OBSOLETE FOR USE IN NAVIGATION* upon
publication of the next edition. See Dates of Latest Editions.

PUBLISHED IN ACCORDANCE WITH INTERAGENCY AIR CARTOGRAPHIC COMMITTEE
SPECIFICATIONS AND AGREEMENTS, APPROVED BY:
DEPARTMENT OF DEFENSE * FEDERAL AVIATION ADMINISTRATION * DEPARTMENT OF COMMERCE

CONTROL TOWER FREQUENCIES ON BOSTON HELICOPTER ROUTE CHART

Airports which have control towers are indicated on this chart by the letters CT followed by the primary VHF local control frequency. Selected transmitting frequencies for each control tower are tabulated in the adjoining spaces, the low or medium transmitting frequency is listed first followed by a VHF local control frequency, and the primary VHF and UHF military frequencies, when these frequencies are available. An asterisk (*) follows the part-time tower frequency remoted to the colocated full-time FSS for use as Airport Advisory Service (AAS) during hours tower is closed. Hours shown are local time. Radio call provided if different from tower name.

Automatic Terminal Information Service (ATIS) frequencies, shown on the face of the chart are normal arriving frequencies, all ATIS frequencies available are tabulated below.

CONTROL TOWER	RADIO CALL	OPERATES	ATIS	FREQ
BEVERLY		0700-2200	118.7	125.2
BOIRE NF	NASHUA	0800-1800 OCT 1-APR 30 0700-2100 MAY 1-SEP 30	125.1	119.7
HANSCOM	HANSCOM	0700-2300	124.6	118.5 236.6
LAWRENCE		0700-2200	126.75	120.0
LOGAN INTL	BOSTON	CONTINUOUS	135.0	119.1 257.8
MOORE AAF		0700-1859 MON-FRI OTHER TIMES AS REQUIRED		119.35 241.0
NORWOOD MEM		0700-2200	119.95	126.0
SOUTH WEYMOUTH NAS-SHEA	NAVY WEYMOUTH	0700-2300 OT BY NOTAMS		126.2 360.2

SPECIAL USE AIRSPACE ON BOSTON HELICOPTER ROUTE CHART

Unless otherwise noted altitudes are MSL and in feet; time is local.
Contact nearest FSS for information.
†Other time by NOTAM contact FSS

The word "TO" an altitude means "To and including."
FL – Flight Level
NO A/G – No air to ground communications

U.S. P-PROHIBITED, R-RESTRICTED, A-ALERT, W-WARNING, MOA-MILITARY OPERATIONS AREA

NUMBER	LOCATION	ALTITUDE	TIME OF USE	CONTROLLING AGENCY
R-4102A	FORT DEVENS, MA	TO BUT NOT INCL 2000	0800-2200 SAT † 24 HRS IN ADVANCE	ZBW CNTR
R-4102B	FORT DEVENS, MA	2000-3995	0800-2200 SAT † 24 HRS IN ADVANCE	ZBW CNTR
W-103	CASCO BAY, ME	TO 2000	INTERMITTENT	ZBW CNTR

ROUTE DESCRIPTIONS

NOTE: Helicopters planning flights to BOSTON and/or within 10.5 NM CONTACT BOSTON AIR TRAFFIC CONTROL TOWER ON FREQ. 121.75

BAY ROUTE

Southern end of Nantasket Beach in Hull, via the coastline to the Long Island Bridge then to the Channel then to the Logan Helipad. NOTE: It is recommended that the Bay Route be used by multi engine and float equipped helicopters due to the low altitudes occasionally imposed.

FENWAY ROUTE

NOTE: Entry Point is within the Norwood Airport Traffic Area. CONTACT NORWOOD TOWER ON FREQ. 126.0.
At the intersection of I95 and I93 and the Conrail Tracks in Norwood. Follow the Conrail Tracks to the Fens (passing over the Fens east of Fenway Park and west of the Prudential Building). Joining the Turnpike Route.

Fig. 4-8. Helicopter route charts are designed primarily to depict helicopter routes in and around major metropolitan areas.

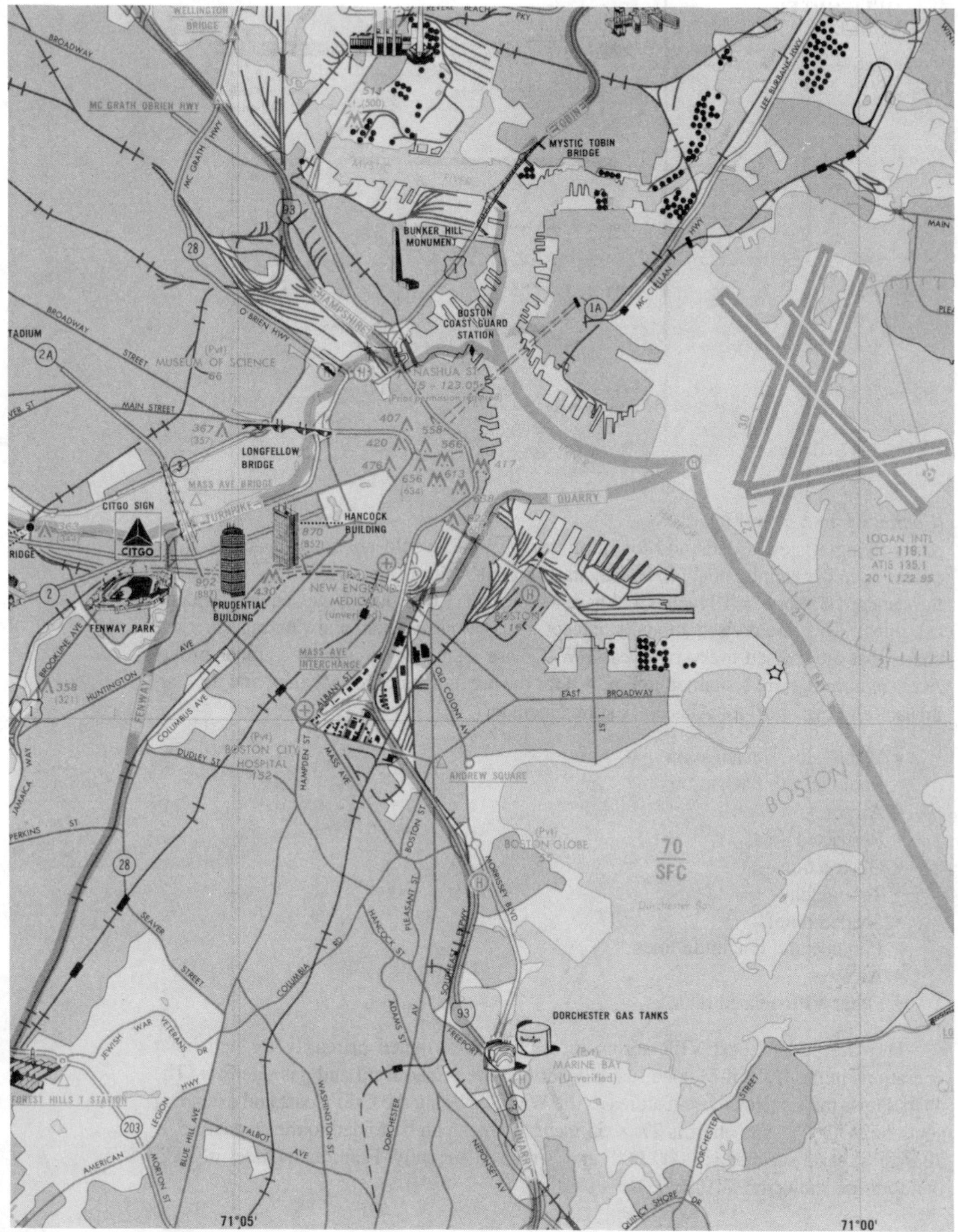

Fig. 4-8. Continued.

HELICOPTER ROUTES
Only the controlled airspace effective below 18,000 feet MSL is shown.

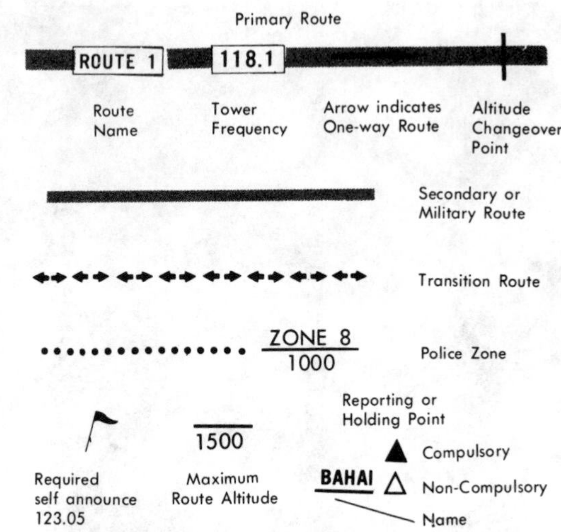

Fig. 4-8. Continued.

(Continued from page 83.)
pilots flying at slow speeds and low altitudes. A WAC would not be satisfactory while operating in the vicinity of a TCA. The charts are 20 × 60 inches, which can be folded to the standard size of 5 × 10 inches and have a scale of 1:1,000,000 (1 inch equals 14 nautical miles). They are revised annually, except for a few in Alaskan and Central American charts that are revised every two years. WACs are identified by a letter-number group. Areas of coverage are contained in Fig. 4-9 for the contiguous U.S., Mexico, and the Caribbean, and Fig. 4-10 for Alaska. Features include:

- Visual aids to navigation
- Radio aids to navigation
- Airports
- Restricted areas
- Obstructions
- Topography
- Shaded relief
- Latitude and longitude lines
- Airways
- Other VFR-related data

The U.S. Gulf Coast VFR aeronautical chart is designed primarily for helicopter operations in the Gulf of Mexico, usually serving the offshore oil and gas interests. The chart shows the same onshore features as the WAC, covering the Gulf Coast and extending south to 26°30′N. The chart is 27 × 55 inches, which can be folded to the standard 5 × 10 inches, has a scale of 1:1,000,000, and is revised annually. Features include, in addition to those shown on WACs:

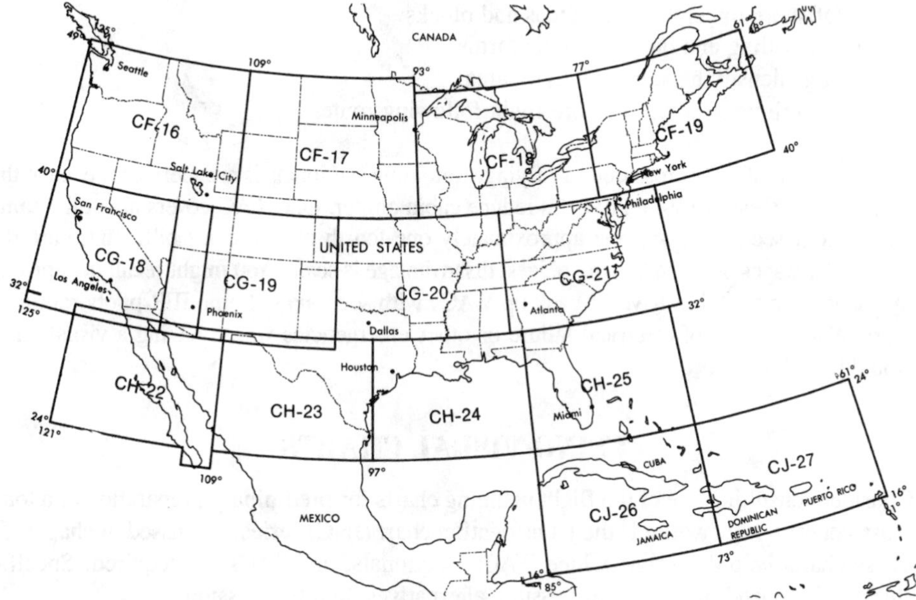

Fig. 4-9. World aeronautical charts are designed for visual navigation of moderate speed aircraft.

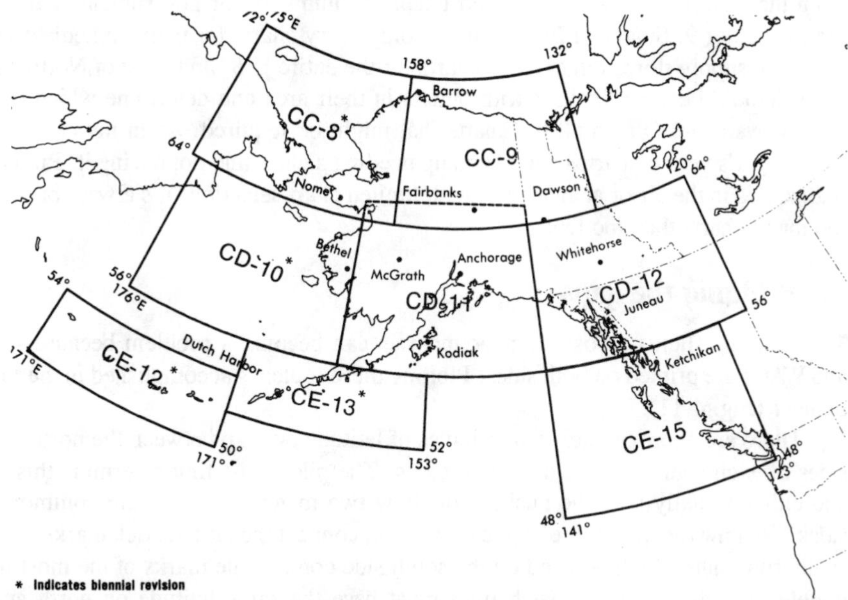

Fig. 4-10. World aeronautical charts are revised annually, except for a few in Alaska and Central America that are revised biennially.

- Offshore mineral leasing areas and blocks
- Oil drilling and production platforms
- High density helicopter activity areas
- Experimental loran offshore flight following routes

I prefer WACs for cross-country flying. Refolding sectionals is cumbersome, even at the speeds of the Cessna 150, and WACs reduce cabin clutter. One WAC covers about the same area as four sectionals, shaving approximately one-fourth off the chart bill. These are the major advantages of the WAC, the biggest disadvantage is scale. You might recall my episode misidentifying the Ohio River. I back up WACs with sectionals. Some IFR pilots routinely carry WACs in case of electrical failure or other emergencies where having a visual chart would be helpful.

USING VISUAL CHARTS

Consider consulting one of the flight planning charts for preliminary preparation of a long cross-country. My favorite is the jet navigation chart (JNC) series, discussed in chapter 5. These charts help determine which WACs, sectionals, and TACs are required. Specific charts will depend on the route, possible alternatives, and the mission.

Ensure chart currency; if a chart is to be revised next month and the proposed flight is six weeks off, wait as long as possible for that new chart to finalize flight planning. Charts can be obtained from many sources; most pilot supply stores and FBOs carry charts, they are available through subscriptions from a number of sources, and they can be obtained from the government. Addresses and telephone numbers for government chart agencies are in chapter 9. Smaller FBOs normally only carry charts for their immediate vicinity.

Pilot supply stores might carry charts for the entire U.S. and most of North America. Pilots should become familiar with outlets in their area and determine which charts are readily available. Obtain all the charts that might be required. As in my case over West Virginia, it's very embarrassing to end up needing a chart and not having it. Pilots usually realize this in the air or at an airport with limited chart selection. It's always better to have too many charts than too few.

Bridging the gaps

Planning flights across chart boundaries can become a problem because sectionals and WACs are printed on both sides. Plotting these routes is accomplished in the following manner (Fig. 4-11).

There are approximately two minutes of latitude overlap between the north and south sides of each chart, greater in certain cases. The pilot must first determine this overlap, and either visually note the position or draw two match lines that are common to both sides. To draw the match line on the north side, connect the latitude tick marks of the most southern minute of latitude and on the south side connect the marks of the most northern minute of latitude. These match lines must have the same latitude on north and south sides.

On the side of the chart having the terminal (departure or destination) nearest the match line, place a sheet of paper so that one edge corresponds to the match line and the

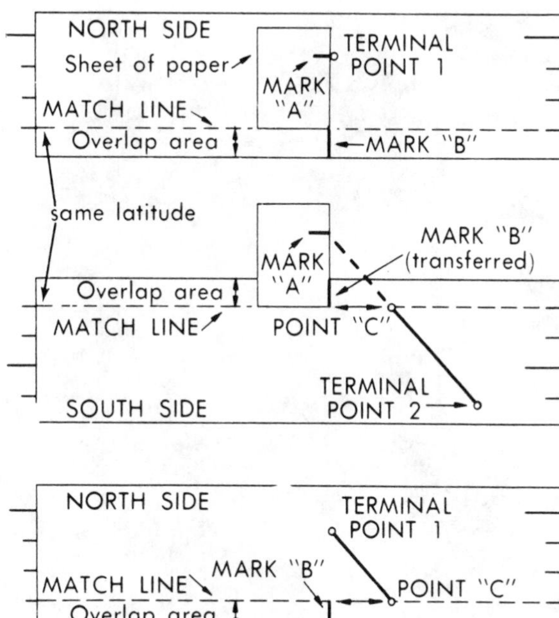

Fig. 4-11. Plotting a course from the front to back side of charts can be cumbersome.

other edge intersects the terminal airport. Mark the edge of the paper at terminal point 1, and label it "Mark A." Then make another mark on the chart extending from the match line to the edge of the chart, and label it "Mark B" as shown in Fig. 4-11.

Turn the chart over and transfer "Mark B" to the other side of the chart, be sure to extend the mark to the match line. Align the sheet of paper to the match line with the corner of the sheet at the transferred "Mark B." With a plotter, or other straightedge, align terminal point 2 with "Mark A," and draw a line from the match line, now called "Point C," to terminal point 2 as shown in Fig. 4-11.

Turn the chart over again and transfer "Point C" to the other side of the chart. This can be done by measuring the distance from "Mark B" to "Point C." With a plotter, draw a line from "Point C" to terminal point 1 as shown in Fig. 4-11.

A direct course now consists of the line segment from terminal point 1 to "Point C" on one side of the chart, and from "Point C" to terminal point 2 on the other side of the chart. Be careful with this procedure. Errors occur from not properly considering the overlap area, incorrectly transferring "Mark B" or "Point C," or including the overlap when measuring total distance. This procedure is a little complex, but can be mastered with a little practice.

Now we're ready to apply the principles and knowledge from chapter 3, and the preceding portion of this chapter. Let's plan a pilotage flight from the Oakland International Airport to Reno's Cannon International. Review the WAC in Fig. 4-12 for initial planning. For relatively short flights WACs might serve as excellent planning charts because of their scale and detail. For our flight there is an added advantage in that we can view the entire route on the WAC; on the sectional, departure and destination airports are on different sides of the chart.

(Continued on page 92.)

Fig. 4-12. WACs are often helpful with initial planning of short trips.

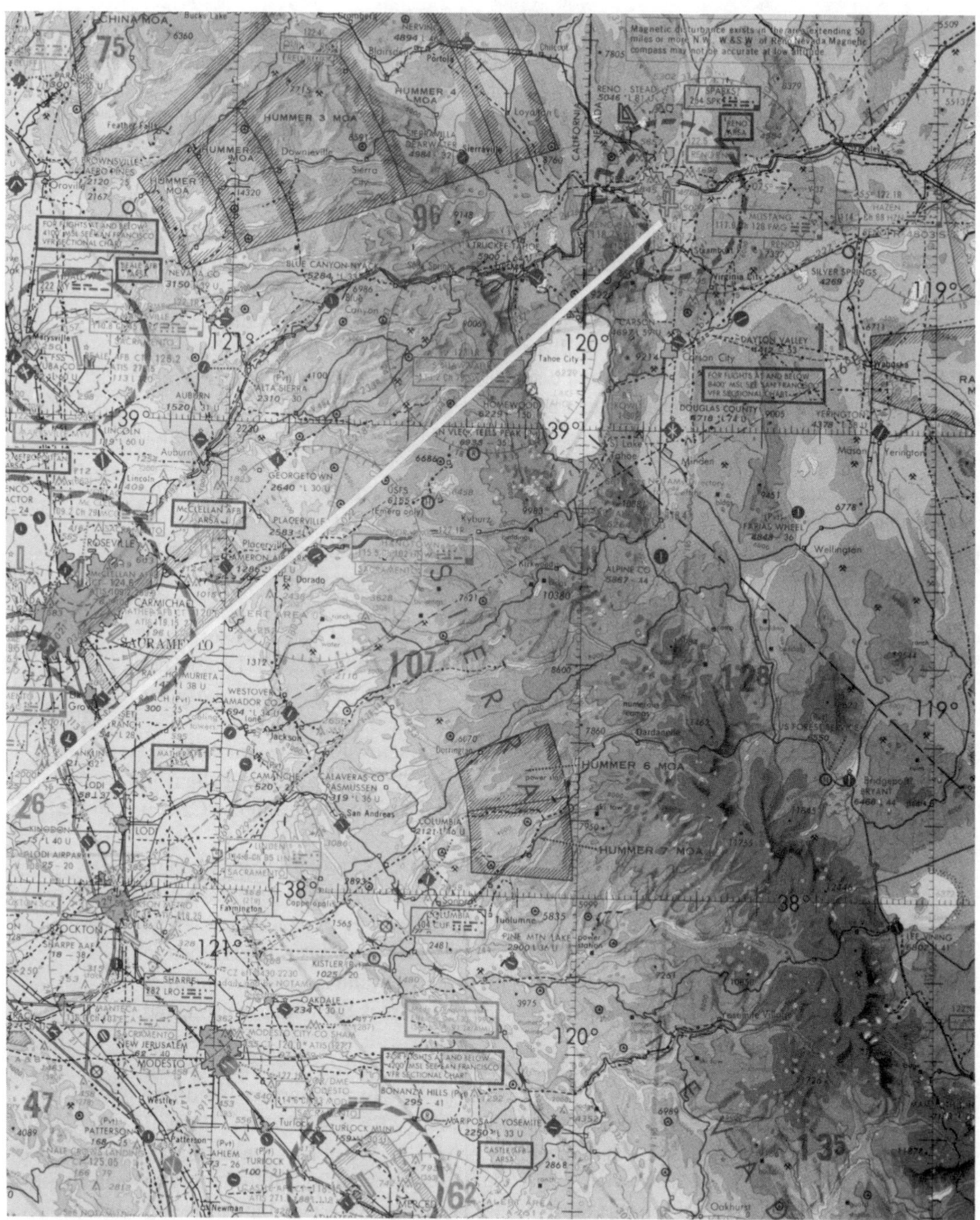

Fig. 4-12. Continued.

(Continued from page 89.)

The Oakland airport is located within an ARSA, under a TCA. Terrain is mostly flat through the Sacramento Valley. We will cut a corner of an ARSA in the Sacramento area, and come close to the Mather AFB Alert Area. Should we choose to verify our route with NAVAIDs, we see that the Sacramento, Hangtown, Squaw Valley, and Mustang VORs are along our course. If you haven't noticed or are new to flying, you might not be aware that in the early days NAVAIDs located in the vicinity of airports usually carried the same name as the associated airport. For example, the VOR just northeast of Reno was called Reno and the Squaw Valley VOR was called Lake Tahoe. To prevent any confusion or misunderstanding in our computerized ATC system a program evolved to change the names of NAVAIDs that are not collocated with the airport of the same name. Policy dictates that names come from the immediate area. A diversion to ward off cross-country boredom might be reviewing airport VOR names and determining the reference; if the reference is not obvious, perhaps a hidden meaning can be explained by a pilot in the area.

For the portion of the route over the Sierra Nevada Mountains, maximum elevation figures (MEF) indicate terrain from 9,600 to 11,100 feet MSL. This might pose a problem for low performance aircraft, especially considering weather and density altitude. Finally, we see that the destination is also located within an ARSA. We might want to consider the note just north of Reno that states, "Magnetic disturbance exists in the area extending 50 miles or more N.W., W, & S.W. of Reno, Nevada. Magnetic compass might not be accurate at low altitude." Apparently gold and silver is in those hills.

From this preliminary view we would also check for any SUA, such as restricted areas and MOAs, or any other high density areas. It is worth noting that MTRs do not appear on WACs. Notes on the TAC reveal that a San Francisco TAC is necessary for flight below 8,000 feet MSL, which is the top of the TCA. Also, the San Francisco Sectional for flight below 4,100 feet in the Sacramento area and below 8,400 feet in the Reno area, which are ARSA ceilings. The WAC would be perfectly acceptable for a flight in good weather from Livermore (east of the San Francisco TCA) to, for example, Carson City (south of Reno).

We could fly east of Livermore and follow the highway to Stockton. The Stockton airport has a control tower so we'll want to stay above that ATA or obtain the required clearance. Proceeding northeast of Stockton toward the large reservoirs, we could pick up the highway and proceed south of the Lake Tahoe Airport, which also has a tower, then to Douglas County, and finally into Carson City. This would be unwise in poor weather because it is too easy to misidentify landmarks from the WAC, pick the wrong canyon, and end up boxed in, it has happened.

Let's move on to the sectional. A direct route has been established on the sectional chart, as illustrated in Fig. 4-13 (south portion) and Fig. 4-14 (north portion). A direct flight will take us out over the coastal mountains where the MEF is 4,200 feet. We will have to negotiate the Oakland ARSA and San Francisco TCA. We review the sectional, end panels, and margin. In good weather the sectional provides enough information to negotiate the ARSA and TCA. From the sectional we see that once out of the ARSA we will have to remain below 4,500 MSL and 6,000 MSL respectively to clear the TCA shelves along our direct route.

Once clear of the TCA we see our course will take us out over the Sacramento River Delta. Just southwest of Franklin are a number of towers that, from the symbols, extend above 1,000 feet. From the chart, these towers extend to 2,001 feet MSL.

From the north portion of the sectional (Fig. 4-14) we see that the top of the Mather AFB ARSA is 4,100 feet; therefore, we'll have to fly above this altitude, around the ARSA, or establish communication with the controlling agency if we wish to penetrate this airspace. The frequency is listed on the chart, or it might be obtained from the end panel that lists: MATHER AFB/127.4 285.6 (SE). The Mather ARSA frequency is 127.4 for the southeast sector. Also, from the margin we determine that Alert Area A-252 serves Mather AFB: "A-252 MATHER AFB, CA TO 5000 0500-2000 MON-FRI NO AG." Alert Area 252 is active from the surface to 5,000 feet, from 5 a.m. until 8 p.m. Monday through Friday, and has no air to ground communications specifically for the alert area.

Terrain starts to rise over the Sierra Foothills. Critical elevations range generally between 8,000 and 11,000 feet. Donner Pass, just west of Truckee, has an elevation of 7,088 feet. The Spooner Summit Pass, on the east central side of Lake Tahoe has an elevation of 7,146 feet. East of the passes it's generally all downhill through the valleys to Reno, which has a field elevation of 4,412 feet.

Notice what appears to be a long runway on the west side of Lake Tahoe next to the Homewood Seaplane Base. A gotcha flight instructor-to-student question is what does this symbol represent? The answer is at the end of the chapter.

Reno's ARSA extends from the surface to 8,400 feet MSL. From the chart or end panel we determine the approach frequency is 120.8. The chart also provides ATIS (124.35) and tower (118.7) frequencies; ground frequency for this airport is 121.9. The only other frequency we might need is the Reno FSS, which the chart indicates is 122.5.

From our review of the sectional we determine there are no military training routes along our proposed course. If there were, we could note the route numbers: four digit numbers for routes flown at or below 1,500 feet and three digit numbers for routes flown above 1,500 feet agl. If our flight altitudes are above 1,500 feet agl we can disregard any routes with four digits. Call an FSS for operational details along any conflicting routes.

Now for the flight from Livermore to Carson City using the sectional, weather is marginal, ceilings between 1,000 and 3,000 feet, visibility 3–5 miles. We'll definitely need the sectional to establish the limits of controlled airspace. We will observe all FARs as they apply to controlled airspace, distances from clouds, and minimum safe altitudes. Departing east of Livermore we could cruise at 1,500 feet MSL, and even lower after crossing into the valley and through the Stockton area. Livermore has a control zone. A transition area based at 700 feet exists to the east, and along the highway to Stockton controlled airspace starts at 1,200 feet agl; therefore, at our altitude a clearance is required to transit the ATA.

We could then proceed northeast of Stockton toward the Comanche and Hogan Reservoirs. Then follow the highway through Carson Pass, which has an elevation of 8,650 feet. The base of controlled airspace through this area is 1,200 feet agl, as indicated by a blue vignette. Where controlled airspace begins at other than this altitude, the chart is labeled in feet above mean sea level. For example, south of Minden, Nevada, the floor of con-

(Continued on page 98.)

Fig. 4-13. The sectional is satisfactory for operating outside of the TCA, but operations close to, or under the TCA require the appropriate terminal area chart.

Fig. 4-13. Continued.

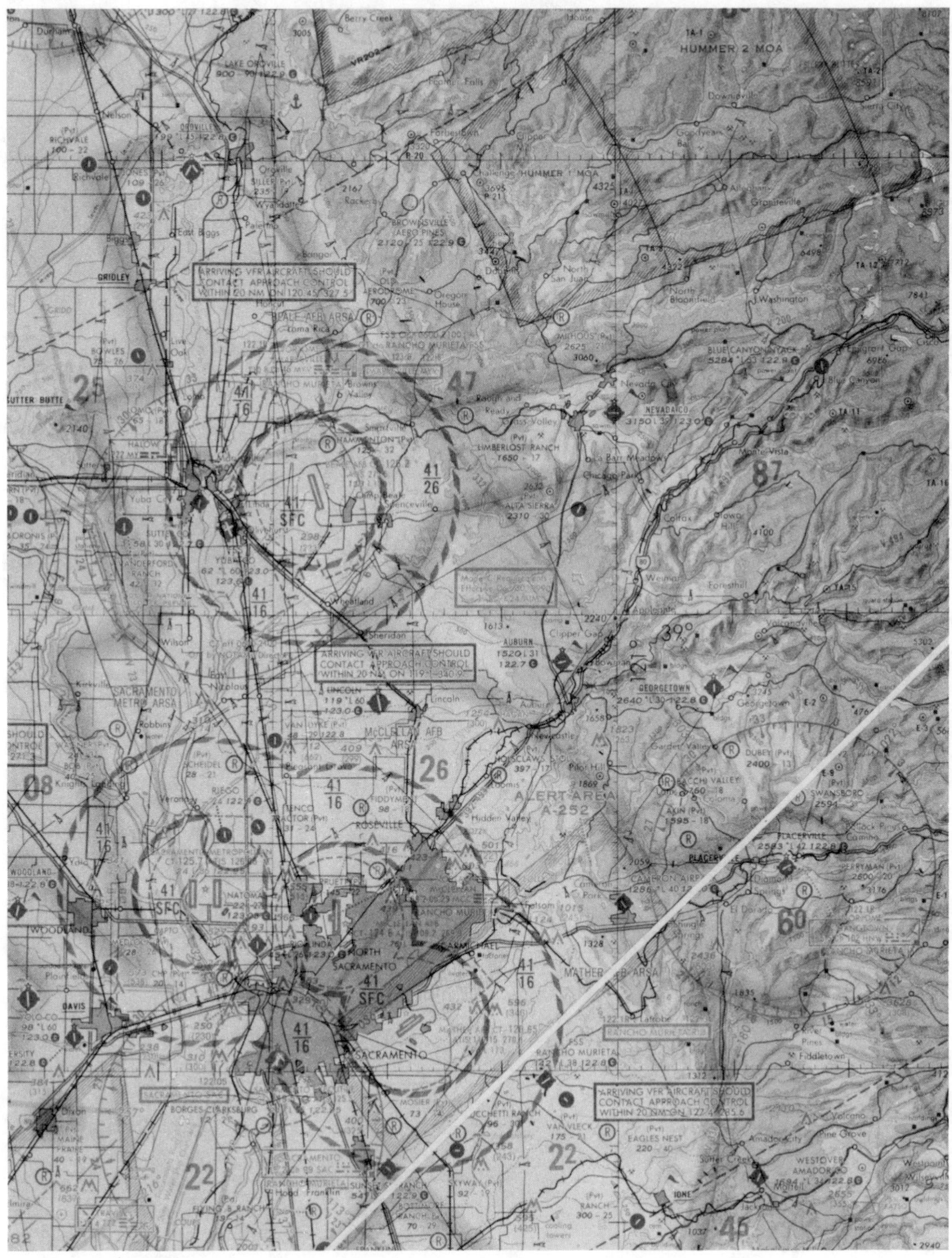

Fig. 4-14. A sectional is required for operating into or in the vicinity of an ARSA.

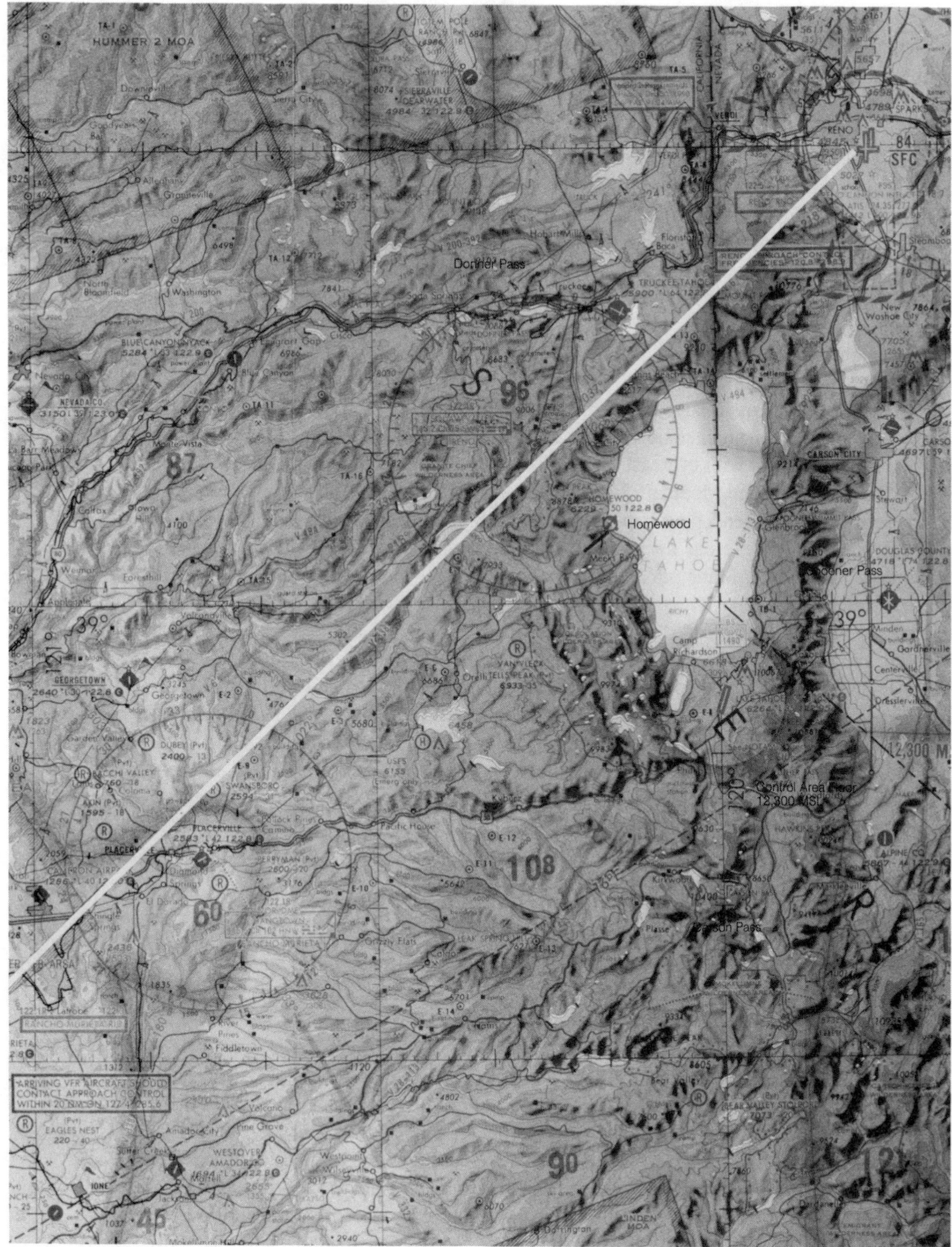

Fig. 4-14. Continued.

(Continued from page 93.)
trolled airspace is 12,300 MSL. From the Carson Pass we could then follow the highway into Carson City.

(This is not a recommendation to fly at low altitudes through mountainous areas. FAA: The decision as to whether a flight can be conducted safety rests solely with the pilot. This depends on a pilot's training and experience.)

Could we fly from Livermore to Stockton with ceilings of 500 to 1,000 feet and visibilities ranging from 1 to fewer than three miles? Technically, yes. This flight would require a special VFR clearance out of the Livermore control zone and into the Stockton control zone. Enroute we would be required to remain clear of clouds below 700 feet agl through the Livermore transition area, and then below 1,200 feet to the boundary of the Stockton control zone. Could this flight be conducted at night under the same conditions? No. At night, even in uncontrolled airspace three miles visibility and standard distance from clouds is required.

When flying in marginal weather, be very careful not to impose on someone's airspace, especially if a preplanned route is altered. (The 1991 Hayward-Bakersfield-Las Vegas Air Race began with overcast ceilings between 1,500 and 2,500 feet agl. The first checkpoint was the Pine Mountain Lake airport in the Sierra foothills at an elevation of 2,900 feet. Most of us ended up bypassing this checkpoint because of the weather. Using pilotage our crew flew to our next plotted checkpoint south of Pine Mountain to resume our preplanned course. Many pilots, after abandoning Pine Mountain, headed straight for the next checkpoint, apparently neglecting to consider, or communicate with the folks controlling, the Castle AFB ARSA.)

San Francisco's terminal area chart will be used to navigate the first part of the flight (Fig. 4-15). The Oakland airport data block reveals ATIS (128.5) and north field tower [118.3* (N)] frequencies; the north tower operates part time, as indicated by the asterisk. FSS at the top of the data indicates a flight service station on the field by the same name. The chart indicates that the south tower frequency operates continuously. This is an example where an airport has two primary local control frequencies. Oakland International is, in effect, two airports. The south field serves air carriers and the north field serves general aviation.

The end panel also notes the ground control frequency (121.9), and times of operation for the north tower. The Oakland VORTAC box indicates that the Oakland FSS has a discrete frequency of 122.5. Appropriate ARSA and TCA frequencies are also contained on the chart and provided in the end panels. It's usually not necessary to obtain departure control frequencies because ATC will normally assign the appropriate frequency upon departure. The navcom portion of the preflight planner and navigation log in Fig. 4-4 has been filled out for the trip.

Take advantage of the detail available on the TAC where they're available, especially in marginal weather. This chart would be useful for our Livermore to Carson City trip. From the chart, east of Livermore we see more clearly the transmission lines that cross the course. We can use these along with the railroad to update progress of the flight. The TAC shows that one of the railroad tracks goes through a tunnel on its second crossing of the highway, which is not seen on the sectional chart. Contours indicate that the approximate elevation of the pass between the Livermore Valley and the San Joaquin Valley is above 1,000 feet, but lower than 1,500 feet.

I obtained a DUAT weather briefing in the morning prior to departure and filed a VFR flight plan through the service. Ground control issued a clearance with heading information and altitude restrictions; the tower provided an ATC frequency for the ARSA and I contacted Bay departure control.

Departure vectored us south of a direct route at 2,500 feet to clear other traffic. Just east of the ARSA we requested and received clearance to climb VFR through the TCA to the initial cruising altitude of 5,500 feet. In the vicinity of Danville, clear of the ARSA and TCA, we terminated radar service and switched to Oakland radio on 122.5 to open the flight plan. We were south of Mt. Diablo, correcting the heading toward the planned flight course, and contacted Oakland flight watch on 122.0 for a weather update over the Sierras.

Checkpoint considerations

A primary checkpoint consists of a topographical feature, or set of features, that cannot be mistaken for any other place in the same general area. Often three or more secondary features can be combined to form a primary checkpoint.

Secondary checkpoints are small towns, streams, a single road or railroad, mountain range, or any other feature that could be mistaken for similar features in the same general area.

A primary checkpoint could be a single feature, such as Arizona's Meteor Crater. It is large and unique, and cannot be mistaken for any other feature in the same area. Several smaller features can be combined to form a primary checkpoint. Unless the features are relatively unique, such as an airport and adjacent town, three features should be used: a town, highway, or railroad. One feature, or even two, can be mistaken, and should if at all possible be avoided.

Pilots should consider the availability of suitable checkpoints during flight planning. This is the time to select an alternate route if you're not comfortable with what's available. If you do get lost, or even think you're lost, call for assistance before a relatively simple flight assist becomes an accident.

Back on course

Returning to our flight we see from the TAC that Mt. Diablo has a critical elevation symbol of 3,849 feet. There is an obstruction on the peak 285 feet agl, 3,865 feet MSL; note the MEF 4,200 feet. Another known feature in this quadrangle apparently requires this higher figure. Mt. Diablo is like a beacon to the bay area, often above the haze and fog. Figure 4-16 shows Mt. Diablo as seen from the east with the haze and bay in the background. Mt. Diablo at one time sported an airway beacon because of its prominence.

When well clear of bay area airspace and traffic congestion, switch from the TAC to the sectional chart for navigation. Still a little south of course, Fig. 4-17 shows the town of Antioch and the Sacramento River Delta as it flows toward the bay. These areas make good primary checkpoints because of the relation of water, towns, and roads. A word of caution, be very careful flying in these areas at low altitudes, especially in low visibilities. The numerous obstruction symbols on the chart represent many power lines that are stretched across the water.

(Continued on page 103.)

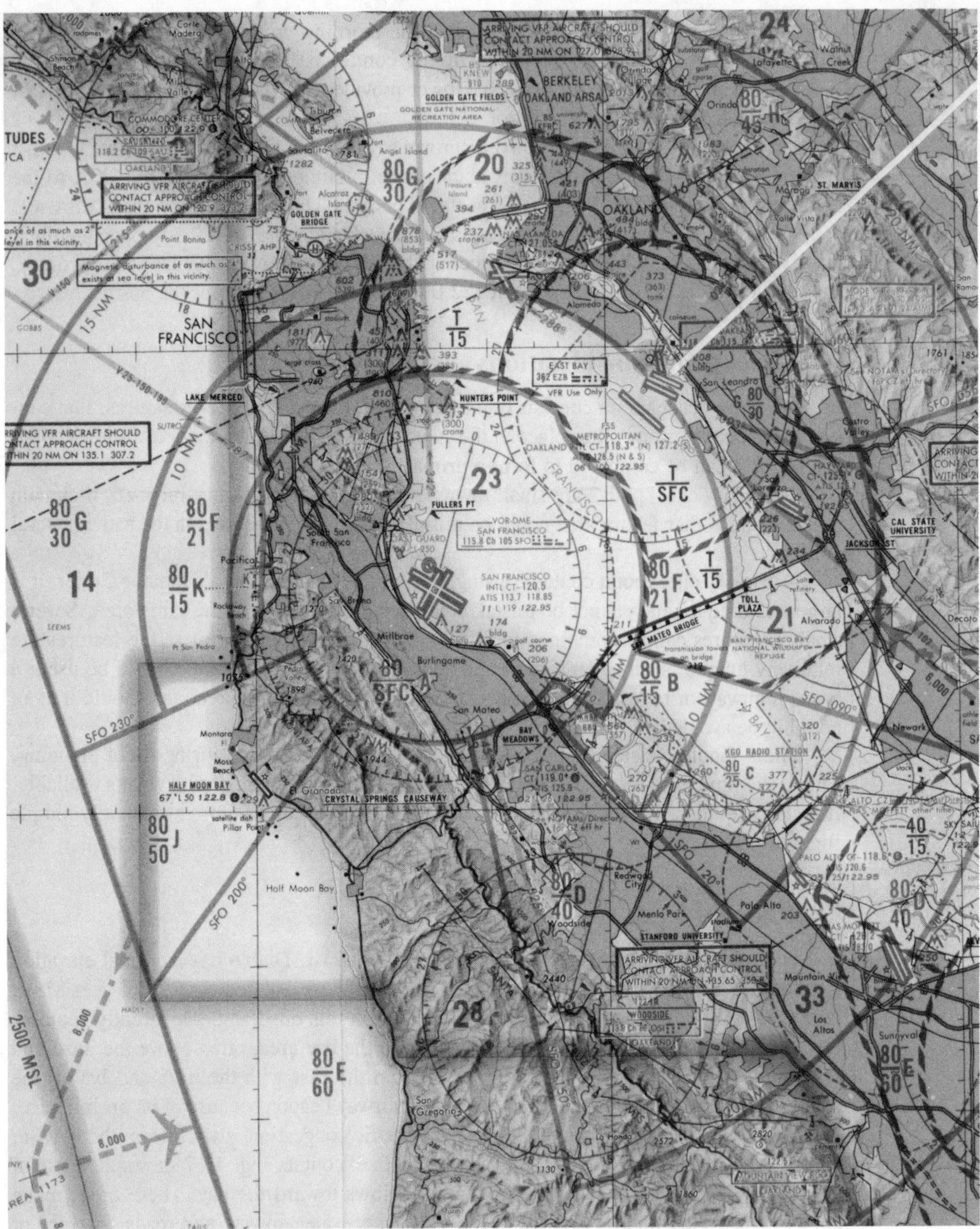

Fig. 4-15. TACs provide the detail necessary for operating in congested terminal airspace.

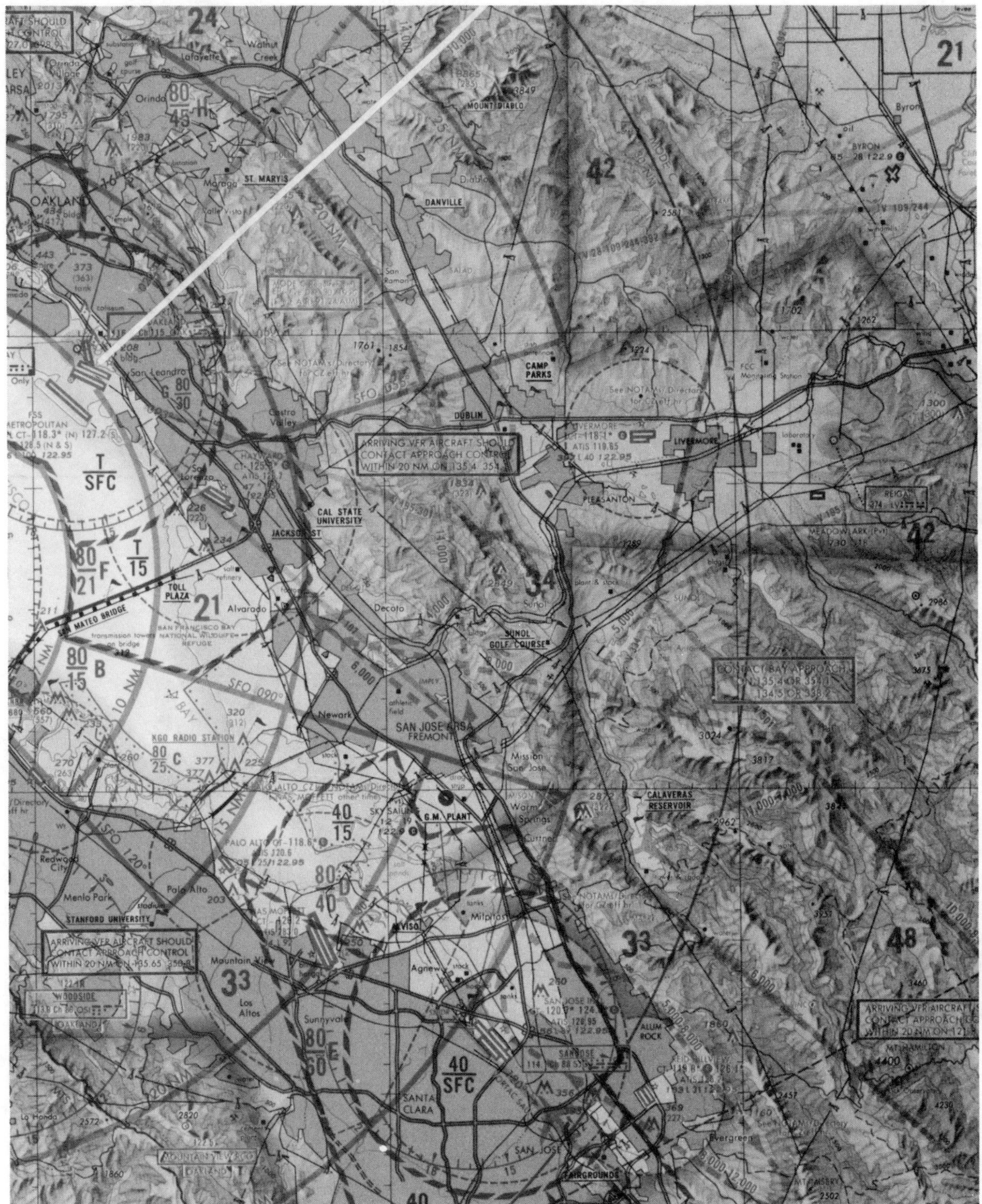

Fig. 4-15. Continued.

Fig. 4-16. Single prominent features can be used as primary checkpoints.

Fig. 4-17. Several secondary features can be combined to make a primary checkpoint.

(Continued from page 99.)

Figure 4-18 shows the town of Rio Vista. This town, waterway, roads, and bridge make an excellent checkpoint. Remember, if at all possible stay away from sparse checkpoints, such as a small town and a road. Almost every small town has a road running through and especially in flat country many of the roads are parallel. Try to select checkpoints with three or more features, such as Fig. 4-18.

Fig. 4-18. The relationship between the town, waterway, roads, and bridge, makes Rio Vista an excellent checkpoint.

An antenna farm is approximately eight miles northeast of Rio Vista. The chart indicates maximum antenna elevation of approximately 2,000 feet MSL, equipped with strobe lights that help but are still hard to see, especially in haze or fog. Realize that large towers usually have guy wires for support, which can easily snag an aircraft flown by an unsuspecting pilot. Flying at low altitudes in low performance aircraft—complying with all FARs, of course—these features are not necessarily hazardous with good visibility, minimum three miles.

Along the northern portion of the sectional in Fig. 4-14, the course runs just south of Sunset Sky Ranch. A four-lane highway and a railroad track crossing, with a town to the north, is another excellent checkpoint. Finding a place where a road, railroad, or river crosses another feature can fix the aircraft position. Figure 4-19 shows the view from the airplane; cruising at 5,500 feet MSL above the Mather ARSA and alert area.

An FSS communication box just south of course reads RANCHO MURIETA RIU. Routine communications, such as position reporting, flight plan updating, or other FSS services are possible on 122.3, the discrete frequency. Also notice that Rancho radio has a receive only on 122.1. We could also contact Rancho radio by transmitting on 122.1 and

Fig. 4-19. Locating the point where a road and railroad cross can fix the aircraft's position.

listening on one of the VOR frequencies in the area; always advise which frequency is being monitored when transmitting on 122.1.

Figure 4-20 shows a view behind the airplane of the Sacramento Valley, and in the distance the coastal range of mountains. Often over flat farmland few landmarks are available; select verifiable checkpoints when planning and keep careful track of your position while flying through the area.

Approaching the Sierra foothills we begin a climb to 9,500 feet. Off to our left is Folsom Lake, shown in Fig. 4-21. These lakes often make excellent checkpoints; the shape of the lake and the dam are easily verifiable with the chart. Compare the presentation of these lakes on the WAC in Fig. 4-12, with the sectional in Fig. 4-14.

A little farther along and to the right is another lake (Fig. 4-22); again, lake and dam shapes are verifiable. The chart shows many small lakes along the Sierra foothills. Care must be exercised selecting them as checkpoints, they are numerous and their shapes are similar.

Approaching the crest of the Sierras, the terrain rises rapidly and the aircraft is buffeted by some turbulence. The lowest terrain is south of a direct course, over the center of Lake Tahoe. As we approach the lake we are greeted by the scene in Fig. 4-23. We can tell that we will clear the crest of the mountains because the terrain beyond appears to be descending in relation to the crest of the mountains. (We were over the lake and decided to continue through Spooner Summit Pass, then over Carson City, and into Reno.)

East of Lake Tahoe we again contact flight watch and file a pilot report regarding conditions over the mountains. Frequent, objective pilot reports cannot be overemphasized, even if the weather and the ride are clear and smooth. Reporting enroute is a timely practice that makes current information available to briefers and ultimately other pilots.

Fig. 4-20. When flying over flat farmlands, few verifiable checkpoints are available.

Fig. 4-21. Lakes and dams make easily verifiable checkpoints.

Fig. 4-22. Be careful about using lakes as checkpoints when many similar lakes are in the same general area.

Fig. 4-23. The airplane is above the crest of the mountains because the terrain behind appears to be descending in relation to the crest.

The VFR flight plan was closed over Carson City via Reno radio prior to switching to another frequency while approaching the destination. The ATIS report is noted prior to contacting approach control for entry into the Reno ARSA. Recall that frequencies are already listed; therefore, we concentrate on flying the airplane, looking for traffic, and navigating, rather than fumbling with the chart trying to find a frequency. We pass Steamboat and follow the highway to the Biggest Little City in the West (Fig. 4-24).

Fig. 4-24. When the destination is in sight, consider closing the flight plan before switching to approach control or the tower.

Coastal cruise

The value of a TAC cannot be overemphasized. Let's take a flight from Livermore to Half Moon Bay, on the coast, southwest of San Francisco (Fig. 4-15). A direct flight will take us through the Oakland ARSA and the San Francisco TCA. Departing Livermore we will probably wish to stay below 4,000 feet because of aircraft inbound to Oakland and Hayward at 4,000 feet, as indicated by the IFR arrival route symbol (blue aircraft on dashed blue line).

The ARSA in the vicinity of Hayward extents from 1,500 feet MSL to the base of the TCA. We can either skirt south of the Oakland ARSA, or contact approach on 135.4 for ARSA services through the area. Don't forget about the Hayward ATA. The San Francisco TCA area C extends from 2,500 feet to the top of the TCA 8,000 feet; area B from 1,500 feet. Crossing the bay we will need to obtain clearance through either the San Carlos or Palo Alto ATAs. This is often easier than trying to obtain clearance through the TCA itself. Tower controllers might assign aircraft specific routes to avoid the airport traffic patterns.

Beyond the ATAs, the base of the TCA in sector D is 4,000 feet. Now we can climb safely over the hills and proceed into Half Moon Bay, where the base of the TCA is 5,000 feet. We need to be very careful of minimum safe altitudes and required cloud clearance and visibility, and keep a sharp eye out for other traffic. Pilots of high performance aircraft might wish to slow down in congested airspace in the vicinity of an ARSA or TCA.

A bit earlier the reader was asked about the symbol adjacent to the Homewood Seaplane Base on the west side of Lake Tahoe. If you look northwest to southeast, from just above the Squaw Valley NAVAID box, to just south of the Lake Tahoe Airport, then south-southwest of the Alpine County Airport, you'll see the large letters: S-I-E-R-R and, if the chart went far enough, A. The so called runway is the "I" in SIERRA. You might want to have some fun and ask your fellow pilots, or better yet an instructor, to decode this symbol.

5
Supplemental visual charts

PILOTS HAVE ACCESS TO A NUMBER OF SUPPLEMENTAL VISUAL CHARTS through the Defense Mapping Agency (DMA), NOS, and other sources, notably Canadian charts published by the Canada Map Office, Department of Energy, Mines and Resources. The DMA maintains a public sales program administered by the DMA Combat Support Center (DMACSC). DMA publishes a catalog that contains descriptions, availability, prices, and ordering procedures for DMA produced aeronautical products, primarily covering foreign regions.

Canada produces charts similar to U.S. counterparts: WACs, VFR navigation charts (sectionals), and VFR terminal area charts. A Canadian chart catalog and products are available by mail or through various authorized dealers located at airports and cities throughout North America. DMA Combat Support Center and Canada Map Office addresses and telephone numbers are in chapter 10.

DMA VISUAL CHARTS

The Defense Mapping Agency produces a series of visual navigation charts, mainly in support of military missions. Some maps may be adapted for civil flight planning and navigation: global navigation charts, loran charts, operational navigation charts, tactical pilotage charts, and joint operations graphic charts.

Global navigational charts

Global navigational charts (GNC) are designed for flight planning, operations over long distances, and enroute navigation in long range, high altitude, high speed aircraft. GNC scale is 1:5,000,000 (1 inch equals 69 nautical miles). Sheet size is approximately

41 × 58 inches. Polar regions use the transverse Mercator projection and other regions, the Lambert conformal conic projection.

The global navigation chart series serves as the base for production of global loran navigation charts (GLCC) and spacecraft tracking charts (NST). Features include:

- Principal cities
- Towns
- Drainage
- Primary roads
- Primary railroads
- Prominent culture
- Shaded relief
- Spot elevations
- NAVAIDs
- Airports
- Restricted areas

Figure 5-1 contains an excerpt from a global navigational chart. Shaded relief contains tints indicating relatively flat areas and those with steep relief, along with spot and critical elevations; however, contours and gradient tints are not included. Cities of strategic or economic importance, major towns, primary road and railroad networks, and other significant cultural features are displayed. Hydrography includes open water vignette, coastlines, and major lakes and rivers.

In Fig. 5-1, a caution note warns that, "Before using this chart, consult the current DMA Aeronautical Chart Updating Manual (CHUM)/CHUM Supplement, and the latest Flight Information Publications (FLIPS) and Notices to Airmen (NOTAMS) for vital updating information." The CHUM and FLIP, in effect the military version of the *Airport/Facility Directory*, are discussed in chapter 10. Figure 5-1 also contains an index to GNC series for the Northern Hemisphere.

These charts can be used for wall display because one sheet covers the United States, Canada, and part of Alaska. They are suitable as planning charts due to relief and major cultural features, for example plotting a flight from San Francisco to Denver. From the chart we can see that a direct route would be over rough, sparsely populated terrain. The lower terrain would be through Reno, Battle Mountain, Elko, Salt Lake City, then through southern Wyoming to Cheyenne, and south to Denver. This was the original airmail route. An alternate, although much longer route would be through Las Vegas, northern Arizona and New Mexico to Las Animas, then north to Denver. This route would generally be over lower terrain, perhaps a preferable route in the winter months.

Loran charts

Loran charts provide a plotting area where ground wave and sky wave correction values have been printed for loran navigation. Loran lines on these charts furnish a constant time difference between signals from a master and slave loran station; however, with most of today's units, which incorporate microprocessors, loran units provide the pilot with direct position readout, along with course, speed, and distance to specified locations. Several loran charts are available from DMA: global Loran-C coastal navigation, Loran-C coastal navigation, and Loran-C navigation.

Global Loran-C coastal navigation charts (GLCC) are selected GNCs modified with Loran-C and Consol/Consolan overprints. Consol/Consolan is a long range radio aid to navigation, the emissions of which, by means of their radio frequency modulation characteristics, enable bearings to be determined. They satisfy high speed, long range navigation requirements over large expanses of water. Chart scale is 1:5,000,000 (1 inch equals 69 nautical miles), sheets are approximately 42 × 58 inches. Polar charts use the transverse Mercator and lower latitudes the Lambert conformal conic projection. The charts include spot elevations, solid land tint, major cities, coastlines, and major lakes and rivers.

Loran-C coastal navigation charts (LCNC) are a series of three charts used for loran navigation for entry into the U.S. when a high degree of accuracy is required to comply with air defense identification and reporting procedures. They are also suitable for celestial navigation. Chart scale is 1:2,000,000 (1 inch equals 27 nautical miles), sheet size is 35 × 57 inches. The chart provides spot elevations only, with land masses portrayed by a light gray tint. Principal cities and towns and international boundaries are shown, along with drainage areas and lake elevations.

Loran-C navigation charts (LCC) are a series of four charts used for precise long range polar loran navigation in support of weather reconnaissance, air search and rescue, and other operations in the Arctic area. The chart uses a transverse Mercator projection with a scale of 1:3,000,000 at the 90°E and 90°W meridian, and is approximately 42 × 58 inches. The charts show spot elevations, major cities, railroads and roads, coastlines, and major lakes and rivers.

Jet navigation charts

Jet navigation charts (JNC) and universal jet navigation charts (JNU) are suitable for long range, high altitude, high speed navigation. Chart scale is 1:2,000,000. Features include:

- Cities
- Major roads
- Railroads
- Drainage
- Contours
- Spot elevations
- Gradient tints
- Restricted areas
- NAVAIDs
- Broadcast stations
- Airports
- Runway patterns

Runway patterns are exaggerated so they can be more readily identified as visual landmarks.

(Continued on page 114.)

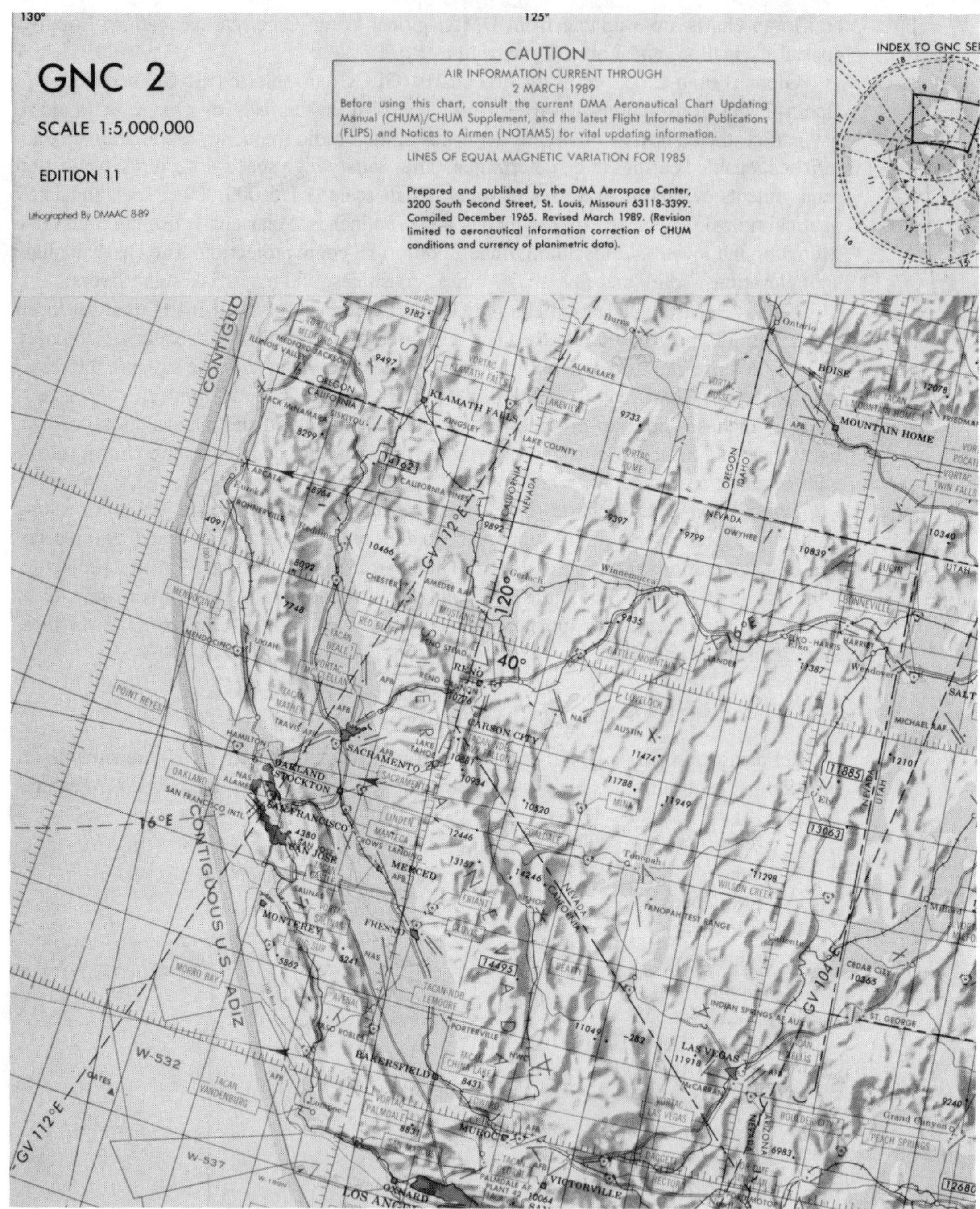

Fig. 5-1. Global navigational charts are designed primarily for flight planning, operations over long distances, and enroute navigation in long range, high speed, high altitude aircraft.

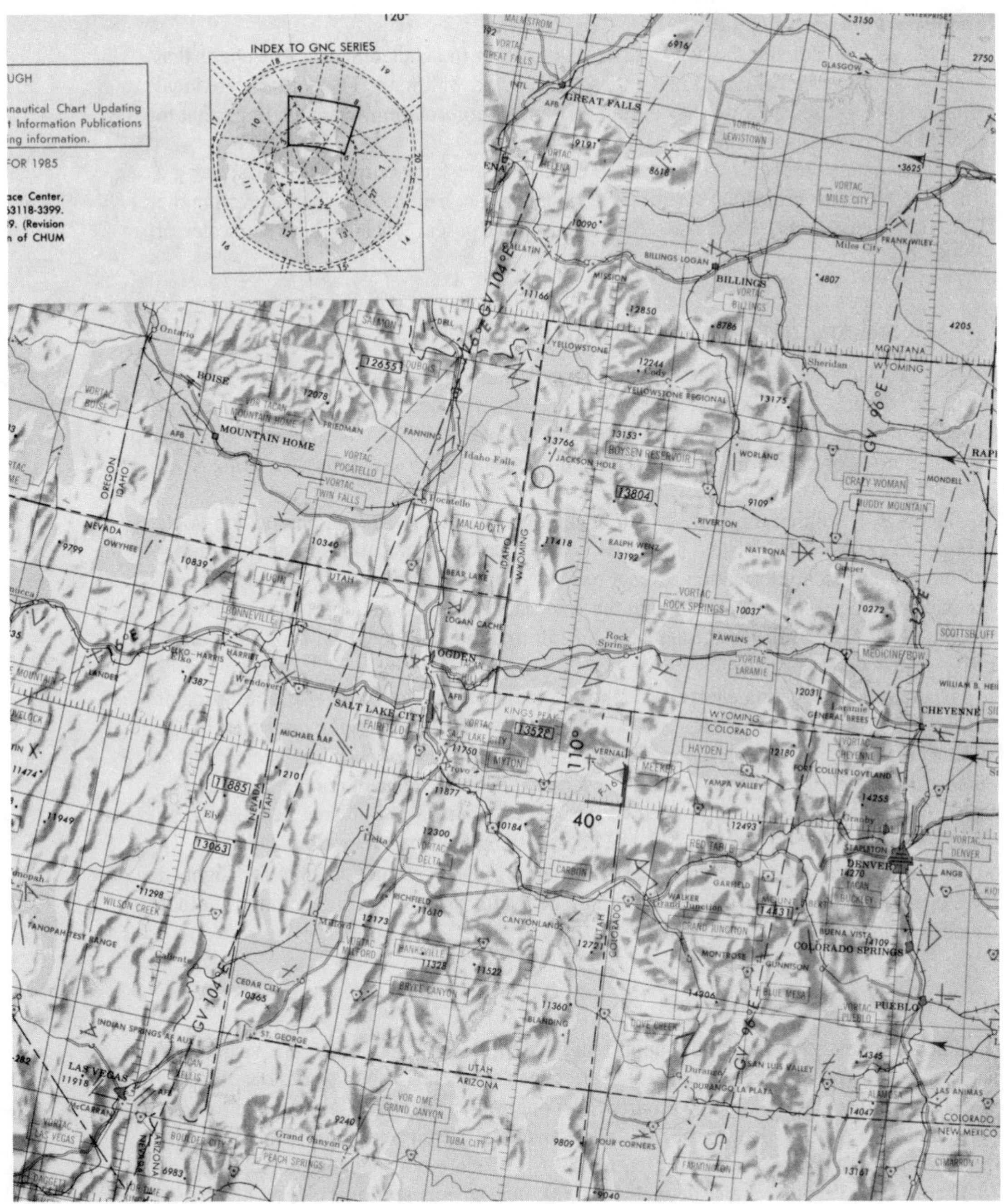

Fig. 5-1. Continued.

(Continued from page 111.)

JNCs are available for the world; three charts cover the United States. The charts that cover the U.S. can be combined into a reasonably sized wall map. These charts are ideal for planning purposes because they have better terrain information, which is important to pilots of low performance aircraft.

Jet navigation charts for the Arctic (JNCA) serve the same purpose as JNCs, using a transverse Mercator projection for Arctic regions, with two additional charts covering the U.S. and Central America, used for training purposes. JNCAs have a scale of 1:3,000,000.

Operational navigation charts

Operational navigation charts (ONC) support high speed radar navigation requirements at medium altitudes. Other uses include visual, celestial, and radio navigation. These charts have a scale of 1:1,000,000, the same WACs. ONCs are available for all of the land masses of the world. Sheet size is approximately 42 × 58 inches, covering 8° of latitude.

Figure 5-2 is an example of an ONC that covers the same general area as the WAC excerpt in Fig. 4-12. Notice that ONCs do not provide communications or airways information, nor airspace information, except restricted, military operations, and alert areas. Air information is only current through the date stated on the chart. Pilots are advised to consult NOTAMs and FLIPs for the latest air information and the CHUM for other chart revision information.

Operational navigation charts are identified in the same manner as WACs. Starting at the North Pole with the letter "A," each successive row of charts uses the next letter of the alphabet through "X," which covers the South Pole. Each row of charts is labeled with a number that generally begins at the Prime Meridian, with subsequent numbers to the east. Because only land masses are charted in this series, chart G-18 will not necessarily be located under chart F-18. In Fig. 5-2, the chart joining G-18 to the north is F-16.

Because of the lack of aeronautical information, ONCs are not suitable for flight in the United States; however, these charts might be useful for pilots planning to fly in other countries where WACs are not available.

Tactical pilotage charts

Tactical pilotage charts (TPC) support high speed, low altitude, radar, and visual navigation of high performance tactical and reconnaissance aircraft at very low through medium altitudes. Tactical pilotage charts cover one-fourth the area of operational navigational charts. They are identified by the respective ONC letter and number and an additional letter representing the TPC (TPC G-18A). TPCs are not available for all areas of ONC coverage. TPCs have a scale of 1:500,000, the same as sectional charts.

Figure 5-3 contains an example of TPC covering the same general area as the sectional in Fig. 4-13. Like ONCs, tactical pilotage charts do not provide communications, airways and fixes, or controlled airspace: airports, shaded relief, and topography are similar to the sectional.

Tactical pilotage charts are not suitable for navigation in the United States because of the lack of aeronautical information; however, these charts might be useful for pilots planning to fly in other countries. Some pilots like to obtain these charts when planning trips outside the U.S. just to get the lay of the land.

Joint operations graphics

Joint operations graphics-air (JOG-A) are suitable for preflight and operational functions. Scale is 1:250,000, the same as TACs. Figure 5-4 contains an example of a JOG-A, which covers the same general area as the TAC in Fig. 4-15. Communications, airways and fixes, and controlled airspace are not indicated.

JOGs are not available for sale outside of the United States. These charts could serve helicopter or other operators where low level navigation is required, and NOS TACs are not available; however, they would have to be used in conjunction with the associated sectional for proper communications and to establish the limits of controlled airspace.

Military training route charts

FAA issued a waiver to the Department of Defense (DOD) in 1967 to conduct various training activities below 10,000 feet MSL at speeds in excess of 250 knots. These activities included low altitude navigation, tactical bombing, aircraft intercepts, air to air combat, ground troop support, and other operations in the interest of national defense. The number and complexity of these routes were to be limited to that considered absolutely necessary. Route widths vary from two to 10 nautical miles. Enroute altitudes will be the minimum necessary for operational requirements but in no case at altitudes less than those specified in FARs for minimum safe altitudes. They range from 500 feet, or lower, to higher than 10,000 feet. Active times vary and are specified for each route, ranging from daylight hours, Monday through Friday, to continuous. Routes are designed to be clear of terminal control areas, control zones, and airport traffic areas. Additionally, to the extent possible, routes remain clear of populated areas, control areas, transition areas, and uncontrolled airports.

Military training routes fall into two categories: IFR military training routes (IR) and VFR military training routes (VR). VRs are only established when an IR route cannot accommodate the mission. IR routes might be flown in all weather conditions. VFR routes are only flown when forecast and encountered weather conditions equal or exceed five miles visibility and a 3,000-foot ceiling.

DMA publishes military training route (MTR) charts. The charts provide a visual depiction of routes, along with a specific route number. Three charts are published for the U.S.: western, central, and eastern. Charts are published every 56 days and are available by single copy or annual subscription. The Department of Defense (DOD) provides these publications to flight service stations for use in preflight pilot briefings.

Pilots should review this information and acquaint themselves with routes located along planned flight paths and in the vicinity of airports from which they operate. Flight instructors, and flight schools especially, should be familiar with the routes that traverse their normal areas of operation. (Obtain the chart and post the information in the flight

(Continued on page 122.)

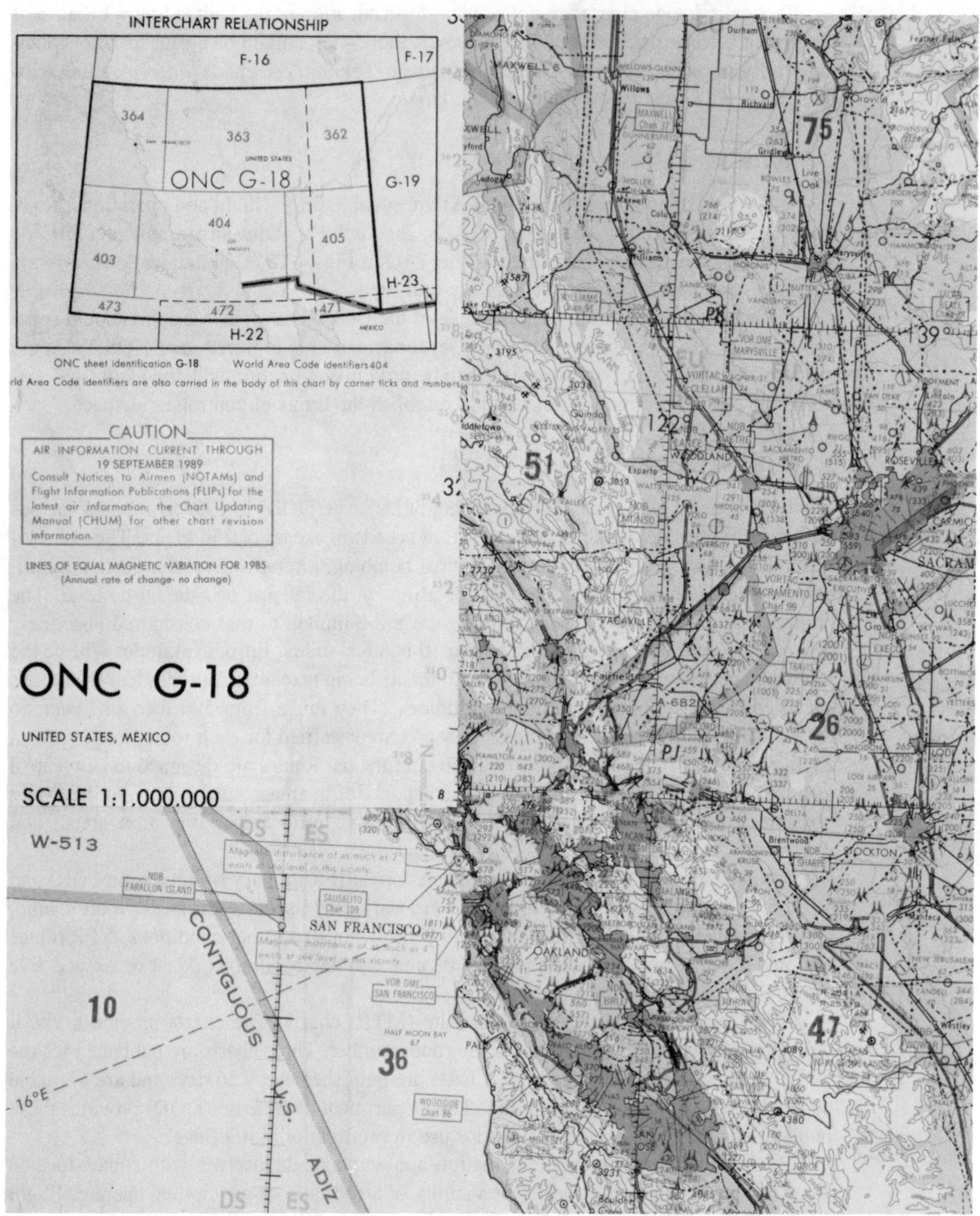

Fig. 5-2. Operational navigational charts support high speed navigational requirements at medium altitudes.

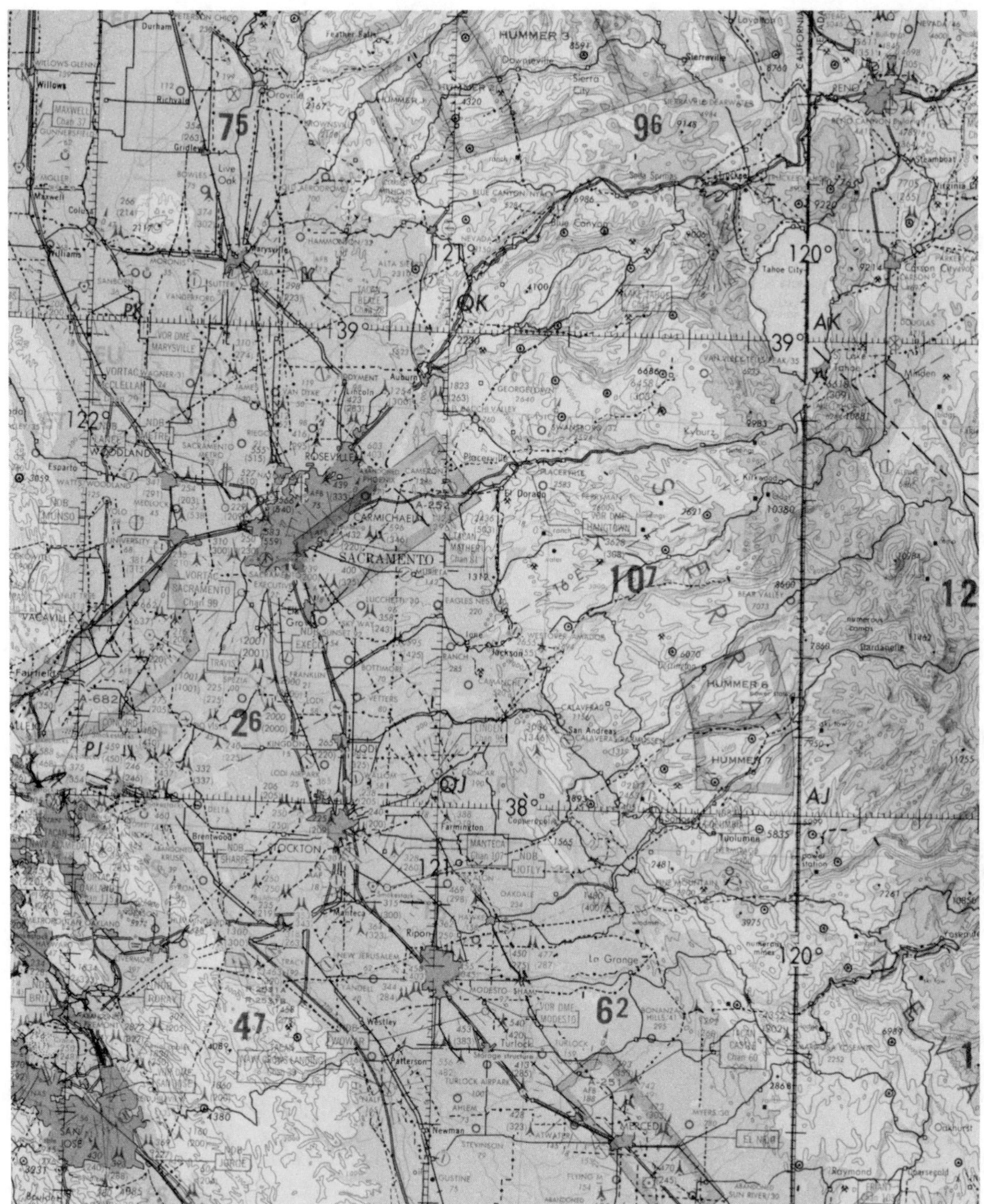

Fig. 5-2. Continued.

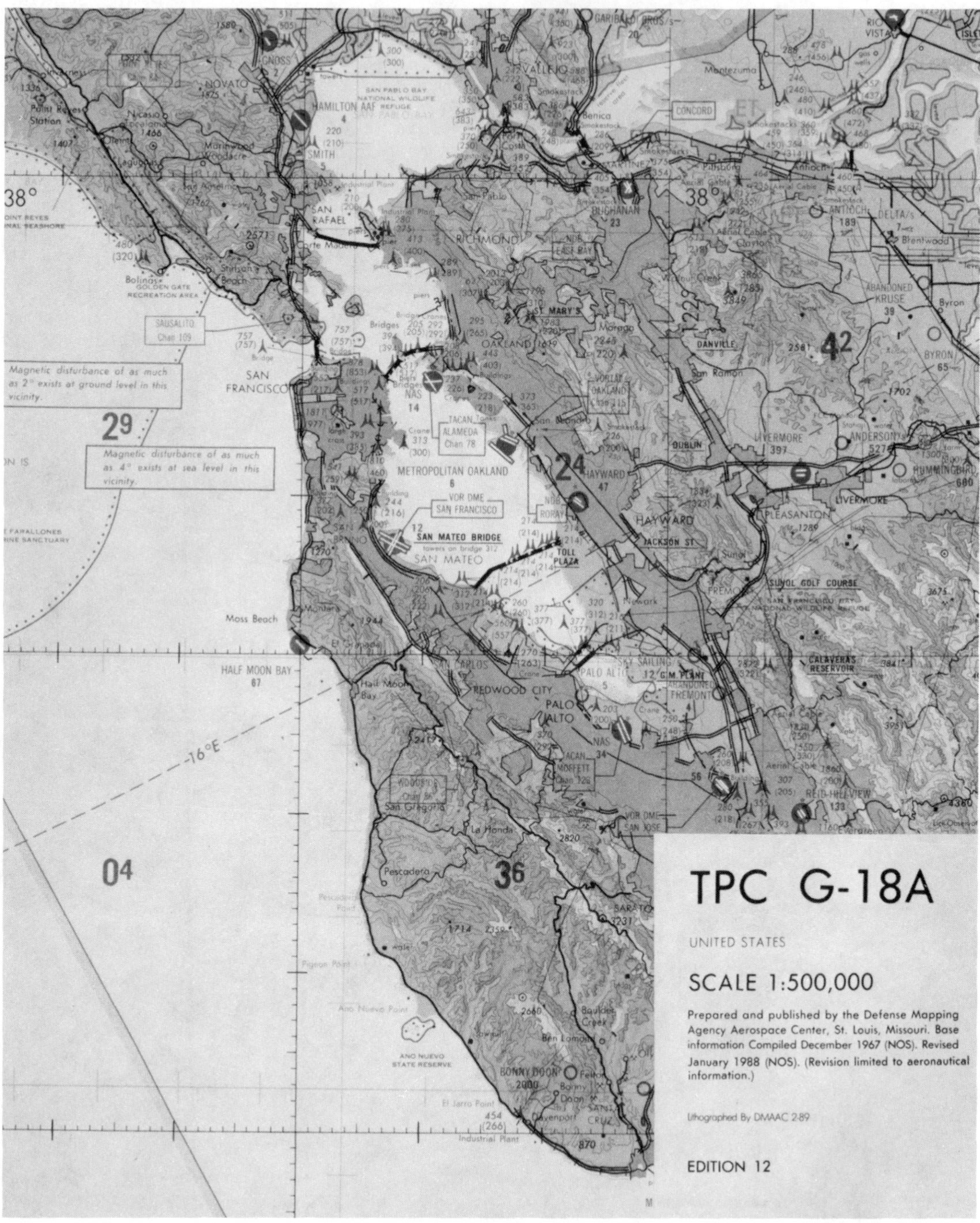

Fig. 5-3. Tactical pilotage charts are designed for high speed, low altitude, radar, and visual navigation.

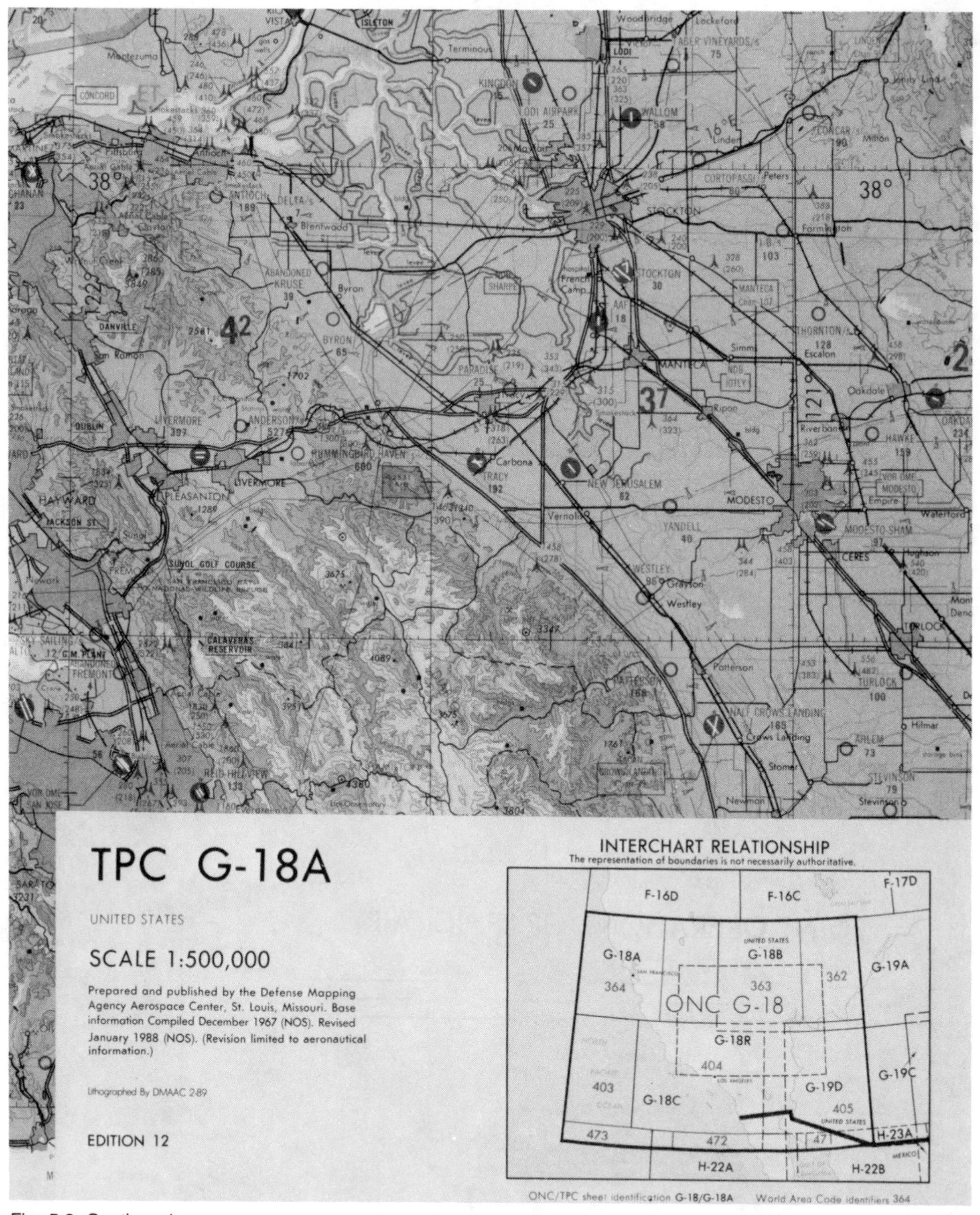

Fig. 5-3. Continued.

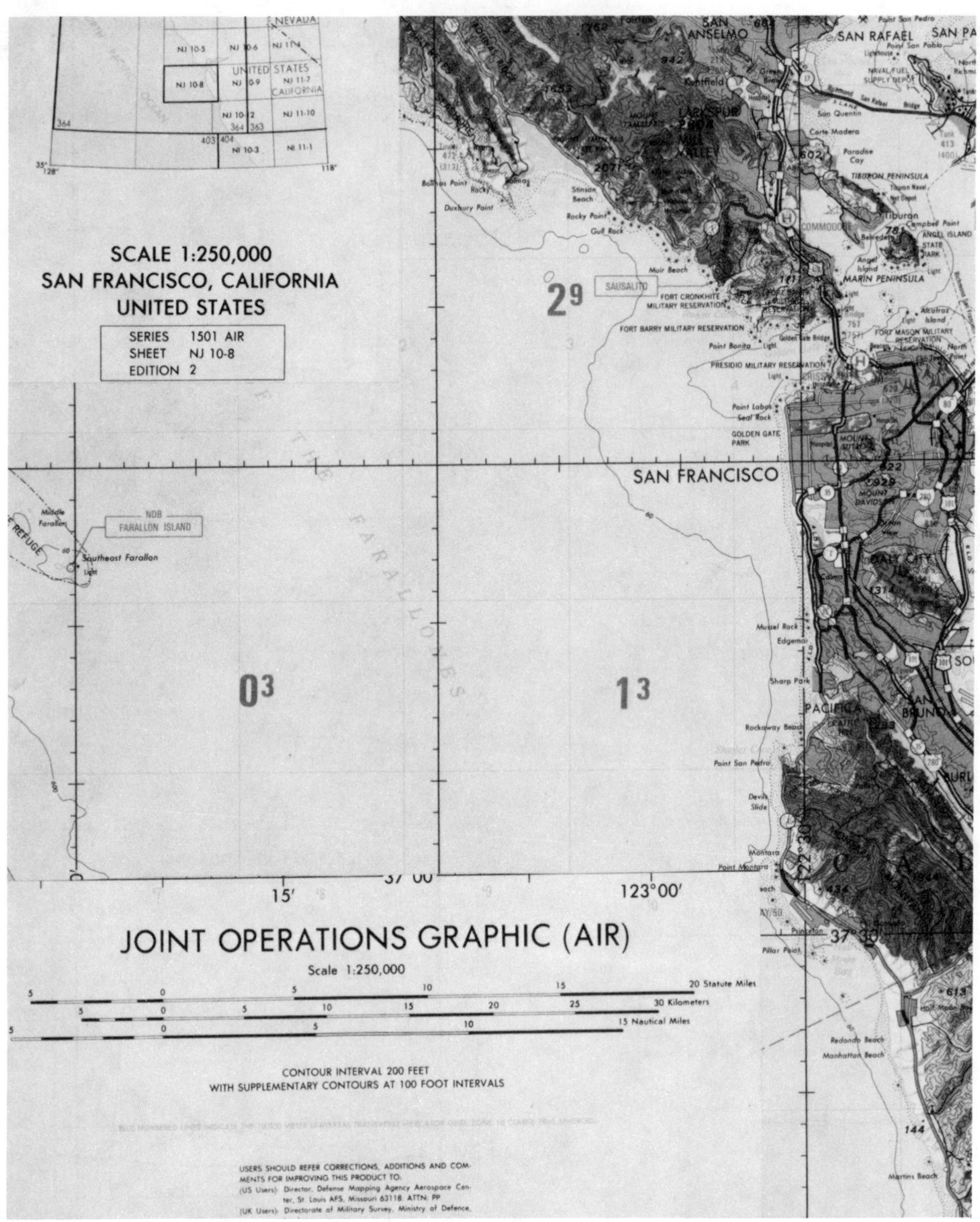

Fig. 5-4. Joint operations graphics are suitable for preflight and operational functions.

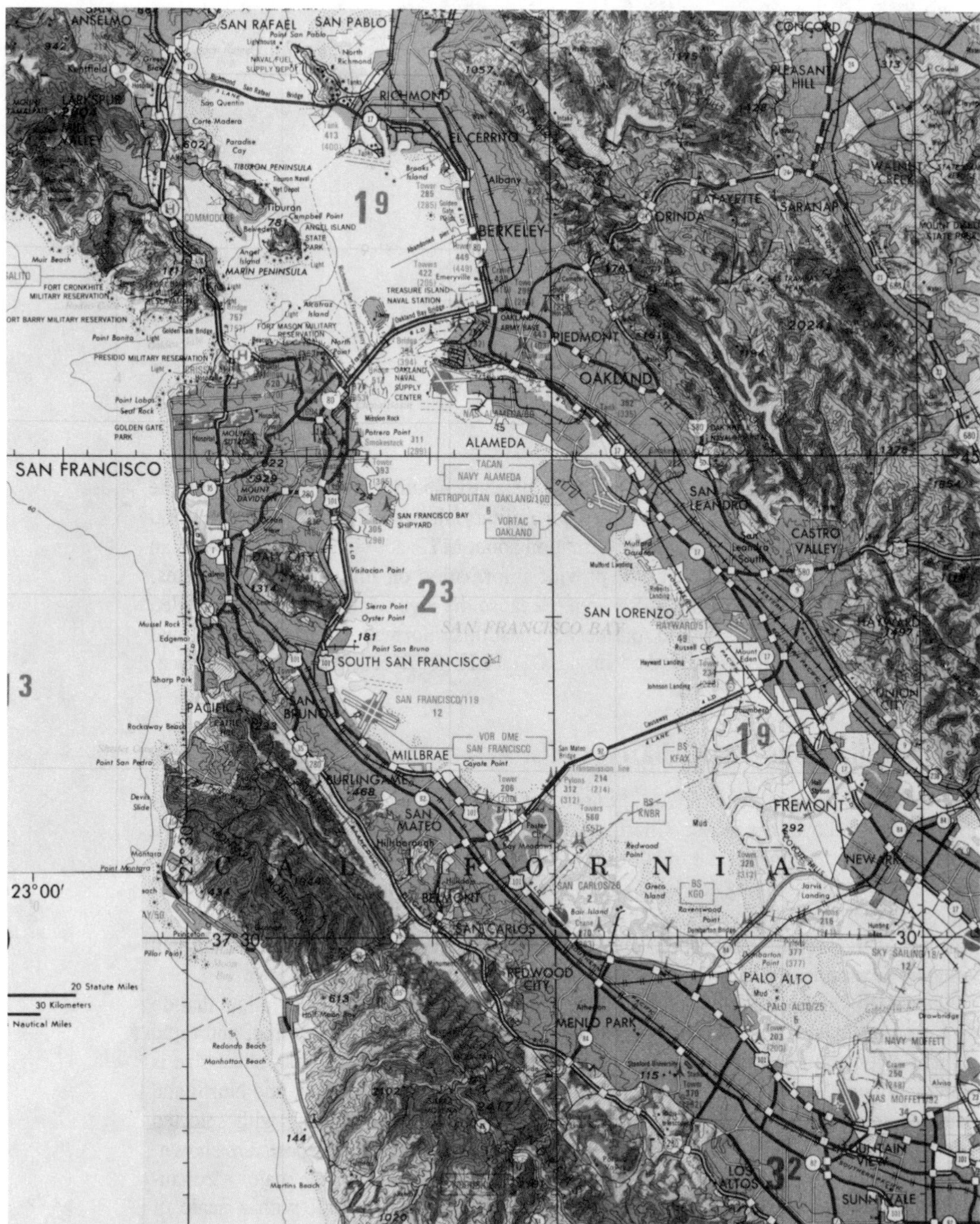

Fig. 5-4. Continued.

(Continued from page 115.)
planning area.) Figure 5-5 contains an excerpt from an area planning, military training routes chart. Features include:

- Major airports
- NAVAIDs
- Flight service stations
- Military training routes
- Route altitudes
- Route hours of operation
- Special use airspace
- Nuclear power plants
- Radioactive waste sites
- VFR helicopter refueling tracks
- Index to tactical pilotage charts

In addition to depicting IR and VR routes, the chart includes slow speed low altitude training routes (SR). These routes are used for military air operations at or below 1,500 feet at speeds of 250 knots or slower. Information about MTR activity is available from an FSS. Upon request, the specialist will provide information on military training routes, normally within 100 miles of the FSS's area. Because the area of MTR activity provided by the FSS is limited, pilots should routinely request this information enroute. Additional information about MTRs is published in a DOD FLIP for North and South America, which is explained in chapter 10.

Miscellaneous DMA charts

Aerospace planning charts (ASC) consist of six charts, at each scale, with various projections that cover the world. Chart scales are 1:9,000,000 and 1:18,000,000, with a sheet size of approximately 58 × 42 or 21 × 29 inches, respectively. These charts are designed for wall mounting, and are useful for general planning, briefings, and studies. Charts do not contain contours or gradient tints, or aeronautical information. Cities of strategic, or economic importance, major towns, transportation networks, international boundaries, prominent landmarks, as well as major lakes and rivers are shown.

Oceanic planning charts (OPC) are designed for transoceanic flights by pilots that do not have a navigator. The charts can be used in preflight and in-flight planning, and rapid in-flight orientation. Chart scales vary from 1:10,000,000 to 1:20,000,000, sheet size is approximately 17 × 11 inches. Charts are available for the North Pacific, and North and South Atlantic. International boundaries and continental outlines, along with selected radio aids, and no wind equal distancc lines between select diversion airports are shown.

Standard index charts (SIC) are graphics with index overprints for the major aeronautical chart series; sheet size is 28 × 48 inches, covering the world, with a scale of 1:35,000,000. SICs are available for the following chart series:

- Global navigational charts
- Jet navigation charts

- Operational navigational charts
- Tactical pilotage charts

OTHER CHARTS

Special purpose and supplementary charts are also available: airport obstruction charts, the Grand Canyon VFR aeronautical chart, Jeppesen VFR Los Angeles Basin charts, and state aeronautical charts.

Airport obstruction charts

NOS publishes an airport obstruction chart (OC), with a scale of 1:12,000, that graphically depicts FAR Part 77, "Objects Affecting Navigable Airspace." OCs provide data for computing maximum takeoff and landing weights of civil aircraft, for establishing instrument approach and landing procedures, and for engineering studies relative to obstruction clearing and improvements in airport facilities. Features include:

- Airport obstruction information
- FAR Part 77 surfaces
- Runway plans and profiles
- Taxiways and ramp areas
- Air navigation facilities
- Selected planimetry

(Planimetry shows manmade and natural features, such as woods and water, but does not include relief.)

Grand Canyon chart

NOS, in coordination with the FAA, is producing a new chart of the Grand Canyon National Park area to promote aviation safety and assist VFR navigation in this popular flight area. The Grand Canyon VFR aeronautical chart has a scale of 1:250,000, same as TACs, and will be revised as needed, probably once a year. The chart covers the procedures and restrictions required by Special Federal Aviation Regulation (SFAR) 50-2. One side of the chart is for noncommercial operations and the other side is for commercial air tour operations. Features include:

- SFAR operations below 14,500 MSL
- Flight free zones, where aircraft operations are prohibited
- Corridors between flight free zones
- VFR checkpoints
- Communications frequencies
- Minimum altitudes
- Navigational data

(Continued on page 126.)

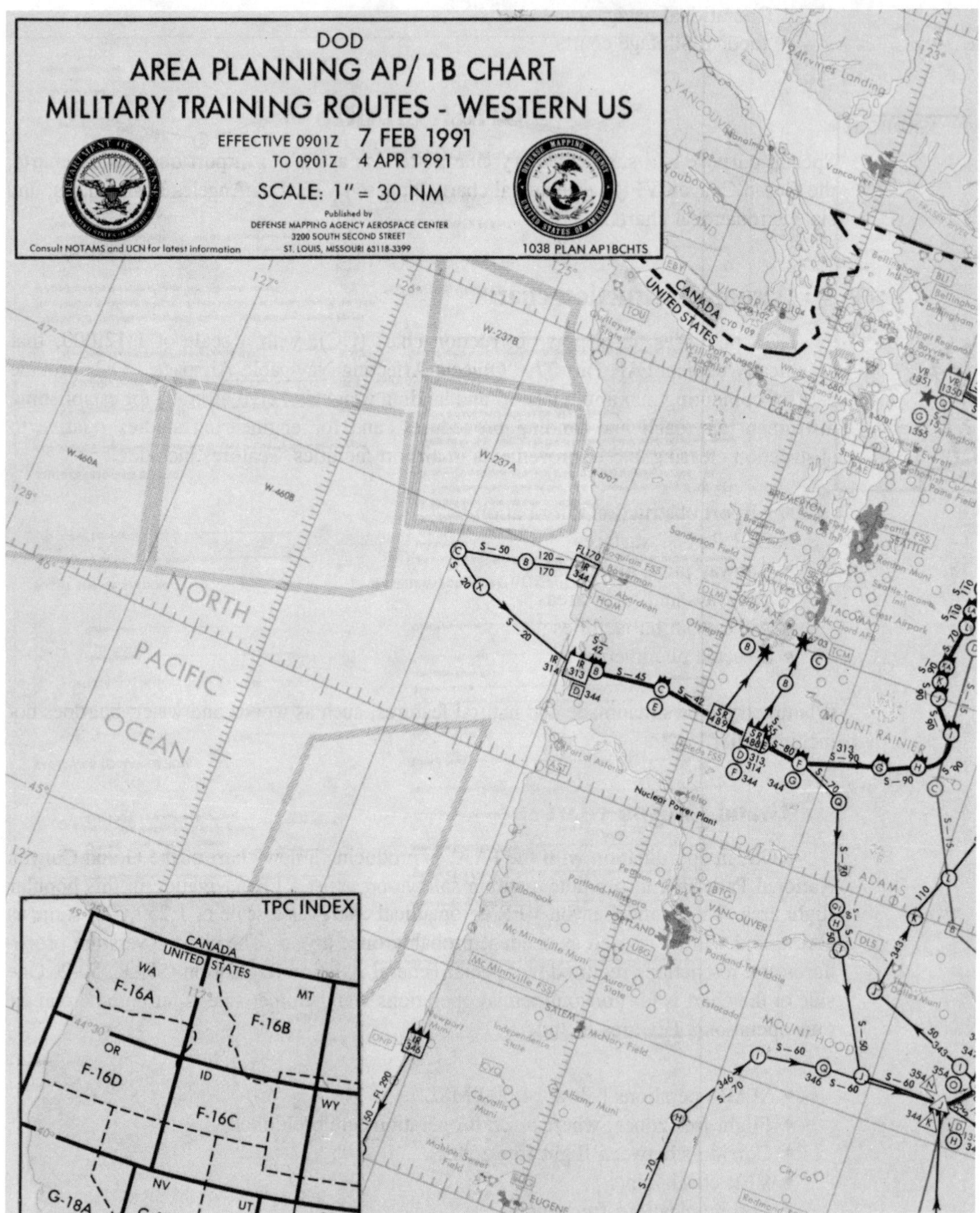

Fig. 5-5. Military training route charts provide a detailed, visual depiction of low level, high speed routes.

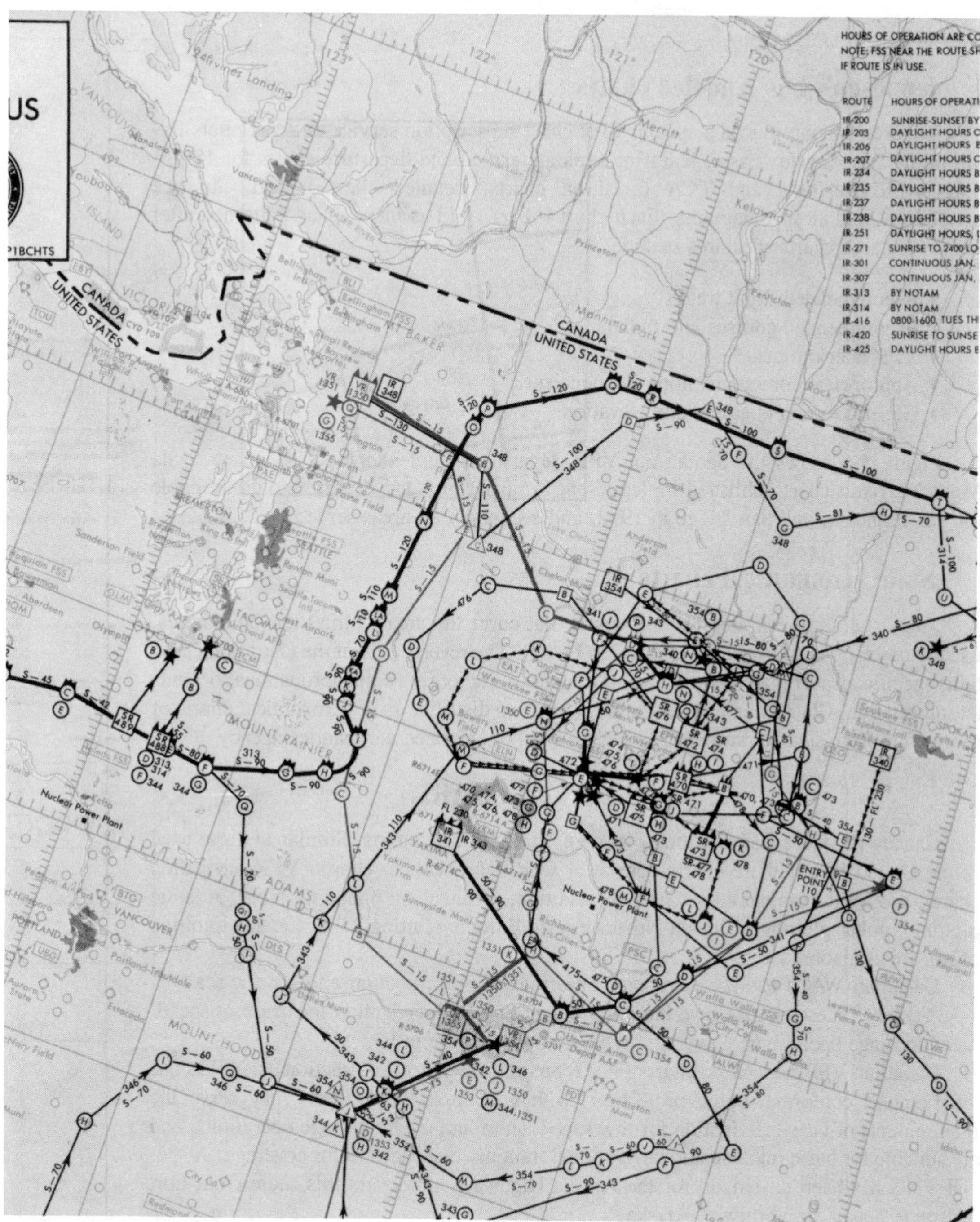

Fig. 5-5. Continued.

(Continued from page 123.)

Jeppesen's Los Angeles charts

Jeppesen has introduced a special VFR chart subscription service to assist pilots flying in the Los Angeles Basin, featuring separate arrival and departure charts for 18 airports with overflight and TCA transition charts, detailed chart legend, airspace boundaries, and airport diagrams. Each chart is $8^1/_2 \times 13^1/_2$ inches, and patterned after Jeppesen's area charts. Features include:

- Recommended flight tracks
- Recommended courses and radials
- VFR checkpoints
- Appropriate communications frequencies
- General terrain contours and elevations

Figure 5-6 shows the Santa Ana VFR departures chart and Fig. 5-7 is the Santa Monica arrivals chart published by Jeppesen. Similar charts are expected to be available for San Francisco and San Diego in 1992, and eventually all areas served by a TCA.

State aeronautical charts

Many states publish aeronautical charts that cover the area within their boundaries, usually based upon the WAC scale of 1:1,000,000. The reverse side of the chart often contains specific airport information, airport diagrams, and other useful aeronautical or tourist information. Charts might be available from local distributors of aeronautical charts or more often from state's transportation department or affiliated aeronautics agency.

CANADIAN VFR CHARTS

Canada produces and distributes its own set of pilotage charts. Similar to those used in the United States, these charts consist of world aeronautical charts, VFR navigation charts, and VFR terminal area charts. In addition, aeronautical planning, North Atlantic plotting, polar plotting, Canada-Northwestern Europe plotting, and Canada plotting charts are available.

Canadian WACs use the Lambert conformal conic projection and have a scale of 1:1,000,000. These charts serve the requirement of visual navigation for medium speed, medium range operations. Coverage, type, and number are in Fig. 5-8.

Canadian VFR navigation charts (VNC) are equivalent to U.S. sectionals. VNCs use the Lambert conformal conic projection, with a scale of 1:500,000. They serve the requirements of visual navigation for low speed, short and medium range operations, and are suitable for basic pilotage and navigational training. *See* Fig. 5-9 for coverage. A special VNC has been developed for the Alaska Highway, and covers this route from Fort Nelson, Canada, to Northway, Alaska.

Canada also produces four VFR terminal area charts (VTA) for Montreal, Toronto, Winnipeg, and Vancouver, respectively, using the transverse Mercator projection, with a scale of 1:250,000, equivalent to U.S. TACs.

Topography and obstructions

Canadian charts display relief as contour lines, shaded relief, and color tints. Green color indicates flat or relatively level terrain, regardless of altitude above sea level. Significant elevations are depicted as spot elevations, critical elevations, and maximum elevation figures; MEF on Canadian charts indicates the highest terrain elevation plus 328 feet, or the highest known obstruction elevation, whichever is higher.

Hydrography and culture symbols are similar to those of U.S. charts. Obstructions are also indicated in the same manner, with obstructions 1,000 feet agl or higher shown with a larger symbol. Obstruction elevation in feet above sea level (ASL) appears above the height in feet agl, which is enclosed with parentheses.

Navigational aids

Approved land airports in Canada having runways 1,500 feet or longer are charted. Airport symbols might be offset for clarity of presentation. Airports with hard-surface runways are depicted using the runway layout. When the use of a particular radio frequency is mandatory, the airport name is followed by the letter "M." The appropriate frequency in the airport data is preceded by the letter "A" (airport traffic frequency). Other airport data and the availability of services is indicated in the same manner as U.S. charts. Airports where customs service is available are indicated by a broken line box around the airport name.

Radio aids to navigation have the same general appearance as those used on U.S. charts. Heavy line boxes indicate services similar to an FSS with standard frequencies of 126.7 and 121.5. Other frequencies are shown above the box. At remote facilities the name of the controlling FSS appears in brackets below the NAVAID box. Control tower frequencies are not shown for all airports, Canadian WACs and VNCs, nor are they available in tabulated form as on U.S. charts. These frequencies must be obtained from other sources, such as VTAs or the *Canada Flight Supplement*, which is discussed in chapter 10. Tower frequencies, along with detailed flight procedures, are contained on Canadian VTAs.

Navigational information

Canada uses the International Civil Aviation Organization (ICAO) category system to describe airspace: A, B, C, D, E, and F.

Class A airspace is controlled airspace within which only IFR flights are permitted, equivalent to U.S. positive control area.

In Canada, controlled high level airspace bases vary from 18,000 feet MSL in the Southern Control Area; from FL230 in the Northern Control Area; and from FL280 in the Arctic Control Area. All controlled high level airspace terminates at FL600.

Class B airspace is controlled airspace within which only IFR and controlled VFR (CVFR) flights are permitted similar to U.S. TCA airspace. It includes all controlled low level airspace above 12,500 feet above sea level or the minimum enroute IFR altitude, whichever is higher. ATC procedures pertinent to IFR flights are applied to CVFR aircraft. Class B airspace terminates at the base of Class A airspace.

Fig. 5-6.

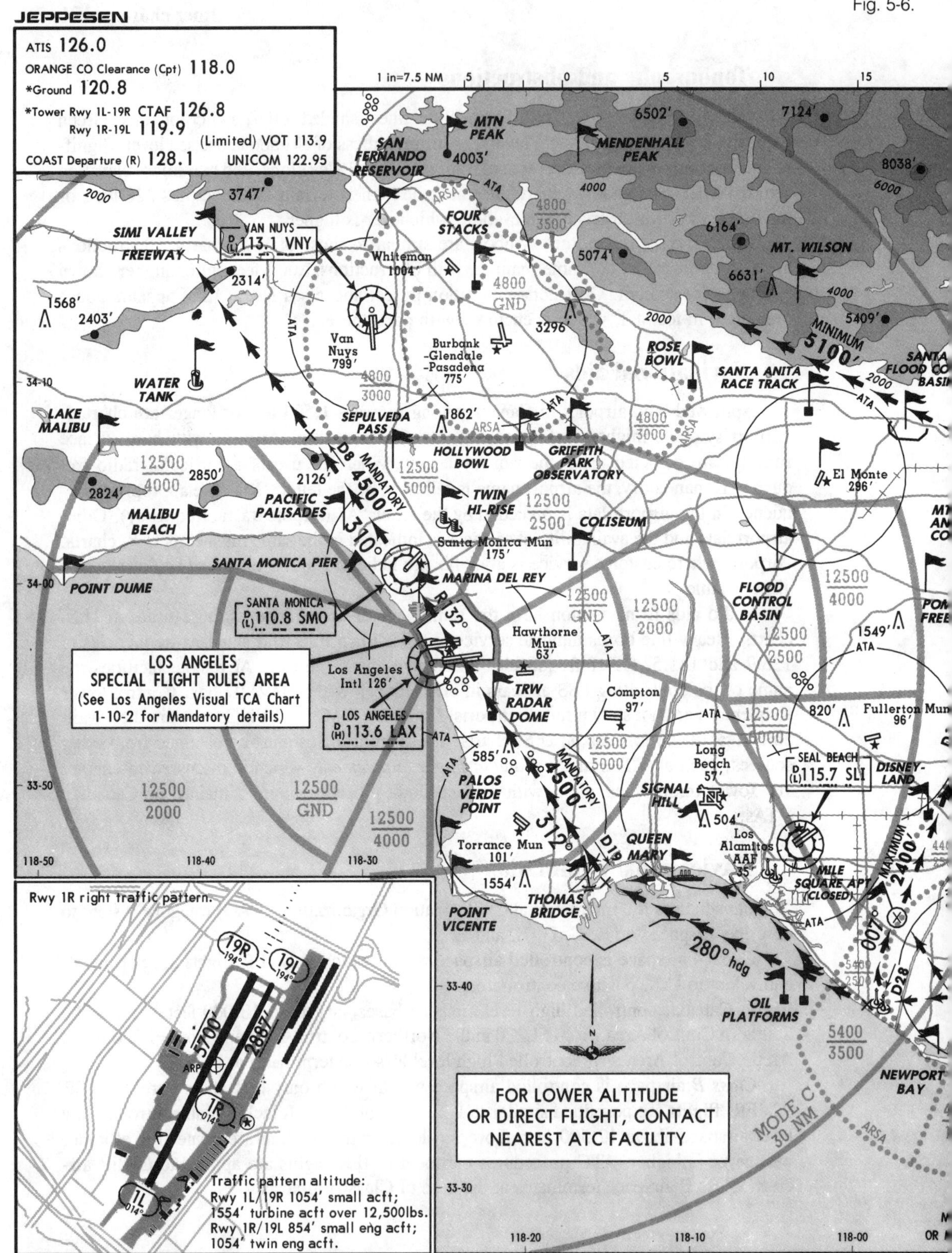
JEPPESEN
ATIS 126.0
ORANGE CO Clearance (Cpt) 118.0
*Ground 120.8
*Tower Rwy 1L-19R CTAF 126.8
Rwy 1R-19L 119.9
(Limited) VOT 113.9
COAST Departure (R) 128.1
UNICOM 122.95
1 in=7.5 NM
LOS ANGELES
SPECIAL FLIGHT RULES AREA
(See Los Angeles Visual TCA Chart 1-10-2 for Mandatory details)
FOR LOWER ALTITUDE
OR DIRECT FLIGHT, CONTACT
NEAREST ATC FACILITY
MINIMUM 5100'
MANDATORY 4500'
310°
MANDATORY 4500'
312°
280° hdg
MAXIMUM 2400'
MODE C 30 NM
VAN NUYS 113.1 VNY
SANTA MONICA 110.8 SMO
LOS ANGELES 113.6 LAX
SEAL BEACH 115.7 SLI
Van Nuys 799'
Burbank -Glendale -Pasadena 775'
Whiteman 1004'
Santa Monica Mun 175'
Los Angeles Intl 126'
Hawthorne Mun 63'
Compton 97'
Torrance Mun 101'
Long Beach 57'
Los Alamitos AAF 35'
El Monte 296'
Fullerton Mun 96'
MILE SQUARE APT (CLOSED)
SAN FERNANDO RESERVOIR
MTN PEAK
MENDENHALL PEAK
MT. WILSON
FOUR STACKS
SIMI VALLEY FREEWAY
WATER TANK
LAKE MALIBU
MALIBU BEACH
POINT DUME
PACIFIC PALISADES
SEPULVEDA PASS
HOLLYWOOD BOWL
GRIFFITH PARK OBSERVATORY
ROSE BOWL
SANTA ANITA RACE TRACK
TWIN HI-RISE
COLISEUM
SANTA MONICA PIER
MARINA DEL REY
FLOOD CONTROL BASIN
TRW RADAR DOME
PALOS VERDE POINT
SIGNAL HILL
QUEEN MARY
THOMAS BRIDGE
POINT VICENTE
OIL PLATFORMS
DISNEY-LAND
NEWPORT BAY
Rwy 1R right traffic pattern.
Traffic pattern altitude:
Rwy 1L/19R 1054' small acft;
1554' turbine acft over 12,500lbs.
Rwy 1R/19L 854' small eng acft;
1054' twin eng acft.
CHANGES: Reissue.

Fig. 5-6.
Continued.

SNA (1-10-2) Eff Nov 14-91

VFR DEPARTURE

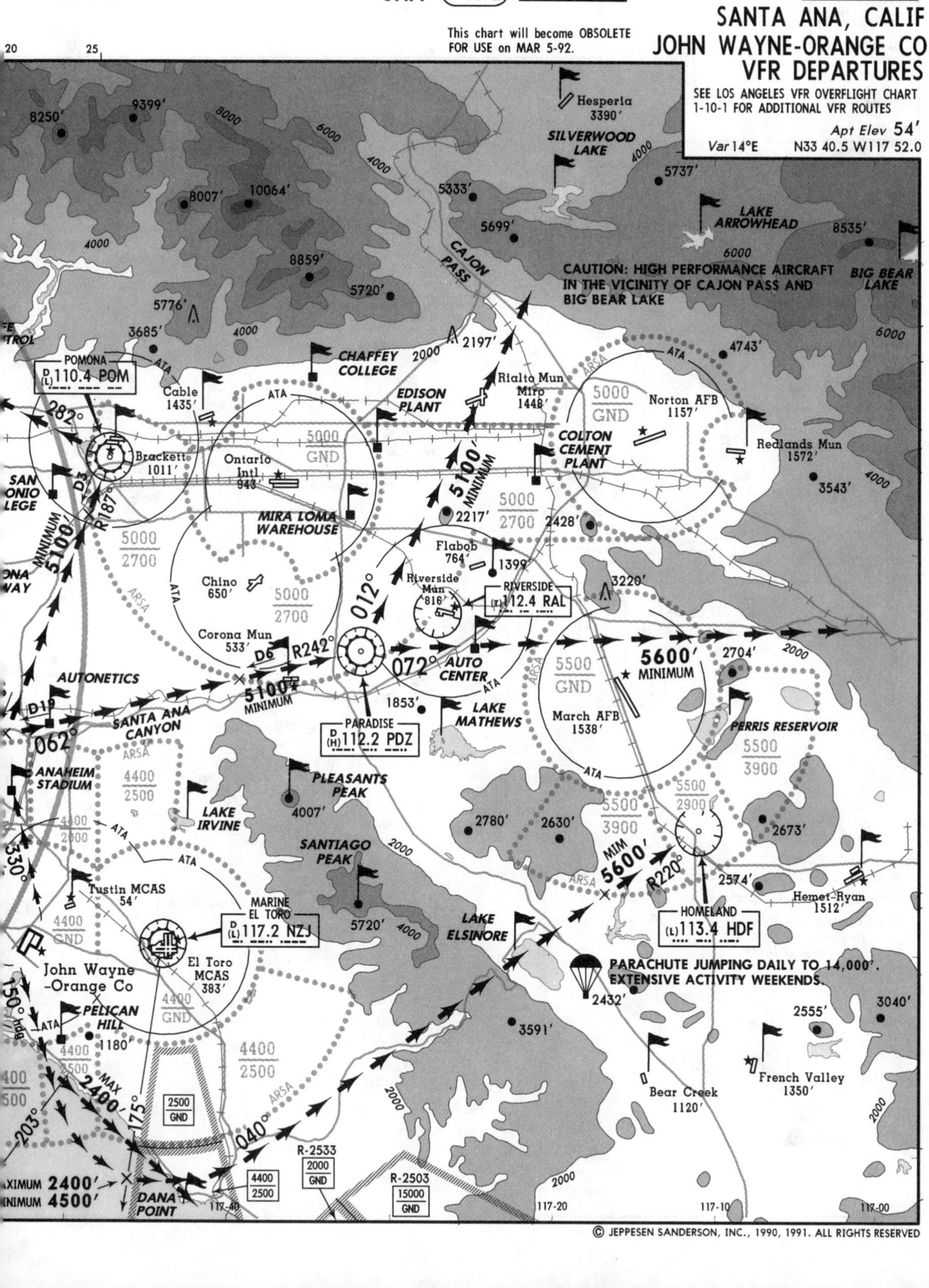

SANTA ANA, CALIF
JOHN WAYNE-ORANGE CO
VFR DEPARTURES

This chart will become OBSOLETE FOR USE on MAR 5-92.

SEE LOS ANGELES VFR OVERFLIGHT CHART 1-10-1 FOR ADDITIONAL VFR ROUTES

Apt Elev 54'
Var 14°E
N33 40.5 W117 52.0

20
25

Hesperia
3390'
SILVERWOOD LAKE
8250'
9399'
8000
6000
4000
5737'
8007'
10064'
5333'
5699'
LAKE ARROWHEAD
8535'
4000
CAJON PASS
8859'
5720'
6000
CAUTION: HIGH PERFORMANCE AIRCRAFT IN THE VICINITY OF CAJON PASS AND BIG BEAR LAKE
BIG BEAR LAKE
5776'
3685'
4000
2197'
4743'
6000
POMONA
D (L) 110.4 POM
CHAFFEY COLLEGE
2000
ATA
ARSA
Cable
1435'
EDISON PLANT
Rialto Mun
Miro
1448'
5000
GND
Norton AFB
1157'
282°
Brackett
1011'
Ontario
Intl
943'
5000
GND
COLTON CEMENT PLANT
Redlands Mun
1572'
3543'
SAN ANTONIO COLLEGE
D3
R187°
5100' MINIMUM
MIRA LOMA WAREHOUSE
2217'
5000
2700
2428'
MINIMUM 5100'
5000
2700
Flabob
764'
1399'
3220'
Chino
650'
5000
2700
012°
Riverside
Mun
816'
RIVERSIDE
(T) 112.4 RAL
Corona Mun
533'
D6
R242°
072°
AUTO CENTER
5600'
MINIMUM
2704'
2000
AUTONETICS
5100' MINIMUM
5500
GND
D19
SANTA ANA CANYON
1853'
LAKE MATHEWS
March AFB
1538'
PERRIS RESERVOIR
062°
PARADISE
D (H) 112.2 PDZ
ARSA
5500
3900
ANAHEIM STADIUM
4400
2500
PLEASANTS PEAK
4007'
5500
2900
LAKE IRVINE
5500
3900
2780'
2630'
2673'
4400
2000
330°
SANTIAGO PEAK
2000
MIM 5600'
R220°
Tustin MCAS
54'
MARINE EL TORO
D (L) 117.2 NZJ
5720'
4000
2574'
Hemet-Ryan
1512'
4400
GND
LAKE ELSINORE
HOMELAND
(L) 113.4 HDF
John Wayne
-Orange Co
El Toro
MCAS
383'
PARACHUTE JUMPING DAILY TO 14,000'. EXTENSIVE ACTIVITY WEEKENDS.
2432'
3040'
150° hdg
PELICAN HILL
1180'
4400
GND
3591'
2555'
4400
2500
4400
2500
MAX 2400'
175°
4400
2500
2000
Bear Creek
1120'
French Valley
1350'
203°
2500
GND
040°
R-2533
2000
GND
2000
MAXIMUM 2400'
MINIMUM 4500'
DANA POINT
4400
2500
R-2503
15000
GND
117-40
117-20
117-10
117-00

Fig. 5-7.

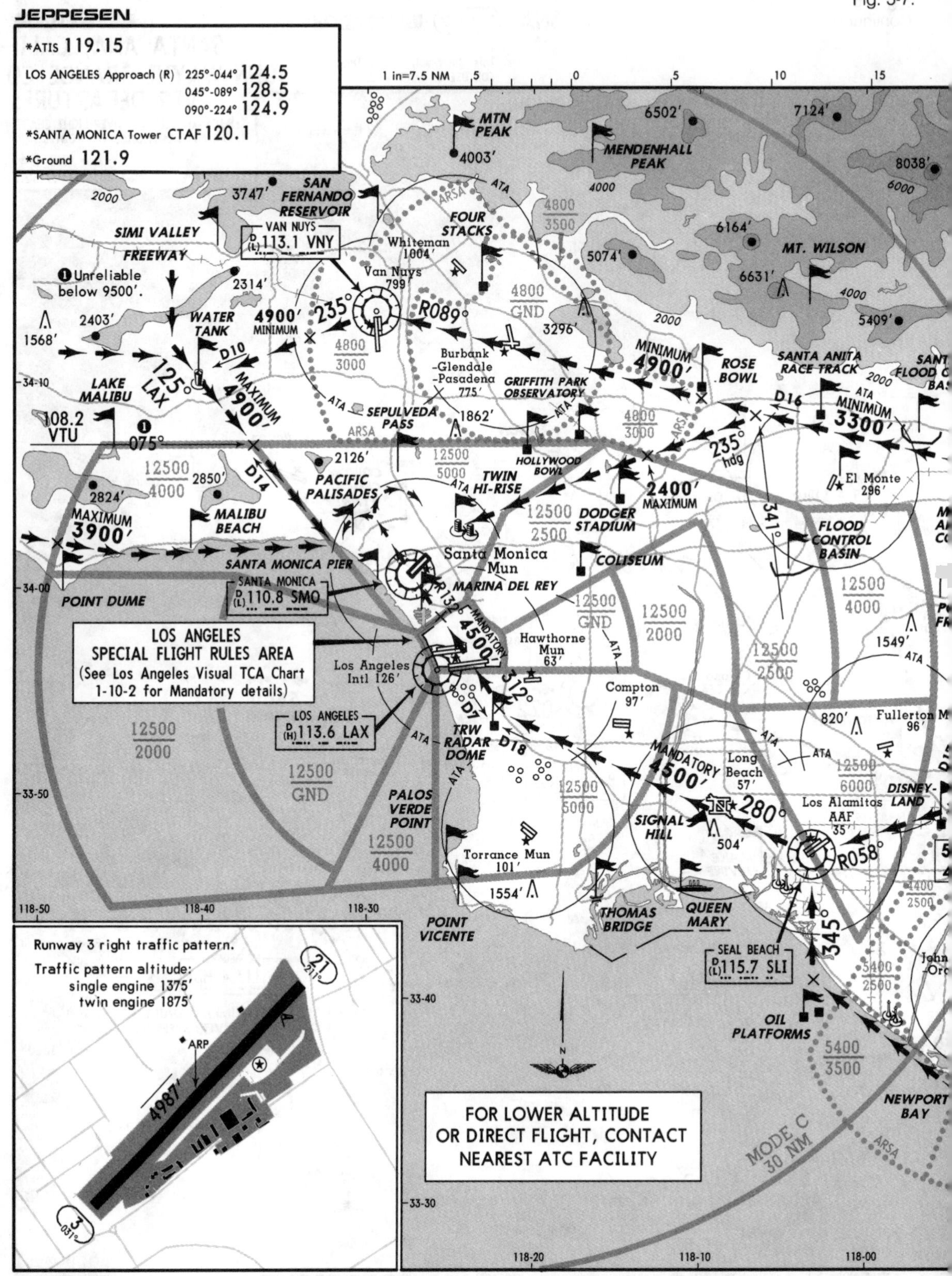

CHANGES: Reissue.

Fig. 5-7.
Continued.

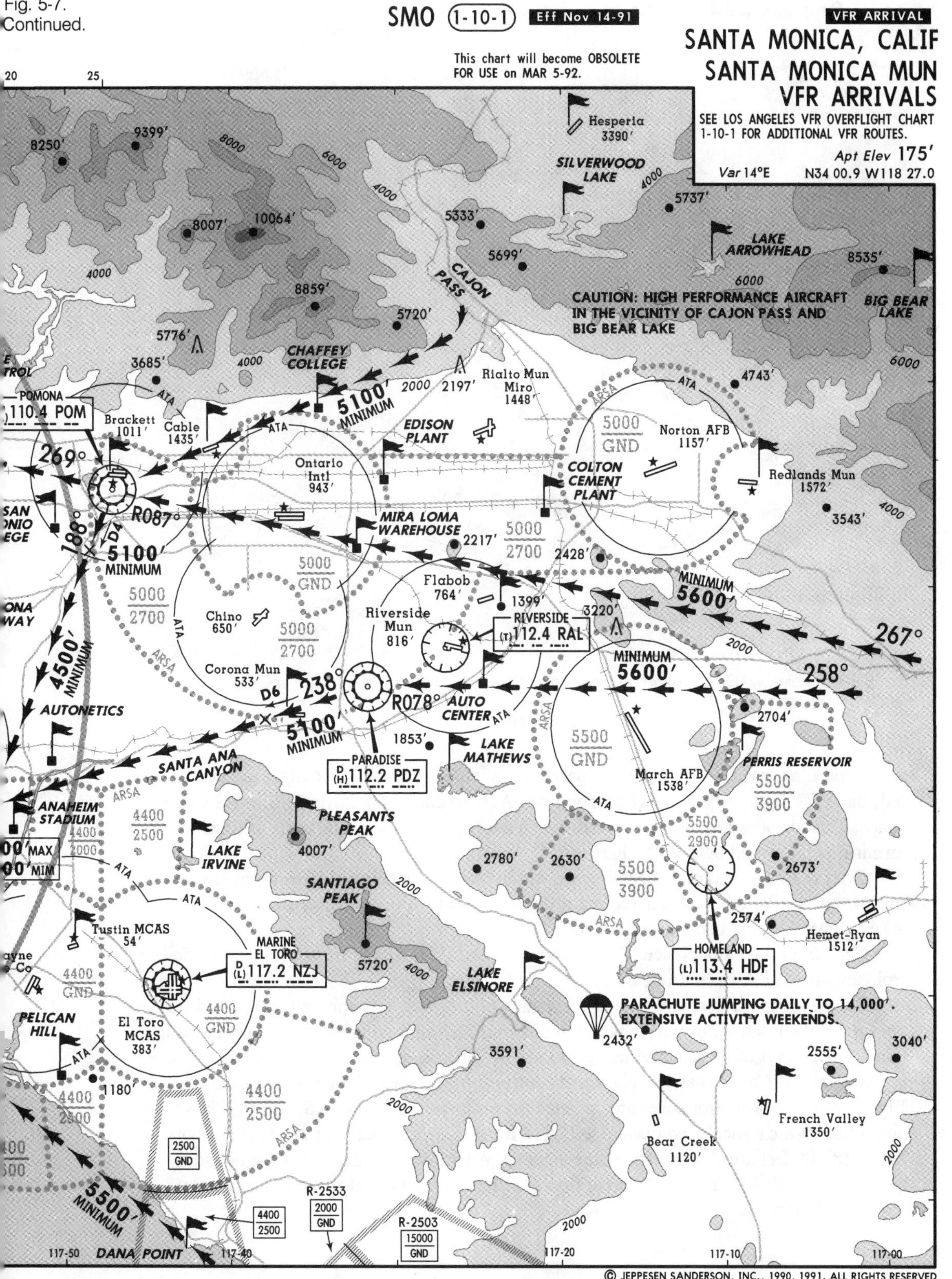

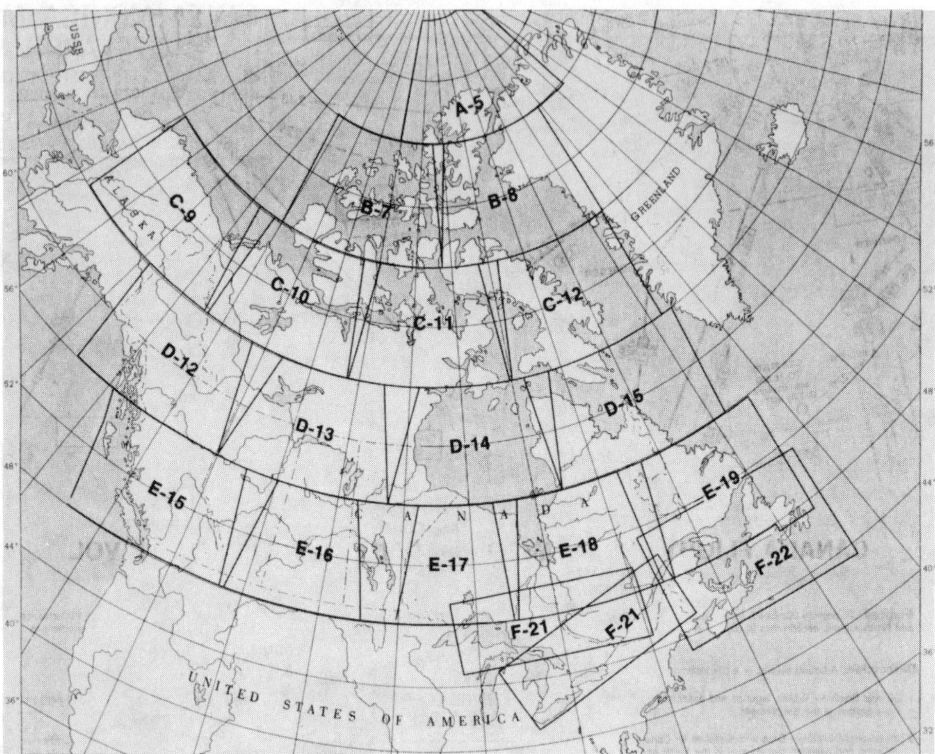

Fig. 5-8. Canadian WACs serve the requirements of visual navigation for medium speed, medium range operations.

Class C airspace is controlled airspace within which IFR and VFR flight are permitted, but VFR flight requires a clearance from ATC to enter, similar to U.S. ARSA airspace. This includes most control zones with a tower in operation, and parts of Canadian terminal areas served by a VTA chart.

Class D airspace is controlled airspace within which both IFR and VFR flight are permitted, but VFR flights do not require a clearance from ATC to enter, similar to U.S. ATAs, control zones, and control areas.

Class E airspace is airspace within which IFR and VFR flights are not subject to control, similar to U.S. uncontrolled airspace.

Class F airspace is of defined dimensions within which activities must be confined because of their nature, or within which limitations are imposed upon aircraft operations that are not a part of those activities, or both: U.S. special use airspace equivalent.

Realize that Canada has two classes of control zones; respective symbols are shown in Fig. 3-13. Class D control zones are depicted as control zones are on U.S. charts; Class C control zones are basically equivalent to U.S. ARSAs. Control area boundaries with floors below 18,000 feet are shown with a blue vignette. Alternate blue lines show the boundary between controlled areas with different floors. Floors are 2,200 feet agl in Canada, unless otherwise indicated.

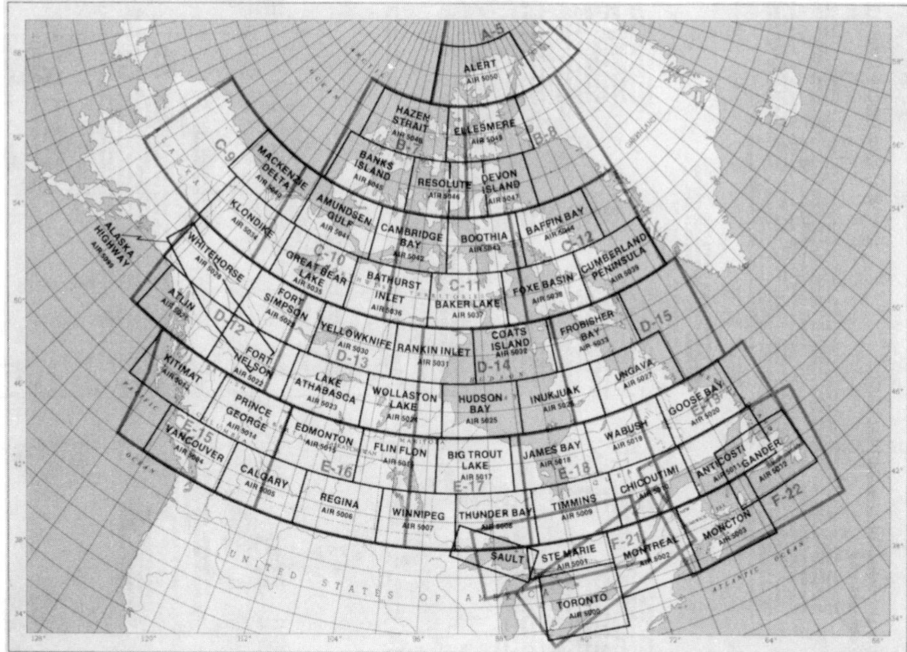

Fig. 5-9. Canadian VFR navigation charts are designed primarily for low speed, short, and medium range operations.

Formal adoption of similar alphabet categories in the United States occurred in late 1991. These changes are to be completed by September 1993. The FAA will provide an education program to help pilots understand and use the new types and classifications of airspace. Figure 5-10 provides a basic comparison between the ICAO and United States airspace designations.

Canadian special use airspace is designated alert (CYA), danger (CYD), and restricted (CYR). Alert area activity is divided into one of the following:

- A Acrobatic
- F Aircraft test
- H Hang gliding
- M Military operations
- P Parachute dropping
- S Soaring
- T Training

Altitudes are inclusive unless otherwise indicated. For example, CYA 125(A) to 5000, indicates an acrobatic flight alert area, active from the surface to 5,000 feet MSL.

Navigational information consists of isogonic lines and values, local magnetic disturbance notes, aeronautical lights, airway intersection depictions, and VFR checkpoints. Most symbols are similar to their U.S. counterparts.

Finally, Canadian charts cost more than double United States' charts.

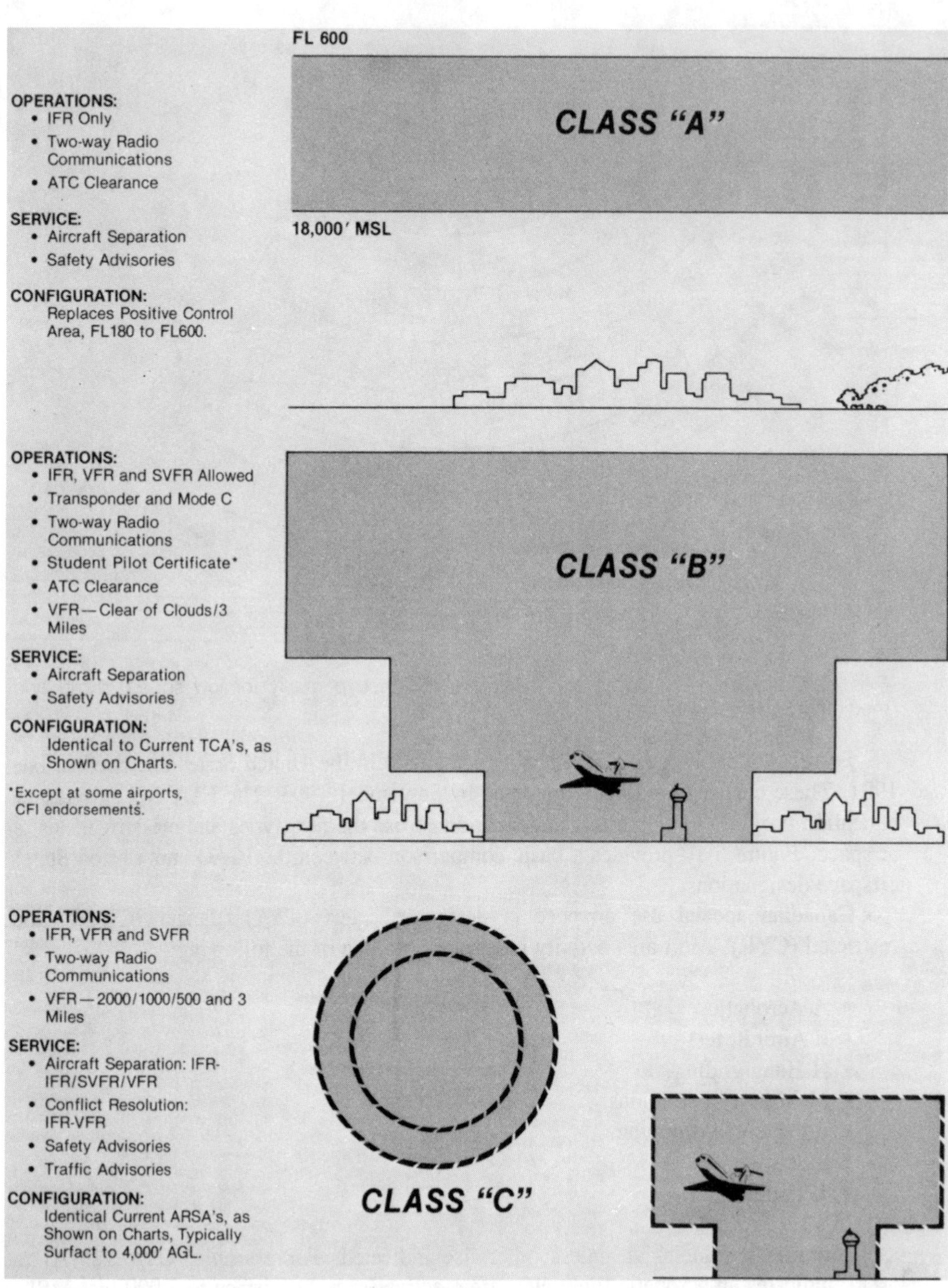
FL 600
OPERATIONS:
• IFR Only
• Two-way Radio Communications
• ATC Clearance
SERVICE:
• Aircraft Separation
• Safety Advisories
CONFIGURATION:
Replaces Positive Control Area, FL180 to FL600.
CLASS "A"
18,000′ MSL
OPERATIONS:
• IFR, VFR and SVFR Allowed
• Transponder and Mode C
• Two-way Radio Communications
• Student Pilot Certificate*
• ATC Clearance
• VFR—Clear of Clouds/3 Miles
SERVICE:
• Aircraft Separation
• Safety Advisories
CONFIGURATION:
Identical to Current TCA's, as Shown on Charts.
*Except at some airports, CFI endorsements.
CLASS "B"
OPERATIONS:
• IFR, VFR and SVFR
• Two-way Radio Communications
• VFR—2000/1000/500 and 3 Miles
SERVICE:
• Aircraft Separation: IFR-IFR/SVFR/VFR
• Conflict Resolution: IFR-VFR
• Safety Advisories
• Traffic Advisories
CONFIGURATION:
Identical Current ARSA's, as Shown on Charts, Typically Surfact to 4,000′ AGL.
CLASS "C"

Fig. 5-10. Canada uses the ICAO category of classes to designate controlled airspace.

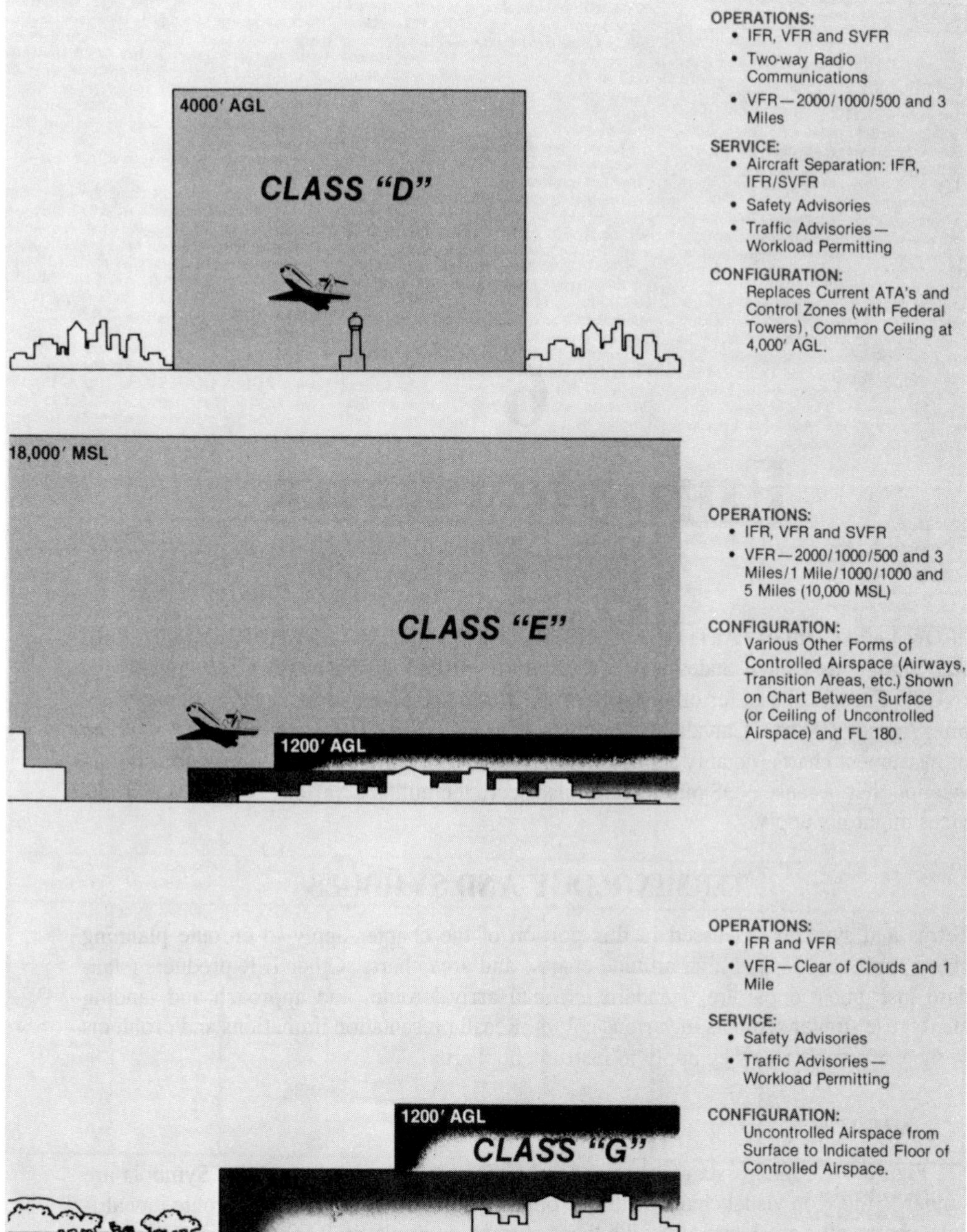

4000′ AGL
CLASS "D"
OPERATIONS:
• IFR, VFR and SVFR
• Two-way Radio Communications
• VFR—2000/1000/500 and 3 Miles
SERVICE:
• Aircraft Separation: IFR, IFR/SVFR
• Safety Advisories
• Traffic Advisories—Workload Permitting
CONFIGURATION:
Replaces Current ATA's and Control Zones (with Federal Towers), Common Ceiling at 4,000′ AGL.
18,000′ MSL
CLASS "E"
1200′ AGL
OPERATIONS:
• IFR, VFR and SVFR
• VFR—2000/1000/500 and 3 Miles/1 Mile/1000/1000 and 5 Miles (10,000 MSL)
CONFIGURATION:
Various Other Forms of Controlled Airspace (Airways, Transition Areas, etc.) Shown on Chart Between Surface (or Ceiling of Uncontrolled Airspace) and FL 180.
1200′ AGL
CLASS "G"
OPERATIONS:
• IFR and VFR
• VFR—Clear of Clouds and 1 Mile
SERVICE:
• Safety Advisories
• Traffic Advisories—Workload Permitting
CONFIGURATION:
Uncontrolled Airspace from Surface to Indicated Floor of Controlled Airspace.

Fig. 5-10. Continued.

6
Enroute charts

LOW AND HIGH ALTITUDE INSTRUMENT ENROUTE CHARTS FOR THE contiguous United States and Alaska are published by the National Ocean Service (NOS). Enroute charts for other parts of the world are available from the Defense Mapping Agency (DMA). Canada also publishes enroute charts as do several private vendors of instrument charts, notably Jeppesen Sanderson. Most of the information is for civilian aviation, but because NOS publications also serve the military, various NAVAIDs and services might not apply.

TERMINOLOGY AND SYMBOLS

Terms and symbols discussed in this portion of the chapter apply to enroute planning charts, enroute low and high altitude charts, and area charts. Other IFR products (standard instrument departure, standard terminal arrival route, and approach and landing charts) use similar symbols in various colors. Recall presentation limitations and problems of the cartographer as they apply to instrument charts.

Airports

Figure 6-1 contains standard enroute chart symbols for landing areas. Symbols are similar to those on visual charts, with the omission of runway layout. On enroute low altitude charts all active airports with hard-surface runways of 3,000 feet or more are depicted, along with all active airports with approved instrument approach procedures regardless of runway length or composition. Airports shown in blue have published instrument approach procedures, those in brown do not.

AIRPORTS			
AIRPORT DATA	Airports/Seaplane Bases shown in BLUE have an approved Low Altitude Instrument Approach Procedure published. Those shown in DARK BLUE have an approved DOD Low Altitude Instrument Approach Procedure and/or DOD RADAR MINIMA published in DOD FLIPS, Alaska Supplement or Alaska Terminal. Airports/Seaplane Bases shown in BROWN do not have a published Instrument Approach Procedure.		
LANDPLANE - CIVIL Refueling and repair facilities for normal traffic.			Douglas Muni
LANDPLANE CIVIL AND MILITARY Refueling and repair facilities for normal traffic.			Charleston AFB/Intl
LANDPLANE-MILITARY Refueling and repair facilities for normal traffic.			MCAF Quantico
SEAPLANE-CIVIL Refueling and repair facilities for normal traffic.			North Bay
SEAPLANE CIVIL AND MILITARY Refueling and repair facilities for normal traffic.			NAS Patuxent River SPB /Trapnell Naples Muni
SEAPLANE-MILITARY Refueling and repair facilities for normal traffic.			NAS Corpus Christi SPB
HELIPORT	Ⓗ	Ⓗ	Ⓗ Allen AHP

Fig. 6-1. Enroute chart airport symbols are similar to those used on visual charts, except that runway layout is omitted.

AIRPORTS	
AIRPORT DEPICTION	Night Landing Capability: Asterisk indicates lighting on request or operating part of night only. Circle indicates Pilot Controlled Lighting. For information consult the Airport/Facility Directory or FLIP IFR Supplement. Airport Elevation — Name 349 * Ⓛ 80 — Longest Landing Runway Length Automatic Terminal Information Service and Frequency — ASR/PAR ATIS ★ 108.5 — Radar Services Availability Indicates less than continuous (Name) 185 – 35s No Runway Lighting Capability — Indicates Soft Surface Parentheses around airport name indicate Military Landing Rights not available. Airport elevation given in feet above or below mean sea level. Length of longest runway given to nearest 100 feet with 70 feet as the dividing point (Add 00). Airport symbol may be offset for enroute navigation aids. Pvt – Private use, not available to general public. A box enclosing the airport name indicates FAR 93 Special Require-ments – See Directory/Supplement.
AIRPORT RELATED FACILITIES	Pilot to Metro Service (PMSV) Continuous Operation Less than Continuous Weather Radar (WXR) PMSV and WXR Combined

Fig. 6-1. Continued.

Airports are identified by their name; in the case of military airports, the abbreviated letters AFB (Air Force base), NAS (Naval air station), NAF (Naval air field), MCAS (Marine Corps air station), or AAF (Army air field) also appear. Airport information is similar to that used on visual charts. Runway length is the length of the longest active runway including displaced thresholds, excluding overruns, and is shown to the nearest 100 feet using 70 feet as the division point: 5,069 is labeled 50 and 5,070 is labeled 51. Runways that are not hard surface have a small letter "s" following the runway length signifying a soft surface. An "L" following the elevation means that runway lights are on during hours of darkness. A circle around the "L" indicates that lighting is less than continuous. In such cases the pilot must consult the approach chart or *Airport/Facility Directory* (A/FD) for light operating procedures.

Airport related facilities, that is pilot to weather services shown in the lower right corner of Fig. 6-1, normally only apply to military bases and military operations. Civil pilots do have access to a similar service, however, known as enroute flight advisory service, flight watch. Because flight watch is available on a common low altitude frequency of 122.0 it is not published on charts; high altitude discrete flight watch frequencies are published on the end panels of enroute high altitude charts. (Always advise flight watch of your approximate position on initial contact.)

Radio aids to navigation

The depiction of radio aids to navigation (NAVAIDs) is similar to visual charts (Fig. 6-2). Very high frequency (VHF) and ultra high frequency (UHF) NAVAIDs (VORs, TACANs, and UHF nondirectional beacons) are shown in blue or black. (UHF NAVAIDs are generally only used by the military.) Low/medium frequency (L/MF) NAVAIDs (compass locators and aeronautical or marine NDBs) are shown in brown. Notice that in addition to the depiction of VORs, VOR/DMEs, and VORTACs there is a TACAN only symbol. TACAN-only facilities are normally located on or in the vicinity of military bases. Without a TACAN receiver civil pilots will normally not be able to utilize these facilities; however, civil pilots would be able to obtain DME information from a TACAN by tuning the DME receiver to the paired VOR frequency for the TACAN. Paired VOR-TACAN frequencies are contained in the *Terminal Procedures Publication* (TPP) and the A/FD. In extremely congested areas, the NAVAID box will only contain the three-letter identifier, the complete NAVAID box will appear in a less congested area of the chart.

VORs, DMEs, and TACANs are classified by their standard service volumes (SSV). This determines the distances and altitude a particular NAVAID can be relied upon for accurate navigational guidance. Nondirectional radio beacons (NDB) are classified according to their intended use. SSVs for NAVAIDs are shown in Table 6-1. This information is of particular importance to pilots using VORTAC area navigation (RNAV). For reliable navigational signals RNAV waypoints must be within the parameters of the SSVs shown in Table 6-1, as well as any restrictions to the NAVAID as published in the A/FD or NOTAMs.

(Continued on page 144.)

RADIO AIDS TO NAVIGATION	
VHF OMNIDIRECTIONAL RADIO RANGE (VOR) DISTANCE MEASURING EQUIPMENT (DME) TACTICAL AIR NAVIGATION (TACAN)	VOR VOR/DME VORTAC TACAN 0 27 9 18 0 27 9 18 COMPASS ROSES oriented to Magnetic North of NAVAID which may not be adjusted to the charted Isogonic Values
NON-DIRECTIONAL RADIO BEACON (NDB) MARINE RADIO BEACON (RBn)	LF/MF Non-directional Radio Beacon or Marine Radio Beacon with Magnetic North Indicator UHF Non-directional Radio Beacon LF/MF Non-directional Radio Beacon/DME
INSTRUMENT LANDING SYSTEM (ILS) SIMPLIFIED DIRECTIONAL FACILITY (SDF) LOCALIZER-TYPE DIRECTIONAL AID (LDA)	BACK COURSE ILS Localizer Course with ATC function Feathered side indicates Blue Sector Published ILS and/or Localizer Procedure available SDF Localizer Course with ATC function Published SDF Procedure available LDA Published LDA Procedure available
COMPONENTS COMPASS LOCATOR MARKER BEACONS	Inner Marker (LIM) Middle Marker (LMM) Outer Marker (LOM)

Fig. 6-2. Enroute chart NAVAID depiction is the same as a visual chart.

BROADCAST STATIONS (BS)	WKBW 1520
WEATHER STATION	Norfolk Weather Radio
REMOTE COMMUNICATIONS OUTLET (RCO)	Flight Service Station (FSS) Remote Communications Outlet (RCO)
FLIGHT SERVICE STATION (FSS) IFR Enroute High Altitude Information	000.0 000.0 NAME 000.0 NME(L) 00 N00°00.00' W00°00.00' Shadow Box indicates FSS and Radio Aid same name (FSS freqs available are 255.4, 122.2 selected discrete freqs and emerg 243.0 and 121.5). The FSS high altitude VHF discrete freq(s) is shown above the box. In Canada a Shadow Box indicates stations with standard group freqs 243.0, 126.7 and 121.5. (L) Frequency Protection Usable range at 18000'-40 NM. "L" category radio aids located off Jet Routes are depicted in black screen. (T) Frequency Protection Usable range at 12000'-25 NM. "T" category radio aids located off Jet Routes are depicted in black screen. Radio Aids to Navigation without Classification are "H" Category.

Fig. 6-2. Continued.

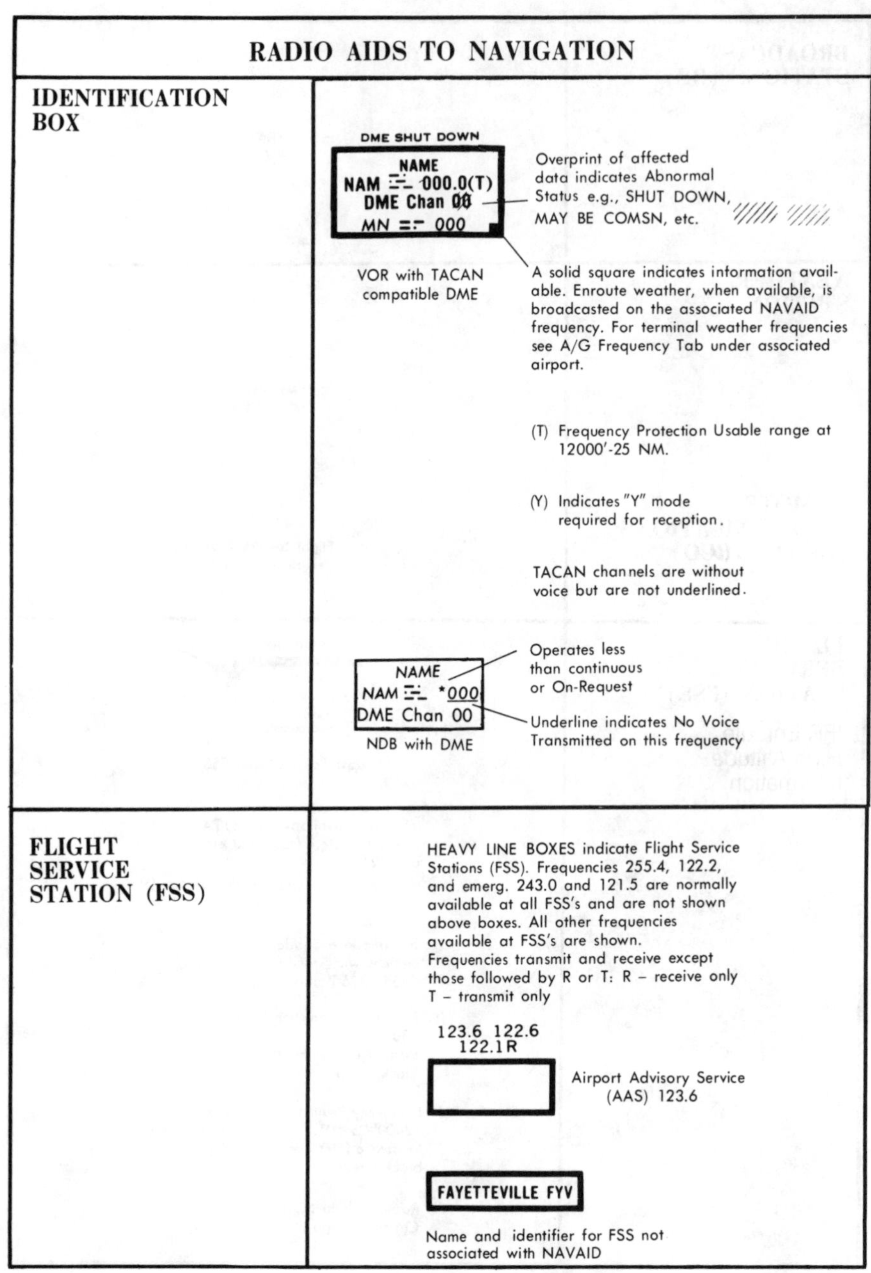

RADIO AIDS TO NAVIGATION
IDENTIFICATION BOX
DME SHUT DOWN
NAME
NAM 000.0(T)
DME Chan 00
MN 000
Overprint of affected data indicates Abnormal Status e.g., SHUT DOWN, MAY BE COMSN, etc.
VOR with TACAN compatible DME
A solid square indicates information available. Enroute weather, when available, is broadcasted on the associated NAVAID frequency. For terminal weather frequencies see A/G Frequency Tab under associated airport.
(T) Frequency Protection Usable range at 12000'-25 NM.
(Y) Indicates "Y" mode required for reception.
TACAN channels are without voice but are not underlined.
NAME
NAM *000
DME Chan 00
NDB with DME
Operates less than continuous or On-Request
Underline indicates No Voice Transmitted on this frequency
FLIGHT SERVICE STATION (FSS)
HEAVY LINE BOXES indicate Flight Service Stations (FSS). Frequencies 255.4, 122.2, and emerg. 243.0 and 121.5 are normally available at all FSS's and are not shown above boxes. All other frequencies available at FSS's are shown.
Frequencies transmit and receive except those followed by R or T: R – receive only T – transmit only
123.6 122.6
122.1R
Airport Advisory Service (AAS) 123.6
FAYETTEVILLE FYV
Name and identifier for FSS not associated with NAVAID

Fig. 6-2. Continued.

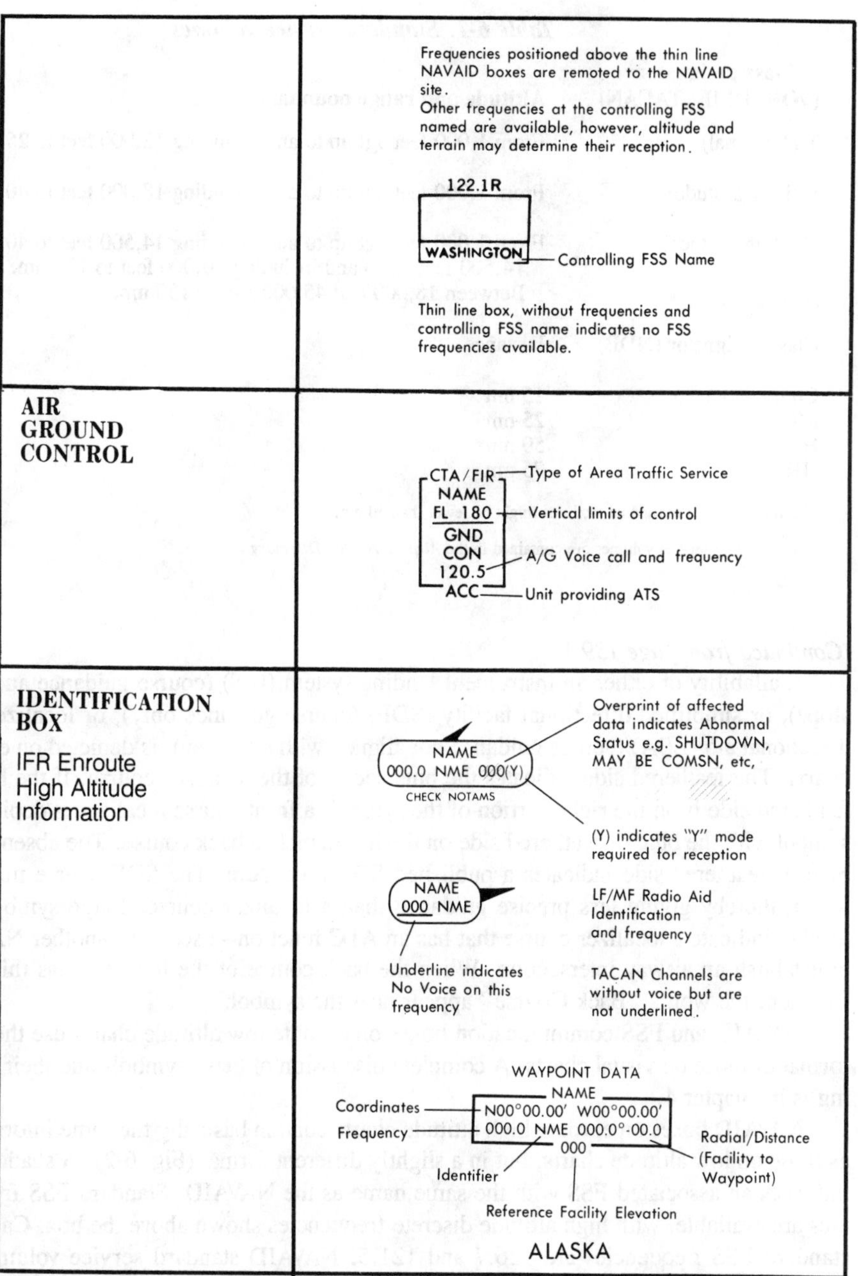

Fig. 6-2. Continued.

Table 6-1. Standard service volumes

Class designator (VOR, DME, TACAN)	Altitude and range boundaries
T (Terminal)	From 1,000 feet agl up to and including 12,000 feet to 25 nm.
L (Low altitude)	From 1,000 feet agl up to and including 18,000 feet to 40 nm.
H (High altitude)	From 1,000 feet agl up to and including 14,500 feet to 40 nm; 14,500 feet up to and including 60,000 feet to 100 nm. Between 18,000 and 45,000 feet to 130 nm.

Class designator (NDB)	Distance
Compass locator	15 nm
MH	25 nm
H	50 nm*
HH	75 nm

*Service ranges of individual facilities might be fewer than 50 nm.

Restrictions to service volumes are contained in the *Airport/Facility DIrectory*.

(Continued from page 139.)

Availability of either an instrument landing system (ILS) (course guidance and glide slope), or simplified directional facility (SDF) (course guidance only), or localizer-type directional aid (LDA) (course guidance not aligned with a runway), is depicted on enroute charts. The feathered side indicates the blue sector of the localizer course. If the blue or feathered side is on the right portion of the symbol, a front course localizer is depicted; a symbol with the blue or feathered side on the left depicts a back course. The absence of a blue or feathered side indicates a published SDF procedure. The SDF course might be wider, thereby giving less precise guidance than a localizer course. Large symbols are used to indicate a localizer course that has an ATC function—used with another NAVAID to establish an airway intersection. When the back course of the localizer has this ATC function, the words "Back Course" appear near the symbol.

NAVAID and FSS communication boxes on enroute low altitude charts use the same format as those on visual charts. A complete discussion of these symbols and their meaning is in chapter 4.

NAVAID boxes on enroute high altitude charts contain basically the same information as those on low altitude charts, but in a slightly different format (Fig. 6-2). A shadow box indicates an associated FSS with the same name as the NAVAID. Standard FSS frequencies are available, with high altitude discrete frequencies shown above the box. Canadian standard FSS frequencies are 126.7 and 121.5. NAVAID standard service volumes are published on these charts. Low (L) and terminal (T) NAVAIDs are indicated by the associated letter following the NAVAID three-letter identifier. The omission of a classification means the NAVAID is class H, high altitude. Overprinted data in the identification box

(Continued on page 146.)

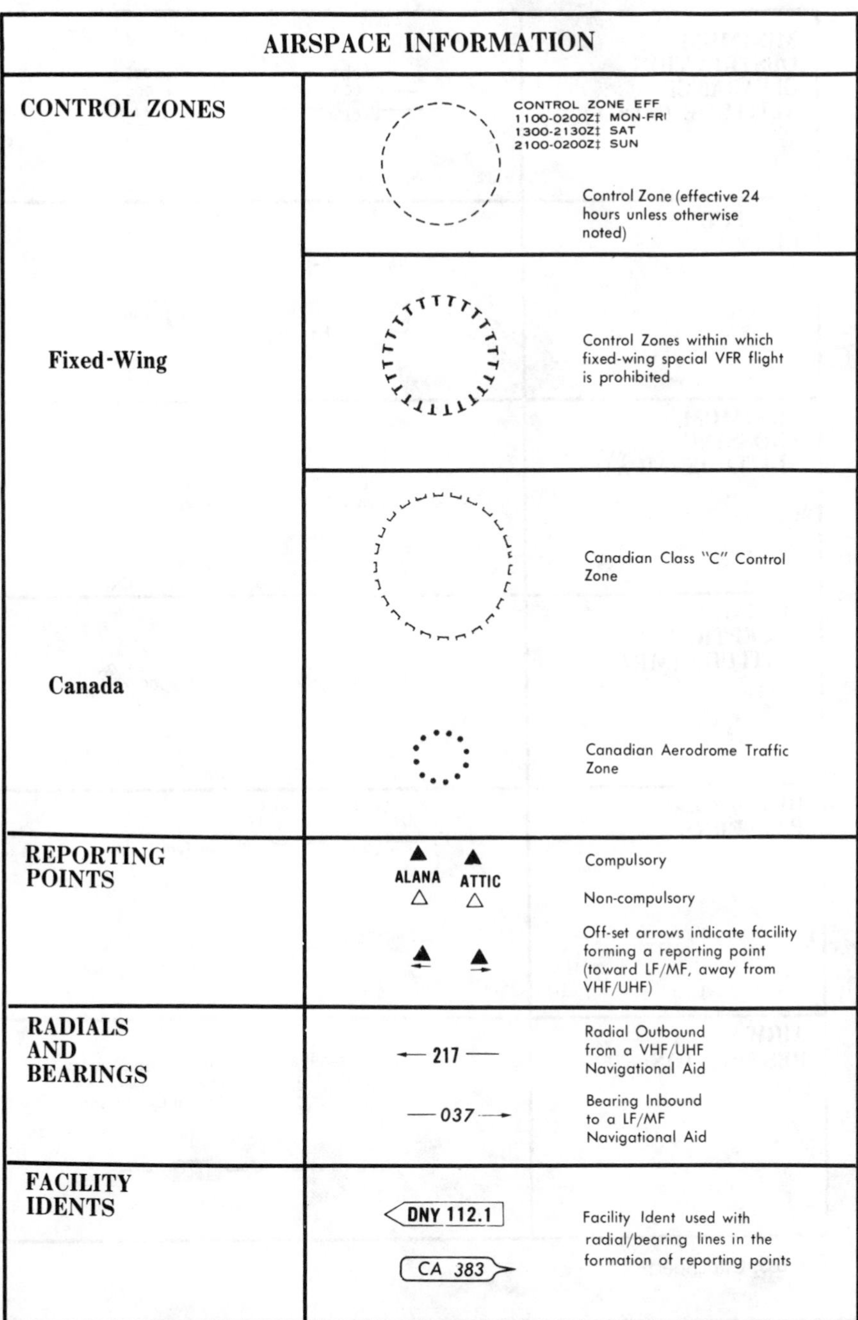

AIRSPACE INFORMATION
CONTROL ZONES
CONTROL ZONE EFF
1100-0200Z‡ MON-FRI
1300-2130Z‡ SAT
2100-0200Z‡ SUN
Control Zone (effective 24 hours unless otherwise noted)
Fixed-Wing
Control Zones within which fixed-wing special VFR flight is prohibited
Canada
Canadian Class "C" Control Zone
Canadian Aerodrome Traffic Zone
REPORTING POINTS
ALANA
ATTIC
Compulsory
Non-compulsory
Off-set arrows indicate facility forming a reporting point (toward LF/MF, away from VHF/UHF)
RADIALS AND BEARINGS
217
Radial Outbound from a VHF/UHF Navigational Aid
037
Bearing Inbound to a LF/MF Navigational Aid
FACILITY IDENTS
DNY 112.1
CA 383
Facility Ident used with radial/bearing lines in the formation of reporting points

Fig. 6-3. Radials and bearings are magnetic, distances nautical, and altitudes feet above mean sea level, unless otherwise noted, on an enroute chart.

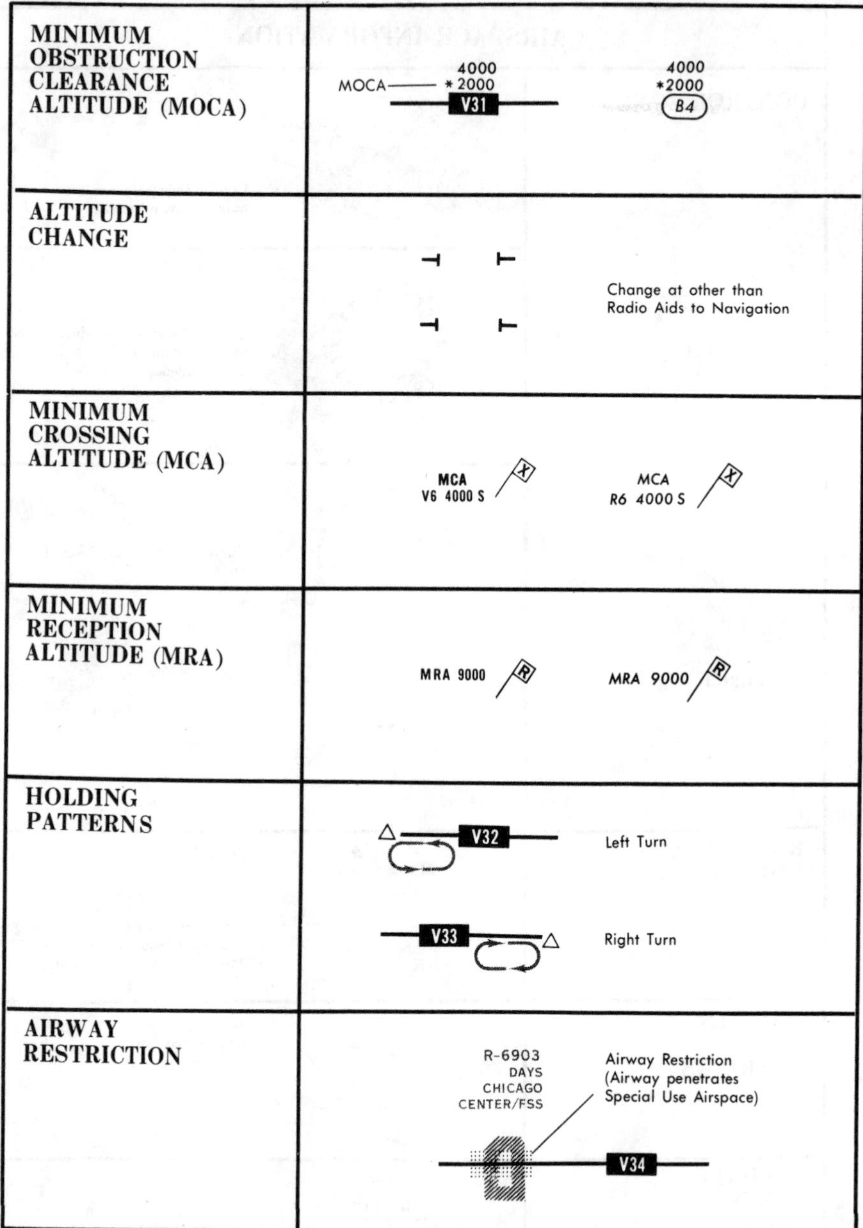

MINIMUM OBSTRUCTION CLEARANCE ALTITUDE (MOCA)
MOCA
4000
*2000
V31
4000
*2000
B4
ALTITUDE CHANGE
Change at other than Radio Aids to Navigation
MINIMUM CROSSING ALTITUDE (MCA)
MCA
V6 4000 S
X
MCA
R6 4000 S
X
MINIMUM RECEPTION ALTITUDE (MRA)
MRA 9000
R
MRA 9000
R
HOLDING PATTERNS
V32
Left Turn
V33
Right Turn
AIRWAY RESTRICTION
R-6903
DAYS
CHICAGO
CENTER/FSS
Airway Restriction
(Airway penetrates
Special Use Airspace)
V34

Fig. 6-3. Continued.

(Continued from page 144.)
indicates the facility might not be operating normally. That is, the facility might be shut down or has not yet been commissioned.

As on low altitude charts, the underlining of a channel means voice communications are not available. Waypoint data is provided in two formats. First the latitude/longitude

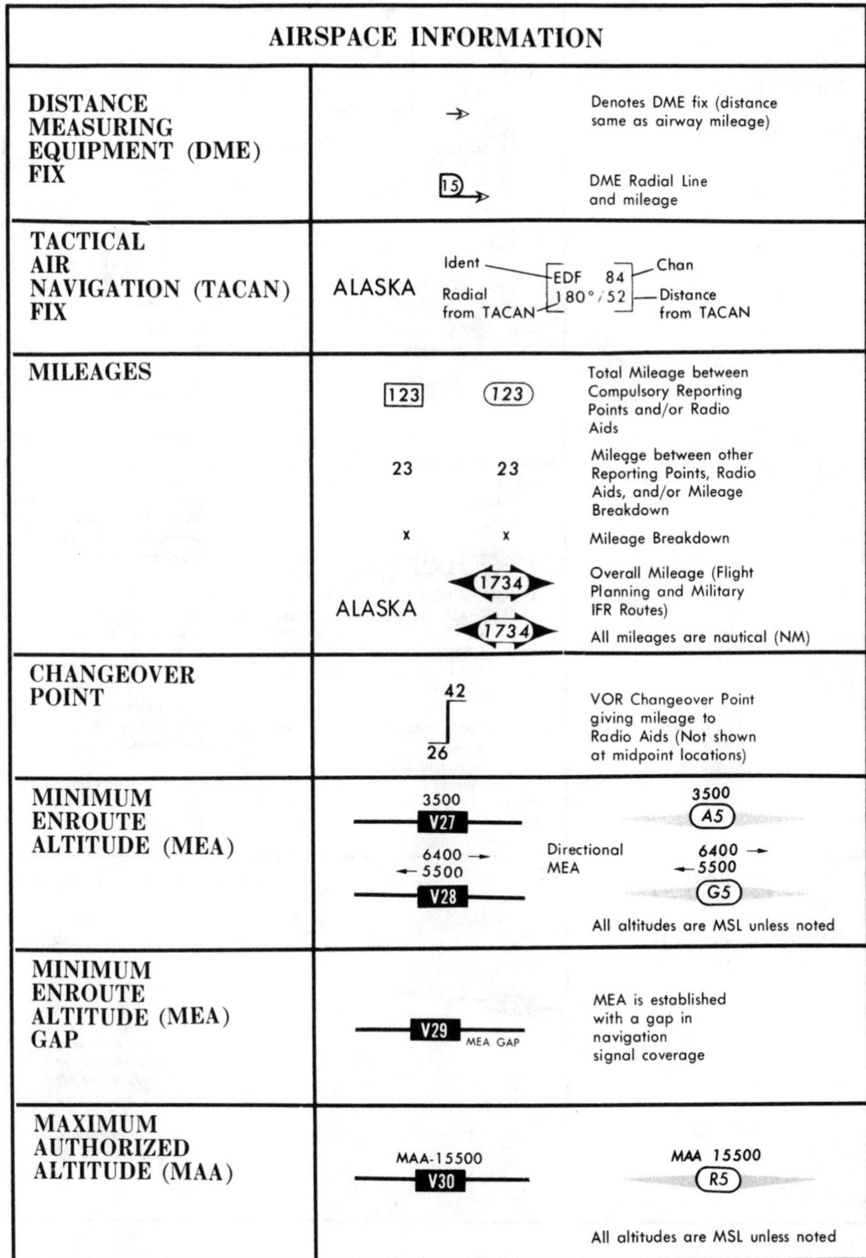

AIRSPACE INFORMATION
DISTANCE MEASURING EQUIPMENT (DME) FIX
Denotes DME fix (distance same as airway mileage)
15
DME Radial Line and mileage
TACTICAL AIR NAVIGATION (TACAN) FIX
ALASKA
Ident
Radial from TACAN
EDF 84
180°/52
Chan
Distance from TACAN
MILEAGES
123
123
Total Mileage between Compulsory Reporting Points and/or Radio Aids
23
23
Mileage between other Reporting Points, Radio Aids, and/or Mileage Breakdown
x
x
Mileage Breakdown
ALASKA
1734
1734
Overall Mileage (Flight Planning and Military IFR Routes)
All mileages are nautical (NM)
CHANGEOVER POINT
42
26
VOR Changeover Point giving mileage to Radio Aids (Not shown at midpoint locations)
MINIMUM ENROUTE ALTITUDE (MEA)
3500
V27
6400 →
← 5500
V28
Directional MEA
3500
A5
6400 →
← 5500
G5
All altitudes are MSL unless noted
MINIMUM ENROUTE ALTITUDE (MEA) GAP
V29
MEA GAP
MEA is established with a gap in navigation signal coverage
MAXIMUM AUTHORIZED ALTITUDE (MAA)
MAA-15500
V30
MAA 15500
R5
All altitudes are MSL unless noted

Fig. 6-3. Continued.

coordinates are provided. These are for users of coordinate area navigation (RNAV) systems (loran, inertial, and the like). VORTAC RNAV waypoints are also identified by the parent facility frequency, identifier, radial, and distance. Elevation of the parent reference facility is below the waypoint data box.

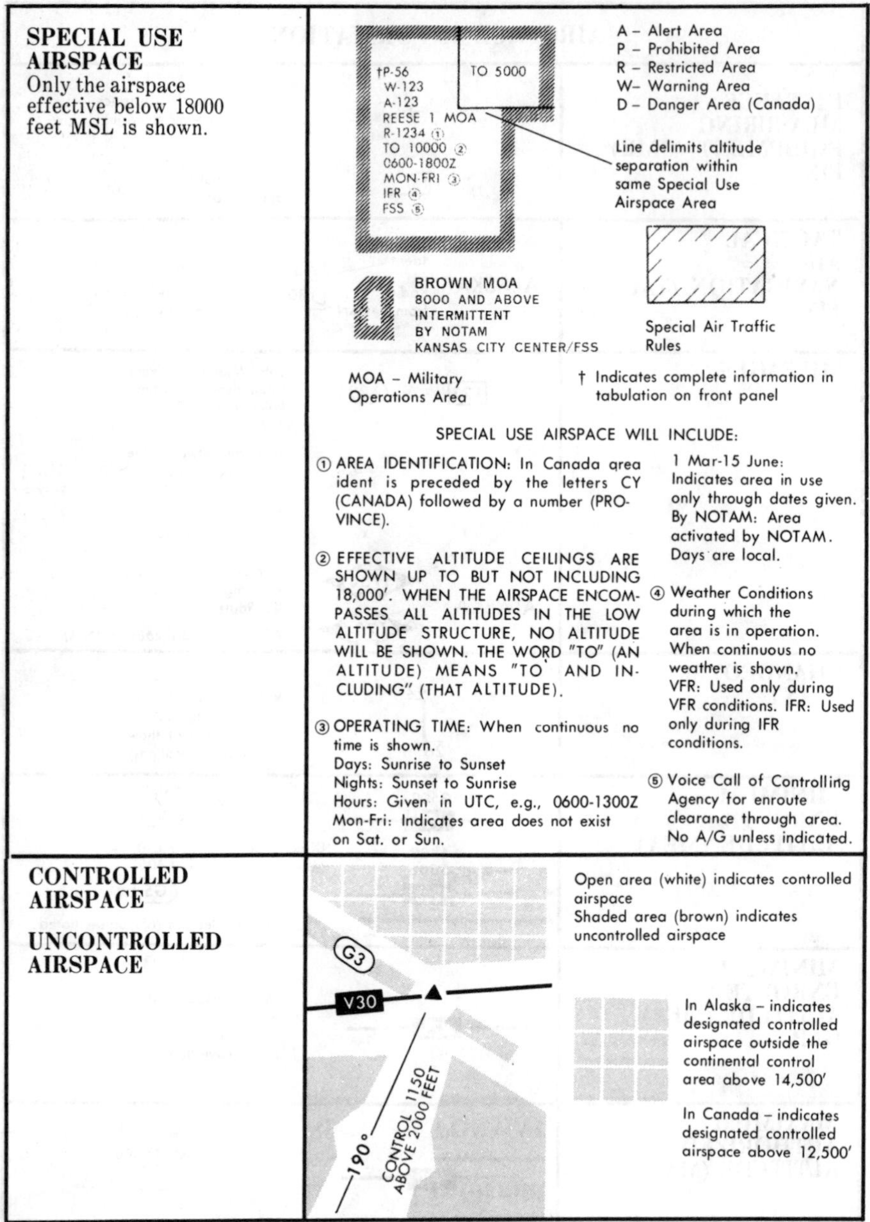

SPECIAL USE AIRSPACE
Only the airspace effective below 18000 feet MSL is shown.
†P-56
W-123
A-123
REESE 1 MOA
R-1234 ①
TO 10000 ②
0600-1800Z
MON-FRI ③
IFR ④
FSS ⑤
TO 5000
A – Alert Area
P – Prohibited Area
R – Restricted Area
W– Warning Area
D – Danger Area (Canada)
Line delimits altitude separation within same Special Use Airspace Area
BROWN MOA
8000 AND ABOVE
INTERMITTENT
BY NOTAM
KANSAS CITY CENTER/FSS
Special Air Traffic Rules
MOA – Military Operations Area
† Indicates complete information in tabulation on front panel
SPECIAL USE AIRSPACE WILL INCLUDE:
① AREA IDENTIFICATION: In Canada area ident is preceded by the letters CY (CANADA) followed by a number (PROVINCE).
② EFFECTIVE ALTITUDE CEILINGS ARE SHOWN UP TO BUT NOT INCLUDING 18,000'. WHEN THE AIRSPACE ENCOMPASSES ALL ALTITUDES IN THE LOW ALTITUDE STRUCTURE, NO ALTITUDE WILL BE SHOWN. THE WORD "TO" (AN ALTITUDE) MEANS "TO AND INCLUDING" (THAT ALTITUDE).
③ OPERATING TIME: When continuous no time is shown.
Days: Sunrise to Sunset
Nights: Sunset to Sunrise
Hours: Given in UTC, e.g., 0600-1300Z
Mon-Fri: Indicates area does not exist on Sat. or Sun.
1 Mar-15 June: Indicates area in use only through dates given.
By NOTAM: Area activated by NOTAM.
Days are local.
④ Weather Conditions during which the area is in operation. When continuous no weather is shown.
VFR: Used only during VFR conditions. IFR: Used only during IFR conditions.
⑤ Voice Call of Controlling Agency for enroute clearance through area. No A/G unless indicated.
CONTROLLED AIRSPACE
UNCONTROLLED AIRSPACE
G3
V30
190°
CONTROL 1150
ABOVE 2000 FEET
Open area (white) indicates controlled airspace
Shaded area (brown) indicates uncontrolled airspace
In Alaska – indicates designated controlled airspace outside the continental control area above 14,500'
In Canada – indicates designated controlled airspace above 12,500'

Fig. 6-3. Continued.

Airspace information

Figure 6-3 and Fig. 6-4 contain airspace information used on enroute charts. Again, many symbols are similar to those used on visual charts.

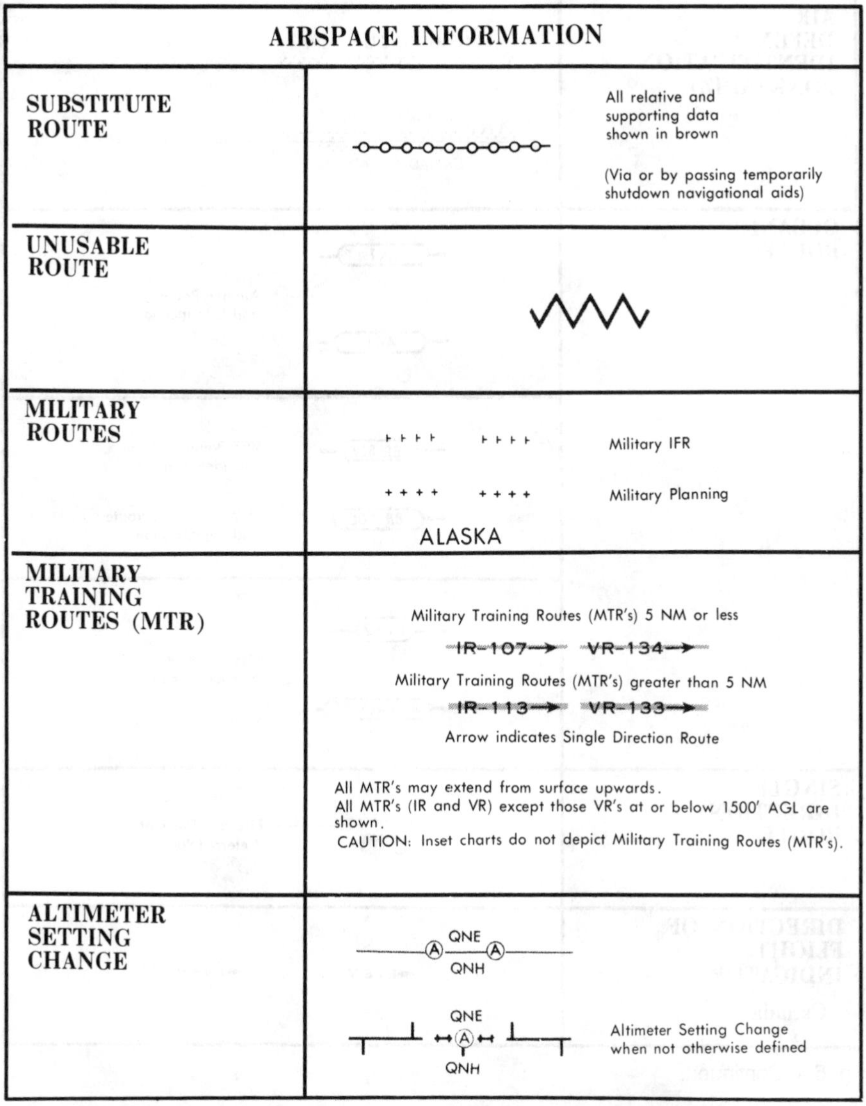

AIRSPACE INFORMATION	
SUBSTITUTE ROUTE	All relative and supporting data shown in brown (Via or by passing temporarily shutdown navigational aids)
UNUSABLE ROUTE	
MILITARY ROUTES	Military IFR Military Planning ALASKA
MILITARY TRAINING ROUTES (MTR)	Military Training Routes (MTR's) 5 NM or less IR-107→ VR-134→ Military Training Routes (MTR's) greater than 5 NM IR-113→ VR-133→ Arrow indicates Single Direction Route All MTR's may extend from surface upwards. All MTR's (IR and VR) except those VR's at or below 1500′ AGL are shown. CAUTION: Inset charts do not depict Military Training Routes (MTR's).
ALTIMETER SETTING CHANGE	QNE (A) (A) QNH QNE (A) QNH Altimeter Setting Change when not otherwise defined

Fig. 6-4. Center remote sites, along with the discrete frequencies, appear on enroute charts.

VOR radials are magnetic, depicted outbound or from the facility; L/MF bearings are magnetic, depicted inbound or the bearing to the facility. All distances are nautical and altitudes in feet above mean sea level (MSL), unless otherwise noted. Airway data, such as identifications, bearings or radials, mileages, and altitudes on low and high enroute charts are shown aligned with the airway and in the same color as the airway.

The minimum enroute altitude (MEA) indicates the minimum published altitude that assures acceptable navigational signal coverage, meets minimum obstruction clearance requirements between fixes, and ensures required radio communications. MEAs are sometimes different for opposite directions along an airway due to rising or lowering ter-

AIR DEFENSE IDENTIFICATION ZONE (ADIZ)	CONTIGUOUS U.S. ADIZ ALASKAN ADIZ CANADIAN ADIZ	Adjoining ADIZ
OCEANIC ROUTES	AR1 AR1	Atlantic Route and Identification
	BR 57V	VHF Bahama Route and Identification
	BR 10L	LF/MF Bahama Route and Identification
	B112 A15 ROUTE	Oceanic Route and Identification
SINGLE DIRECTION ROUTE	1000-0600Z V5	Effective Times of Preferred Route
DIRECTION OF FLIGHT INDICATOR **Canada**	←EVEN	←EVEN

Fig. 6-4. Continued.

rain. The primary enroute obstacle clearance area extends from one navigational facility to the next. The minimum obstacle clearance over areas not designated as mountainous under Federal Aviation Regulation (FAR) 95, "IFR Altitudes," will be 1,000 feet above the highest obstacle. Because of the Bernoulli effect and atmospheric eddies, vortices, waves, and other phenomena associated with strong winds over mountains, steep horizontal pressure gradients develop in these regions. Downdrafts and turbulence are also prevalent under these conditions, which create significant hazards to air navigation; therefore, minimum obstacle clearance over terrain designated as mountainous in FAR 95 will be 2,000 feet.

VHF airways have a width of four nautical miles (nm) each side of centerline to 51

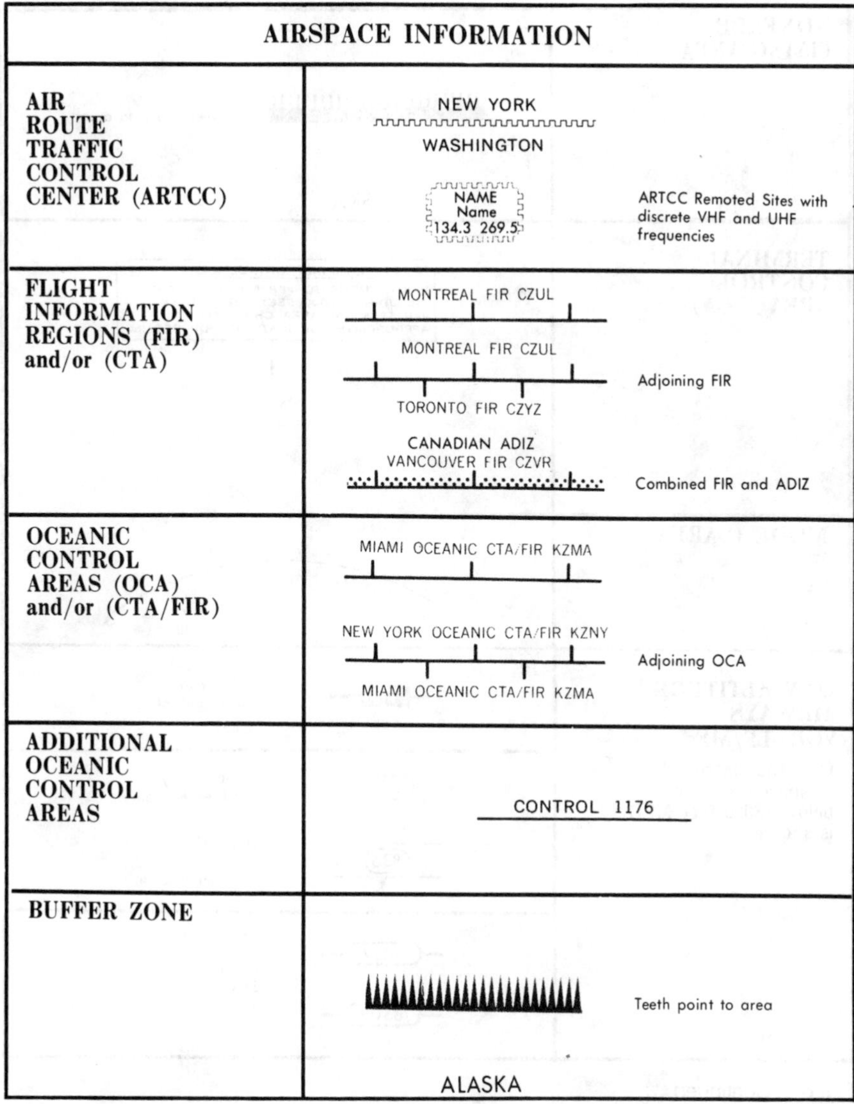

Fig. 6-4. Continued.

nautical miles (Fig. 6-5). If the distance from the facility to the changeover point is more than 51 nautical miles, the outer boundary of the primary area extends beyond four nautical miles diverging at an angle of $4^1/_2$° as shown in Fig. 6-5. This area provides obstruction protection based on system accuracy. That is, where the NAVAID is within tolerance, the pilot is flying the centerline, and the navigation receiver is within FAR limits.

Changeover points (COP) tell the pilot where to change from one navigational facility to the next. These COPs assure continuous reception of reliable navigation signal at the prescribed MEA. Where frequency interference or other navigation signal problems exist, the COP will be at the optimum location, taking into consideration signal strength, align-

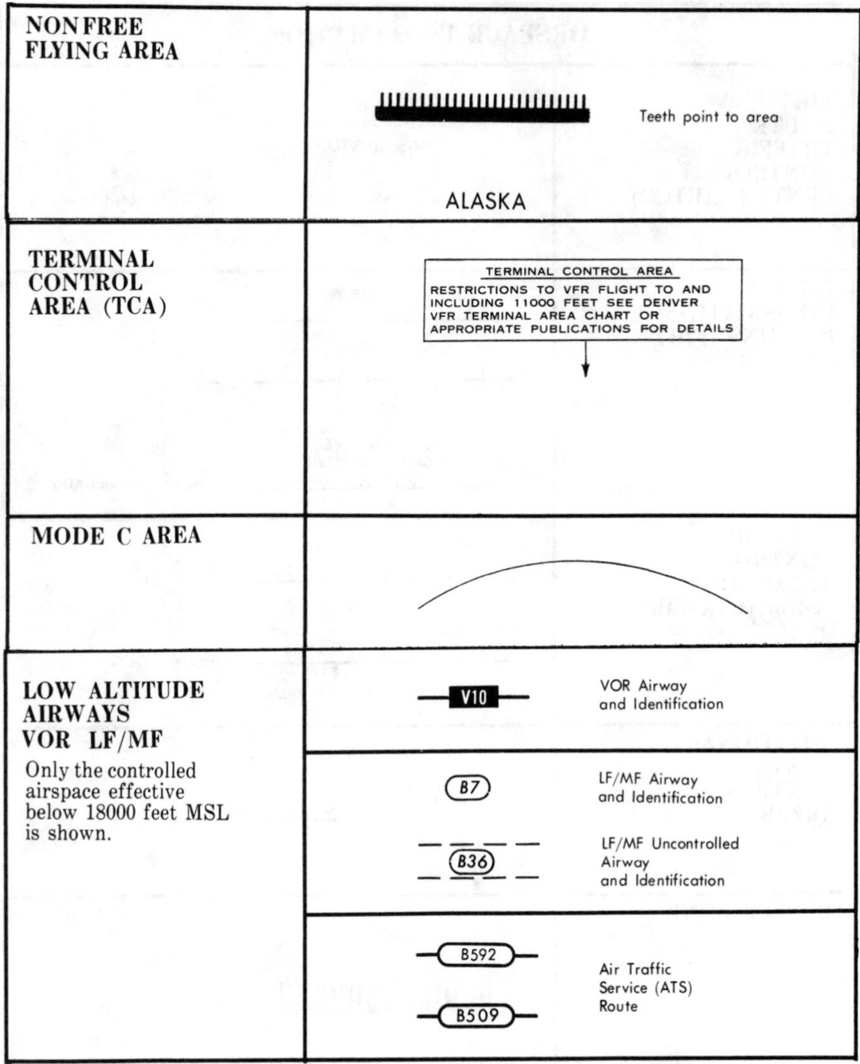

NONFREE FLYING AREA
Teeth point to area
ALASKA
TERMINAL CONTROL AREA (TCA)
TERMINAL CONTROL AREA
RESTRICTIONS TO VFR FLIGHT TO AND INCLUDING 11000 FEET SEE DENVER VFR TERMINAL AREA CHART OR APPROPRIATE PUBLICATIONS FOR DETAILS
MODE C AREA
LOW ALTITUDE AIRWAYS
VOR LF/MF
Only the controlled airspace effective below 18000 feet MSL is shown.
V10
VOR Airway and Identification
B7
LF/MF Airway and Identification
B36
LF/MF Uncontrolled Airway and Identification
B592
B509
Air Traffic Service (ATS) Route

Fig. 6-4. Continued.

ment error, or any other known condition that affects reception. Where signal coverage overlaps, the COP will normally be designated at the midpoint. The COP symbol will be omitted when the COP is at the midpoint of an airway segment.

Some airways have an MEA gap indicating that NAVAID signal coverage will be lost for a portion of the flight. Specific criteria exists for an airway or route segment to be designated with a navigational gap. The gap cannot exceed a distance that varies directly with altitude from zero at sea level to a maximum of 65 nautical miles at 45,000 feet. Not more than one gap can exist and the gap will not normally occur at a turning point. When a gap occurs, it will be identified by distances from the navigation facilities.

It is the pilot's responsibility to select altitudes that comply with obstruction clearance

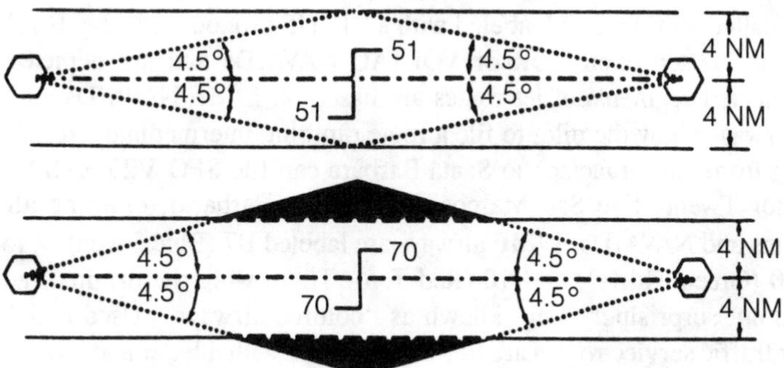

Fig. 6-5. Airway primary obstacle clearance areas extend four nautical miles each side of centerline to 51 nautical miles, then the area increases at an angle of 4.5 degrees.

requirements for instrument flight along routes not in controlled airspace, or routes for which no specified minimum IFR altitude has been established.

Minimum crossing altitude (MCA) points out an associated NAVAID or intersection that cannot be crossed below a specific altitude. MCAs are established where obstacles prevent a pilot from maintaining obstacle clearance during a normal climb to a higher MEA after the aircraft has passed the point where a higher MEA applies. Standards for determining MCA are based on the following climb rates, and computed from the flight altitude:

- Sea level through 5,000 feet: 150 feet per nautical mile
- 5,000 through 10,000 feet: 120 feet per nautical mile
- Above 10,000 feet: 100 feet per nautical mile

Some fixes have more than one MCA depending upon the direction of flight.

Minimum obstruction clearance altitude (MOCA) meets obstruction clearance criteria between fixes, but only assures navigational signal coverage within 22 nautical miles of the NAVAID. A MOCA is shown directly below the MEA and is identified by an asterisk. The designation of a MOCA indicates that a higher MEA has been established for that particular airway or segment because of signal reception requirements. When no MOCA appears, the MEA and MOCA are considered to be the same.

Maximum authorized altitude (MAA) is the highest altitude, for which an MEA is designated, where adequate NAVAID signal coverage is assured. For example, an MAA will be established for route segments where interference from VOR signals on the same frequency prevent reliable navigation.

Minimum reception altitude (MRA) denotes the lowest altitude required to receive adequate signals to determine specific fixes. Often DME can be used to identify the fix, in which case the MRA would not apply because the off-airway facility would not be needed to establish the intersection. Pilots must use caution, if the aircraft is not DME equipped, the pilot might not be able to establish a required fix.

VOR routes on low altitude charts are designated victor airways, labeled with a "V" (V31 pronounced "Victor thirty-one"), and depicted in blue; except in Mexican airspace on NOS charts where L/MF airways are charted in blue. VOR airways on high altitude charts are called jet routes and labeled with a "J" (J5 pronounced "Jay Five" not jet five). Jet routes are based upon VOR or VORTAC NAVAIDs and are depicted in black. In Alaska, selected segments of jet routes are based on L/MF NAVAIDs and are shown in brown. Airways allow the pilot to file a route omitting intermediate fixes. For example, a pilot filing from San Francisco to Santa Barbara can file SFO V25 RZS SBA (San Francisco Victor Twenty-five San Marcos direct Santa Barbara), omitting all intermediate intersections and NAVAIDs. L/MF airways are labeled B7 (Blue Seven), A15 (Amber Fifteen), G30 (Green Thirty), or R10 (Red Ten). These were the original radio navigation routes and not surprisingly were known as "colored airways." Used mainly for oceanic routes, air traffic service routes are normally labeled with a letter and a number, similar to L/MF airways (R464, B592, and the like). VHF/UHF oceanic airways are depicted in blue, L/MF airways in brown.

Special use airspace (SUA) below 18,000 feet MSL is shown on low altitude charts; high altitude charts show SUA above 18,000 feet. Open areas (white) indicate controlled airspace, shaded areas (brown) indicate uncontrolled airspace on low and high altitude charts.

Altimeter settings to be used are designated by the code letters QNH and QNE. QNH is the altitude above mean sea level displayed on the altimeter when the altimeter setting window is set to the local altimeter setting. This is the setting used in the U.S. when flying below 18,000 feet. QNE is pressure altitude. This is the altitude shown on the altimeter with the altimeter set to 29.92 inches (1013.2 millibars). This is the setting used in the U.S. when flying at or above 18,000 feet.

Air route traffic control center (ARTCC) remote sector discrete frequencies are depicted within a communications box. The name of the center appears at the top, with the center's sector name in the middle. Specific frequencies are displayed at the bottom of the box (VHF for civilian use; UHF for the military).

Figure 6-6 illustrates miscellaneous culture, hydrography, and navigational and procedural information symbols. Cultural symbols identify international, convention or mandate line, and dateline boundaries. Isogonic lines and values, time zones, and a translation of Morse code signals are provided. Shorelines are indicated by a water vignette; smaller scale area charts indicate a shoreline with a broken blue line. Match marks show chart overlap.

IFR ENROUTE CHARTS

IFR enroute charts consist of planning charts, low and high altitude charts, and area charts. IFR planning charts provide the IFR pilot with essentially the same information as VFR planning charts provide the VFR pilot. Low altitude charts can be compared with sectionals, high altitude charts with WACs, and area charts with TACs. Each is designed to provide enough detail to allow the pilot to operate safely in that particular environment.

Planning charts

IFR planning charts consist of the IFR wall planning chart, the flight case planning chart, and the IFR enroute high altitude planning chart-U.S. Additionally, NOS publishes North Atlantic and North Pacific route charts. The IFR wall planning chart is the reverse

CULTURE	
BOUNDARIES **International**	— — — — International Boundary (Omitted when coincident with ARTCC or FIR)
Convention or Mandate Line	— — — — U.S. – Russia Convention Line of 1867 ALASKA
Date Line	INTERNATIONAL DATE LINE MONDAY •• SUNDAY ALASKA
NAVIGATIONAL AND PROCEDURAL INFORMATION	
ISOGONIC LINE AND VALUE Isogonic lines and values shall be based on the five year epoch chart.	7°E
TIME ZONE	Central Std +6=UTC ⋮ Eastern Std +5=UTC ‡DURING PERIODS OF DAYLIGHT SAVING TIME (DT) EFFECTIVE HOURS WILL BE ONE HOUR EARLIER THAN SHOWN. ALL STATES OBSERVE DT EXCEPT ARIZONA AND THAT PORTION OF INDIANA IN THE EASTERN TIME ZONE. All time is Coordinated Universal (Standard) Time (UTC)
MORSE CODE	A .– F ..–. K –.– P .––. B –... G ––. L .–.. Q ––.– C –.–. H .... M –– R .–. D –.. I .. N –. S ... E . J .––– O ––– T – U ..– 1 .–––– 6 –.... V ...– 2 ..––– 7 ––... W .–– 3 ...–– 8 –––.. X –..– 4 ....– 9 ––––. Y –.–– 5 ..... 0 ––––– Z ––..

Fig. 6-6. Enroute charts only depict cultural features significant to IFR operations, such as international, convention or mandate, and dateline boundaries.

HYDROGRAPHY	
SHORELINES	Water Vignette
NAVIGATIONAL AND PROCEDURAL INFORMATION	
CRUISING ALTITUDES	0° → 179°M: IFR ODD Thousands; VFR or ON TOP ODD Thousands Plus 500′ 180° → 359°M: IFR EVEN Thousands; VFR or ON TOP EVEN Thousands Plus 500′ VFR above 3000′ AGL unless otherwise authorized by ATC IFR outside controlled airspace IFR within controlled airspace as assigned by ATC All courses are magnetic
ENLARGEMENT AREA	DETROIT AREA CHART A-1
MATCH MARKS	ALASKA
NOTES	FAA AIR TRAFFIC SERVICE OUTSIDE US AIRSPACE IS PROVIDED IN ACCORDANCE WITH ARTICLE 12 AND ANNEX 11 OF ICAO CONVENTION. ICAO CONVENTION NOT APPLICABLE TO STATE AIRCRAFT BUT COMPLIANCE WITH ICAO STANDARDS AND PRACTICES IS ENCOURAGED.
Warning	WARNING UNLISTED RADIO EMISSIONS FROM THIS AREA MAY CONSTITUTE A NAVIGATION HAZARD OR RESULT IN BORDER OVERFLIGHT UNLESS UNUSUAL PRECAUTION IS EXERCISED.

Fig. 6-6. Continued.

side of the VFR wall planning chart discussed in chapter 4. In addition to the features of the VFR wall planning chart, the IFR wall planning chart contains ARTCC boundaries and an index of enroute low altitude charts. Indexes, in light gray, show coverage of associated IFR enroute low altitude charts, Caribbean and S. American charts, and Canadian charts. The flight case planning chart, also discussed in chapter 4, contains all the features of the IFR wall planning chart.

Symbols used on the IFR wall planning chart are standard, with the following exceptions. Airports shown have a minimum 5,000 foot hard-surface runway. Those in green have approved instrument approach procedures. Airports in blue, in addition to a low altitude instrument approach, have a high altitude or jet penetration procedure for military aircraft. VHF/UHF NAVAIDs are depicted using standard symbols. Low altitude NAVAIDs are shown with a black screen, which makes the facility appear light gray. NDBs are depicted as a dot within a circle, UHF beacons in black, and L/MF beacons in brown. Airways with an MEA of 10,000 feet or higher are shown as a wide line, as opposed to those with an MEA lower than 10,000 feet depicted with a thin line.

Symbols used on the flight case planning chart are standard, with exceptions; airports with an approved instrument approach procedure are annotated with a "(P)" and a red underline indicates a flight service station located at the airport.

The IFR enroute high altitude planning chart-U.S. is the same size and scale as the flight case planning chart. It is designed for preflight and inflight planning of IFR flights higher than 18,000 feet. This chart is the first in NOAA's new series of multicolored enroute charts. The chart is revised every 56 days. Features include:

- Published jet routes
- Mileages
- Radio aids to navigation, high and low frequency
- NAVAIDs
- Airports with a minimum of 5,000 feet of hard-surface runway
- Special use airspace above 18,000 feet
- ARTCC boundaries
- Isogonic lines
- Time zone boundaries
- NAVAID location index
- Special use airspace tables
- Indexes of some related chart products

Figure 6-7 contains airspace information unique to the IFR enroute high altitude planning chart. This consists of high altitude, oceanic, single direction, and bypass routes, and control area boundary symbols. Note the bypass route symbol. It indicates a route that bypasses a facility that is not part of that specific route. This could be important in flight planning. Should a pilot file the route and include the bypassed facility, the ARTCC computer will reject the flight plan.

North Atlantic and North Pacific route charts are available from NOS. These multicolored charts are designed for the monitoring of oceanic flights by air traffic controllers. They may also be used by pilots for planning transoceanic flights. Charts are revised every 24 weeks.

AIRSPACE INFORMATION		
HIGH ALTITUDE ROUTES Only the controlled airspace effective at and above 18,000 feet MSL is shown.	J520	Jet Route and Identification
	J591 / J592	Canadian Jet Route and Identification
OCEANIC ROUTES	AR8	Atlantic Route and Identification
	BR22V	VHF Bahama Route and Identification
	BR18L	LF/MF Bahama Route and Identification
	B724 / UB 724	VHF/UHF Oceanic Route and Identification
	A23	LF/MF Oceanic Route and Identification
SINGLE DIRECTION ROUTE		Preferred Route
BYPASS ROUTE		Jet Route centerline bypassing a facility which is not part of that specific route
UPPER INFORMATION REGIONS (UIR)	MONTERREY UTA/UIR MMTY	
	MONTERREY UTA/UIR MMTY / MAZATLAN UTA/UIR MMZT	Adjoining UTA/UIR
UPPER CONTROL AREAS (UTA)	MAZATLAN FIR/UIR MMZT	Combined FIR and UIR

Fig. 6-7. Airspace information unique to the enroute high altitude planning chart consists of high altitude, oceanic, single direction, and bypass routes, and control area boundary symbols.

The North Pacific route chart is a multicolored composite chart covering the North Pacific area. The chart has a scale of 1:12,000,000 (1 inch equals 165 nautical miles), sheet size is approximately 60 × 43 inches unfolded. Four larger scale area charts are in this series. The area charts cover four quadrants of the North Pacific for flight planning. Area charts have a scale of 1:7,000,000 (1 inch equals 96 nautical miles), and unfolded are the same size as the composite chart. Figure 6-8 shows an excerpt from the North Pacific route chart composite depicting the routes from the west coast to Hawaii. Features include:

- Selected ATS routes
- NAVAIDs and reporting points with geographic coordinates
- International boundaries
- USSR and USSR dominated areas
- U.S.-Russia Convention Line of 1867
- Buffer zones and nonfree flying areas
- Air defense identification zones
- International dateline
- Aerial refueling tracks
- Special use airspace
- Airports of entry
- Mileage circles

The North Atlantic route chart is available in two sizes. Full size the scale is 1:5,500,000 (1 inch equals 75 nautical miles). The chart is 58 × 41 inches unfolded. A half-size version is also available with a scale of 1:11,000,000 (1 inch equals 151 nautical miles): unfolded 29 × 21 inches and folded standard 5 × 10 inches. Features include:

- Selected ATS routes
- Oceanic control areas
- NAVAIDs and reporting points, with geographic coordinates
- North Atlantic/minimum navigation performance specifications area
- Air defense identification zones
- Airports of entry
- Flight information region (FIR) boundaries
- Shorelines
- International boundaries
- Special use airspace

Enroute low altitude charts

Enroute low altitude charts are designed to provide navigation information for IFR flights below 18,000 feet MSL. Contiguous U.S. scale varies from 1:583,307 (1 inch equals 8 nautical miles) to 1:1,458,267 (1 inch equals 20 nautical miles). The scale of the individual charts depends on the amount of information to be displayed. In Alaska, due to the large area and relatively small amount of information depicted, charts have a scale of 1:2,187,402 (1 inch equals 30 nautical miles). All charts can be folded to the standard 5 × 10 inches. Revision cycle is 56 days. Low altitude enroute charts are labeled "L" with a

(Continued on page 162.)

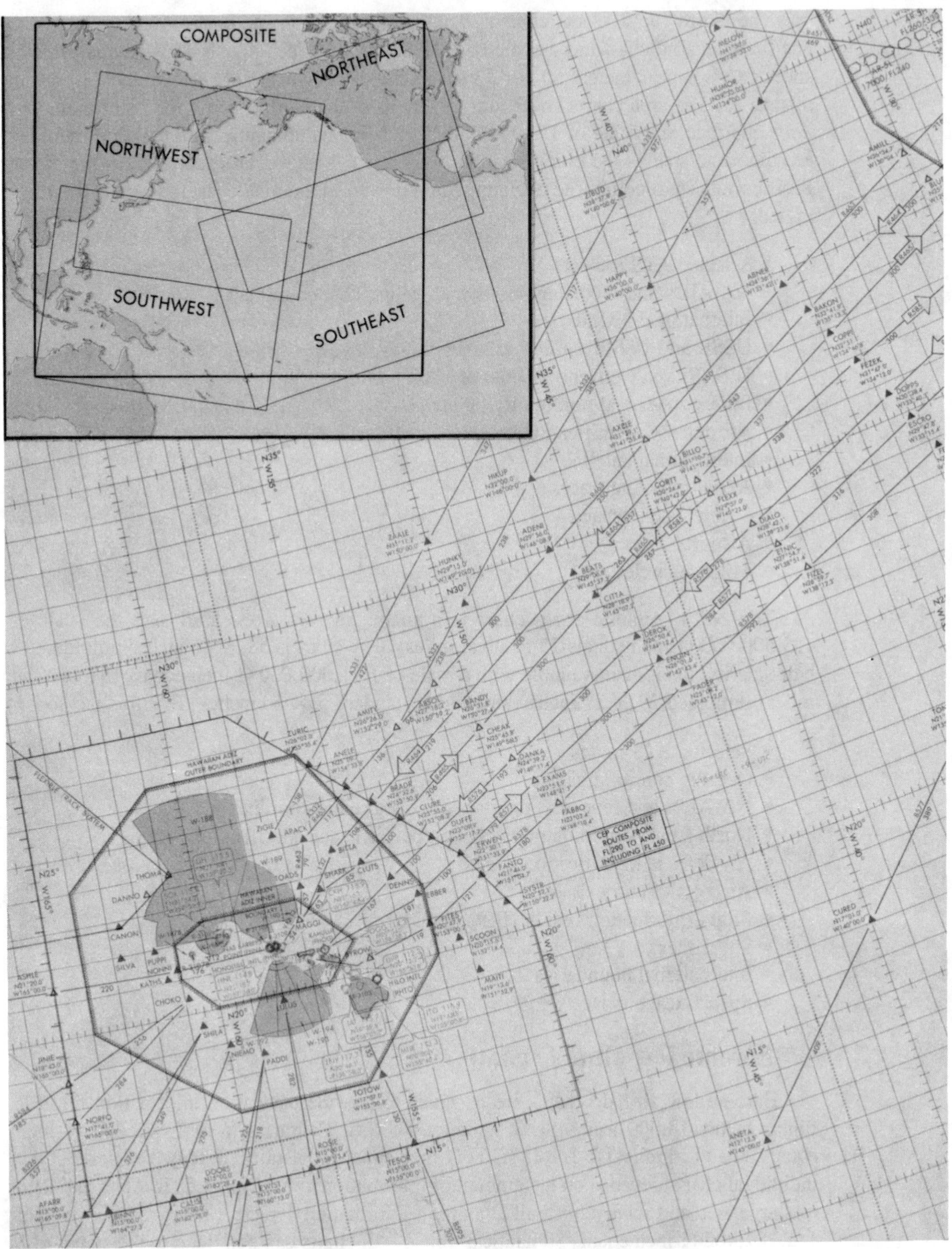

Fig. 6-8. North Pacific and North Atlantic route charts are designed for the monitoring of oceanic flights by ATC and may be used by pilots for planning transoceanic operations.

Fig. 6-8. Continued.

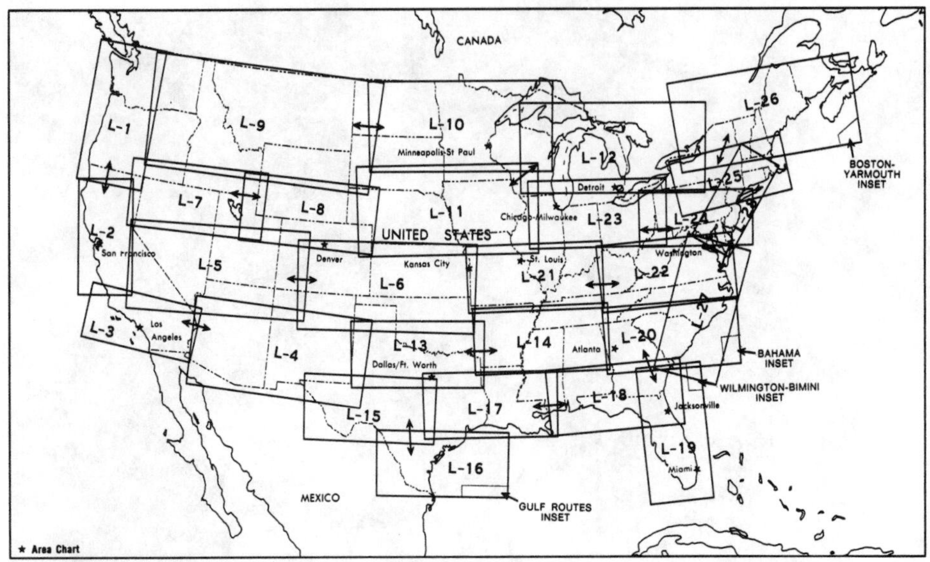

Fig. 6-9. The contiguous Unites States is covered by a series of 28 enroute low altitude charts, printed on 14 sheets.

(Continued from page 159.)
number (L-13). Coverage for the contiguous U.S. is contained in Fig. 6-9. Coverage for Alaska is represented in Fig. 6-10. Features include:

- Low altitude airways
- Limits of controlled airspace
- Radio aids to navigation
- Selected airports
- Route altitude descriptions (MEAs, MOCAs, MCAs)
- Airway distances
- Reporting points
- Special use airspace
- Military training routes
- Operational notes
- Adjoining chart numbers
- Outlines of area chart coverage

Area charts are larger-scale representations of congested terminal areas. Scale varies from 1:364,567 (1 inch equals 5 nautical miles) to 1:583,307 (1 inch equals 8 nautical miles). NOS area charts are distributed on a single chart sheet. The chart folds to the standard 5 × 10 inches, with a revision cycle of 56 days. End panels contain the standard air to ground communication frequencies listing for the areas covered. Area charts are available for the following locations, which are also indicated by the star symbols with associated city names in Fig. 6-9:

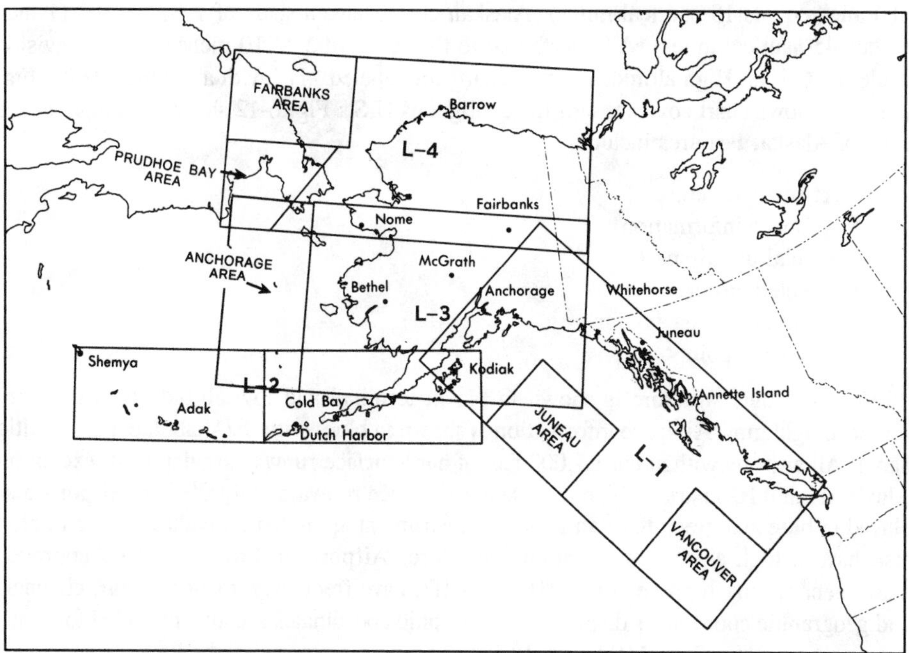

Fig. 6-10. Due to the large area and relatively small amount of information related to Alaska, all four charts have the same scale.

- San Francisco
- Los Angeles
- Denver
- Minneapolis/St. Paul
- Dallas/Fort Worth
- Kansas City
- Chicago/Milwaukee
- Detroit
- St. Louis
- Atlanta
- Jacksonville
- Washington
- Miami

Enroute high altitude charts

Enroute high altitude charts are designed to provide navigation information for IFR flights at and above 18,000 feet MSL. Except to effect transition within or between route structures, the altitude limit for victor airways should not be exceeded (below 18,000 feet). Pilots planning flights at or above 18,000 feet should file appropriate direct or jet routes for operations within this stratum. Charts for the contiguous U.S. have a scale of 1:2,187,402 (1 inch equals 30 nautical miles, except for the H-6 chart, which has a scale of 1 inch equals 18 nautical miles). Alaskan charts have a scale of 1:3,281,102 (1 inch equals 45 nautical miles). All charts fold to the standard 5 × 10 inches, with a revision cycle of 56 days. High altitude enroute charts are labeled "H" with a number (H-4). Figure 6-11 shows chart coverages for the contiguous U.S.; Fig. 6-12 shows coverage for the state of Alaska. Features include:

- Jet route structure
- Airspace information
- Special use airspace
- Selected airports
- Radio aids to navigation
- Reporting points

The jet route structure is shown in black; terminal and low altitude NAVAIDs are screened light gray. Airspace information is shown in blue, with SUA tabulated on the title panel. All airports with at least 5,000 feet of hard-surface runway are depicted, except on Alaska H1 and H2 charts, which include hard-surface runways of 4,000 feet. Airports displayed in blue and green have an approved instrument approach procedure, those in blue also have a high altitude penetration procedure. Airports in brown have no approved instrument approach procedure. VHF NAVAIDs have frequency, identification, channel, and geographic coordinates displayed. Geographic coordinates are also provided for compulsory reporting points. MEAs are 18,000 feet, unless otherwise shown.

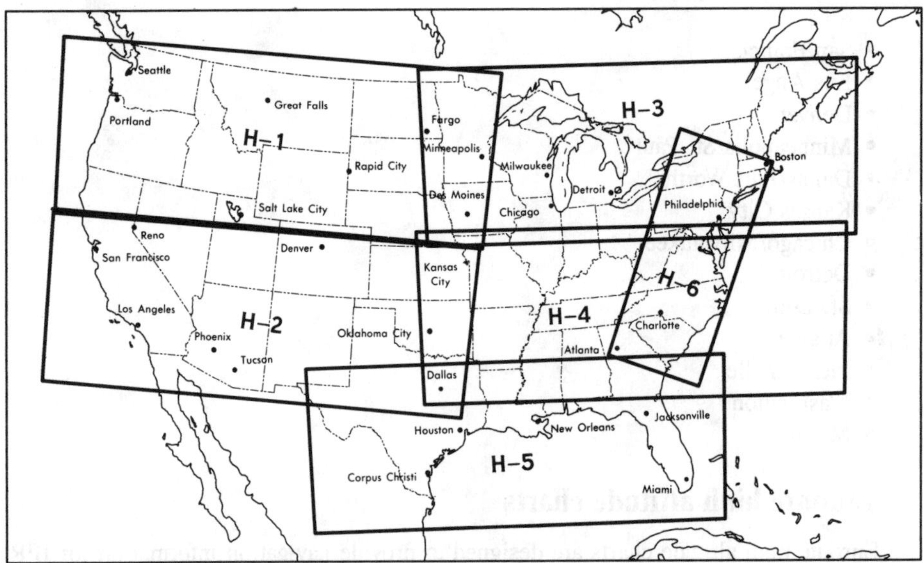

Fig. 6-11. The contiguous United States is covered by a series of six enroute high altitude charts, printed on three sheets.

USING ENROUTE IFR CHARTS

A pilot's first task is to ensure currency. NOS IFR charts are published every 56 days. Changes that occur between publication cycles are distributed in the form of a change

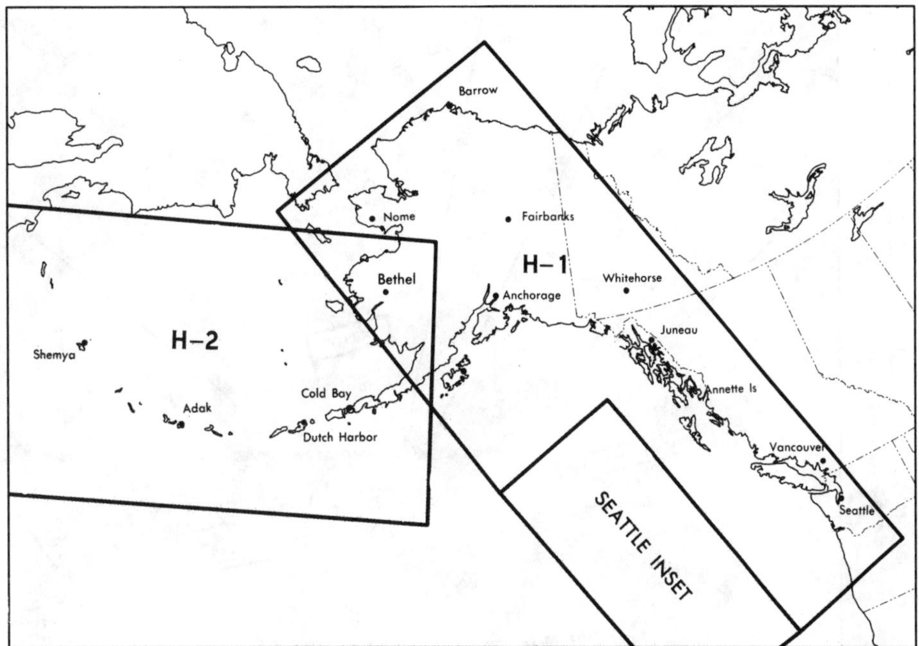

Fig. 6-12. Alaskan enroute high altitude charts provide coverage for the state, West Coast of Canada, and Northwest United States.

notice volume, NOTAM publication, and National Flight Data Center (NFDC) NOTAM. NAVAID and airport restriction, changes, or outages, might also appear in these publications, or be published as NOTAM (D)s. Publication, sources, and availability of these products is discussed in chapter 2. Pilots, especially those using NOS products, must be aware of these limitations. It might be a good idea to review the currency portion of chapter 2 before proceeding.

Enroute low and high altitude chart cover panels identify the chart, provide effective times and date, and contain other useful information, such as applicable MTR tables, and SUA data. Low altitude charts provide a tabulation of airport and terminal communication frequencies. High altitude chart cover panels contain discrete high altitude flight watch frequencies for their area of coverage. Inside panels contain a brief chart legend. Chart margins provide a mileage scale, panel labels (panel E, F, G, and the like) with associated major city name, and pertinent remarks (the name of the next fix off the chart, or "Overlaps chart Nr L-5"). The A/FD references airport and NAVAID locations to their associated low and high altitude chart and panel. For example, the directory for the Santa Rosa, California, airport lists: H2G, L2G. This airport is located on the enroute high altitude chart number 2, panel G (H2G), and the enroute low altitude chart number 2, panel G (L2G).

Figure 6-13 contains a portion of the enroute low altitude L-2 chart. The margin identifies this as the "G" panel associated with the city of Sacramento, California. A mileage scale is provided and the remark: "Overlaps DOD PAA chart Nr-1." The L-2 overlaps Department of Defense, enroute Pacific, Australasia and Antarctica chart number 1.

(Continued on page 168.)

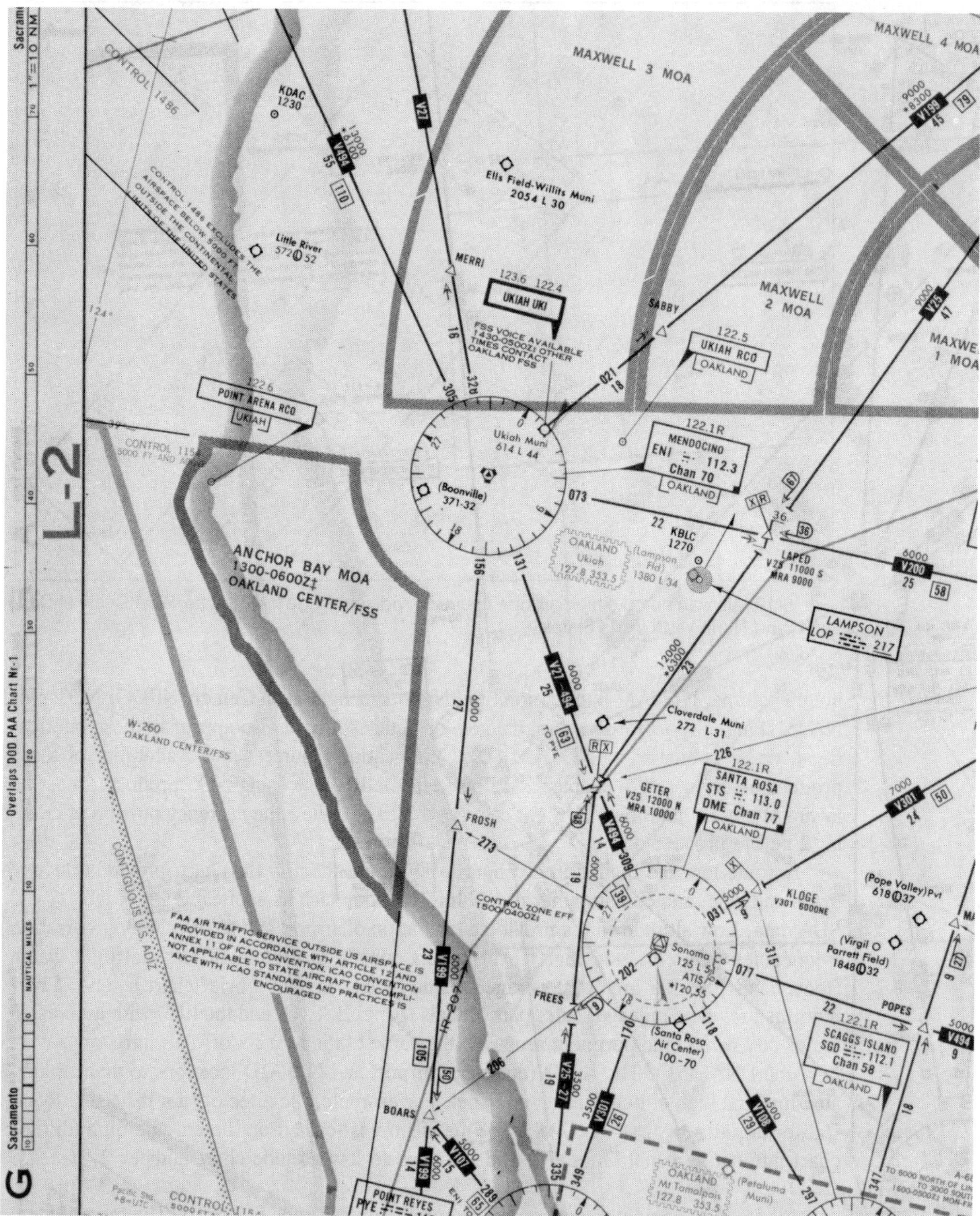

Fig. 6-13. Airport symbols on enroute low altitude charts depict facilities, including the availability of approach procedures.

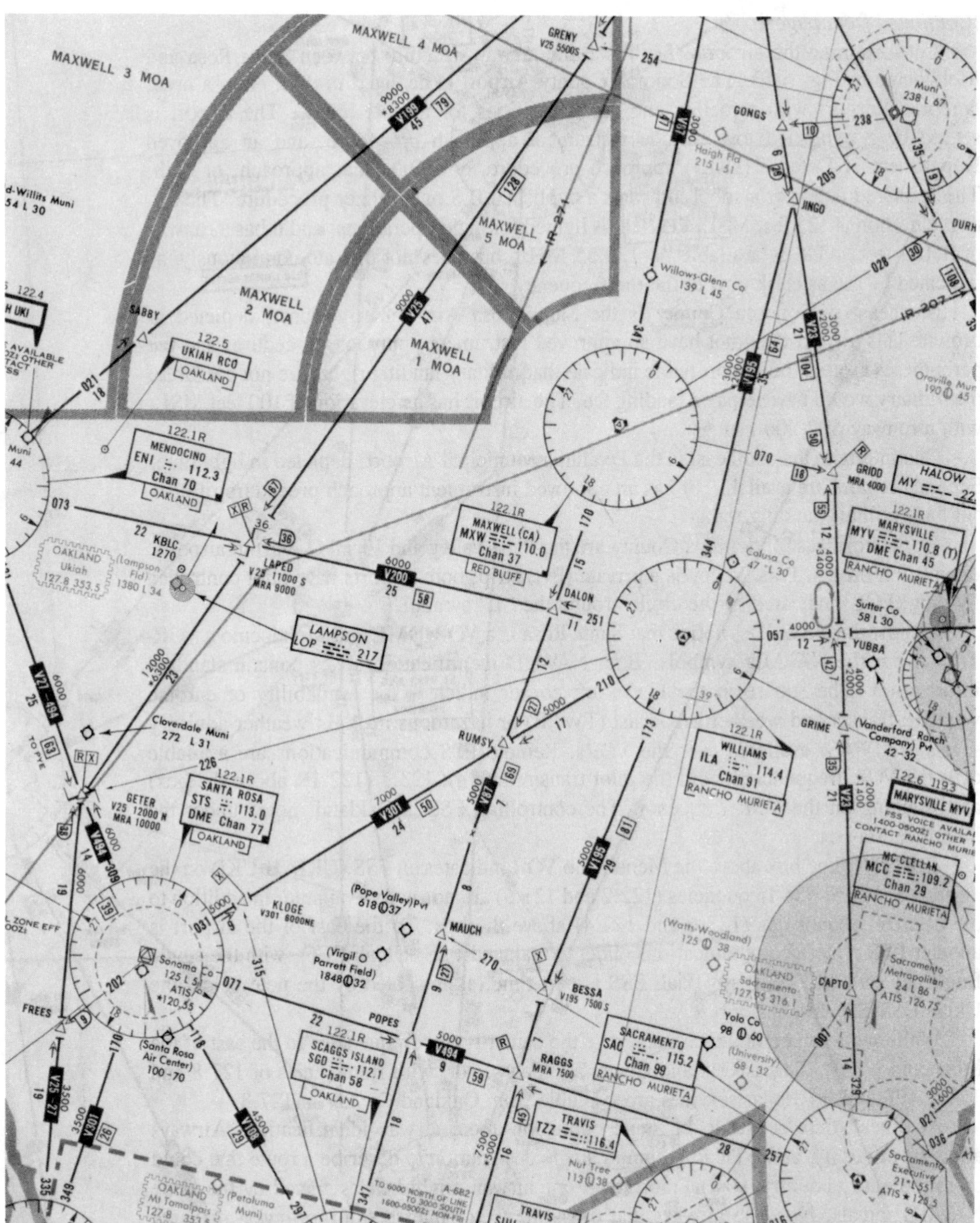

Fig. 6-13. Continued.

(Continued from page 165.)

Let's examine the airports, facilities, and airway structure between Santa Rosa and Mendocino in Fig. 6-13. The Sonoma County Airport is depicted in dark blue, a civil, landplane airport with refueling and repair facilities for normal traffic. The airport is served by a published low altitude instrument approach procedure, and an approved Department of Defense (DOD) approach procedure, or DOD radar approach, or both. The feathered localizer symbol indicates a published ILS or localizer procedure. The airport elevation is 125 feet MSL, the field is lighted for night operations, and it has a runway of 5,100 feet. ATIS is available on 120.55 MHz, but does not operate continuously, as indicated by the asterisk next to the the frequency.

Southeast of Sonoma County is the Santa Rosa Air Center Airport, depicted in brown. This airport does not have an approved instrument approach procedure, and the parentheses around the airport name indicate that military landing rights are not available, the military would have to pay a landing fee. The airport has an elevation of 100 feet MSL, with a runway of 7,000 feet.

Continuing to the southeast is the Petaluma Municipal Airport, depicted in light blue. Fuel and repairs are available; it has an approved instrument approach procedure; it does not have military landing rights.

To the northeast of Sonoma County are the Pope Valley and Virgil O. Parrett airports, depicted in brown. Pope Valley is a private (Pvt) strip; both airports have pilot controlled lighting (PCL) indicated by the circle around the "L" symbol.

Amidst the NAVAIDs, notice that Santa Rosa is a VOR/DME and Mendocino a VORTAC, from the NAVAID symbols. Both NAVAID identification boxes contain standard information. The square in the lower left corner indicates the availability of enroute weather: transcribed weather broadcast (TWEB) or hazardous in-flight weather advisory service (HIWAS) available over the VORs. Remote FSS communications are available over the VOR frequencies, with the pilot transmitting on 122.1 (122.1R above the box) and listening on the VOR frequency. The controlling FSS is Oakland, noted under the identification box.

The heavy line box above the Mendocino VOR indicates an FSS (UKIAH UKI) on the airport. Standard FSS frequencies (122.2 and 121.5) are normally available, in addition to the discrete frequencies (123.6 and 122.4) above the box. To the east of the airport is another FSS remote communications outlet to Oakland FSS, Ukiah RCO, with the single frequency of 122.5 because Ukiah FSS is part-time, as indicated by the note below the Ukiah UKI FSS frequency box.

While we're on communications, note the center sector frequencies to the east of the Mendocino VOR. This is Oakland Center's Ukiah sector, with frequencies of 127.8 and 353.5. IFR or VFR radar services are available from Oakland Center on 127.8.

At times, there seems to be some confusion about airway identification. Airways allow a pilot to fly, and ATC to assign, a single designation to describe a route that could cross the entire country. Often, more than one airway might be designated by the same NAVAID radials. Note the airways that cross the Geter intersection northwest of Santa Rosa. The airway between Santa Rosa (STS) and Mendocino (ENI) is designated V494. After the Geter intersection, V27 also overlies the same radials as V494 (V27-494). Where did V27 come from? Airways V25 and V27 come from the south to the Geter intersection. At Geter, V25 goes off to the north and V27 to Mendocino. Northwest of Mendocino, V27 and V494 separate. This is important during flight planning, especially for pilots using DUATs. The airway between STS and ENI is V494, not V27 or V27-494. Neither V27 nor V27-494 will be accepted during computer flight planning.

Airway information interpretation is straightforward. The airway from STS to Geter is made up of the STS 309 radial, MEA is 6,000 feet, and the distance is 14 nautical miles. Geter is the COP, as indicated by the COP symbol at the intersection. The airway from Geter to ENI is the ENI 131 radial, MEA is the same, and distance is 25 nautical miles. The airway contains an MEA change bar at Geter on V25, in this case the change, represented by the T is quite significant, from 6,000 feet to 12,000 feet. The chart also provides total distances between NAVAIDs. The number 39 in a box within the STS VOR compass rose represents the distance between STS and ENI. What about the 63 in a box just to the northwest of Geter? As the note states, 63 miles is the distance between ENI and the PYE (Point Reyes) VOR, or the total distance between NAVAIDs on V27. Total distance between NAVAIDs on V25 is found along V25 north away from Geter in the box showing 128 nautical miles. Also note the MEA north of Geter on V25 is 12,000 feet. The number below, with the asterisk, 6,300 feet is the MOCA. Do not expect navigational signal coverage at the MOCA, this portion of V25 is beyond 22 nautical miles from the NAVAID.

Geter intersection has MCA and MRA flags. Geter has an MCA on V25 of 12,000 feet northbound. Using VORs only, the Geter intersection is designated as any one of the four airway radials and the Williams 226 radial. The MRA flag warns that the lowest altitude to establish the intersection using VOR radials is 10,000 feet. How else could a pilot establish this intersection? Geter can be determined using DME from STS (14 nautical miles), ENI (25 nautical miles), or PYE (38 nautical miles)— the arrows on either side of Geter, or the DME arrow box south of the intersection. How would a pilot establish Geter without DME, below 10,000 feet? The pilot's only option would be timing, and although not as accurate as DME or VOR, this would be acceptable for determining Geter as the COP. A pilot flying with VOR receivers only could not hold, nor would ATC assign a hold, at the Geter intersection, below 10,000 feet.

A portion of enroute low altitude chart L-17 offers additional examples of chart elements (Fig. 6-14). Airspace in the Houston area is dominated by a TCA and its 30-nautical-mile mode C veil. Houston Intercontinental (IAH) and Hobby (HOU), as well as many other airports in the area are served by ILSs; none of the precision landing systems establish airway intersections.

Note that the VORTAC at IAH is named Humble and the VORTAC at HOU is named Hobby, but with the identifier HUB. Recall that many NAVAIDs in the vicinity of airports were often given the same name and identifier as the nearby airport. Computerization has impacted this practice. To ensure accurate flight data processing and eliminate ambiguity, NAVAIDs that are not collocated with airports will have different identifiers. Some still remain, but all will eventually be changed.

Above the Hobby NAVAID box is the frequency 122.35. Because the suffix "R" is missing, the associated FSS (Montgomery County) has transmit and receive capability on this frequency. Voice receive capability in the aircraft is not available on the VOR frequency because the VOR frequency is underlined, indicating no communication through the VOR. Refer to the Trinity VOR/DME NAVAID box east of Houston. There is no frequency above the box, but Montgomery County appears below the box. The Montgomery

(Continued on page 172.)

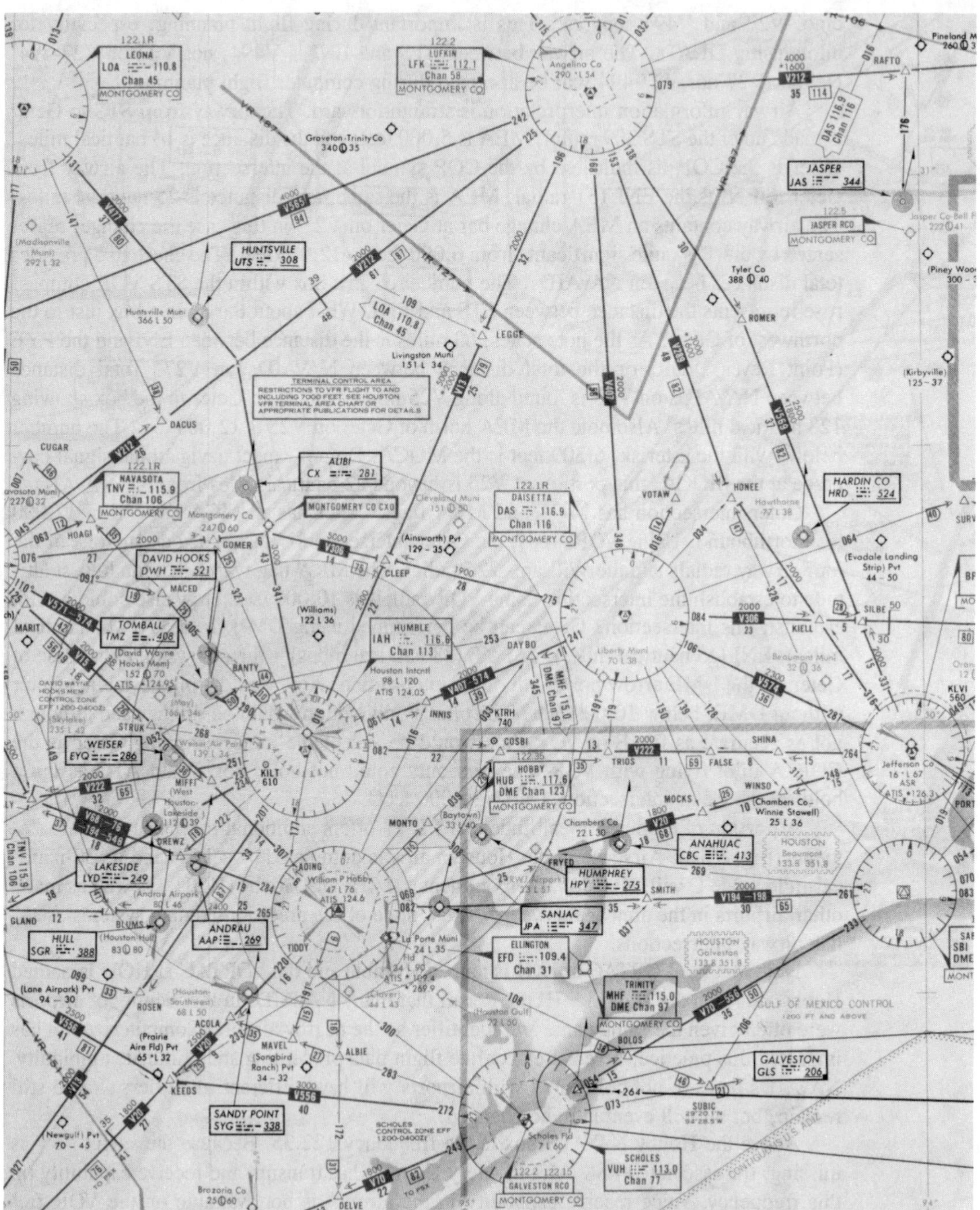

Fig. 6-14. Enroute low altitude charts provide navigational information for flights below 18,000 feet.

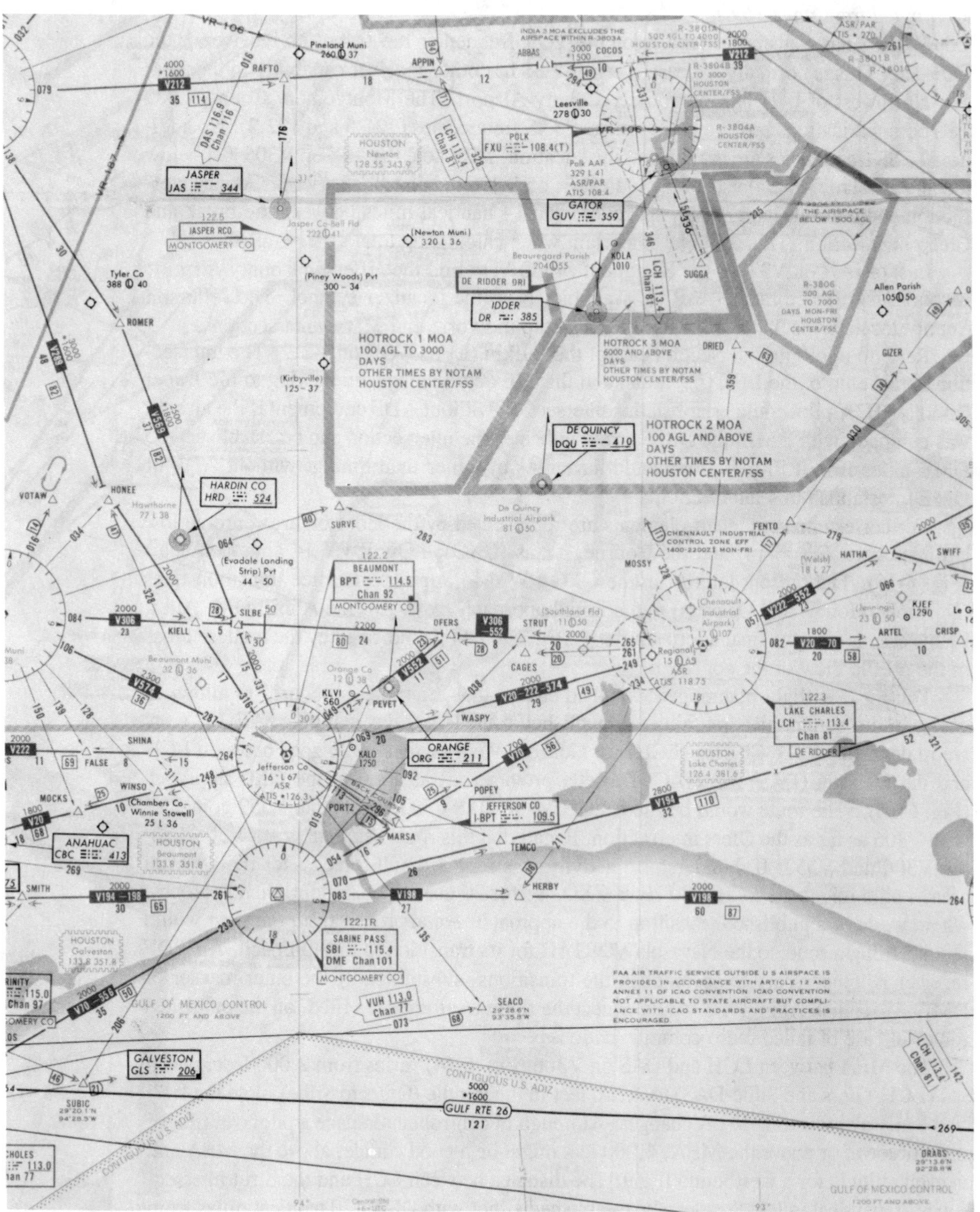

Fig. 6-14. Continued.

(Continued from page 169.)
County FSS has transmit capability on the VOR frequency 115.0, but no receiver at the VOR site. This is normally the case when an FSS has other receiver capability in the area.

Northwest of IAH is Montgomery County Airport. The Montgomery County CXO FSS communications box indicates standard FSS frequencies (121.5 and 122.2). No additional FSS frequencies are available at this location. Also note the "x" on V306 just below the airport. This is a mileage break indicator; on V306 the distance between Gomer intersection and the mileage break is six miles, and 14 nautical miles between the break and Cleep intersection. This could be important when calculating total airway mileage.

East of IAH on V222 is the Beaumont (BPT) VOR and the Jefferson County Airport. Airport surveillance radar (ASR) is available. Note the localizer symbol. The Jefferson County localizer back course is used to make up the Portz and Marsa intersections.

Refer to the Rafto intersection east of the Lufkin (LFK) VOR on V212. The intersection is made up of the LFK 079 radial and the 176 degree magnetic bearing to the Jasper (JAS) NDB. A pilot could establish this intersection without ADF equipment if the aircraft was equipped with DME; the DME arrow indicates the intersection can be established 35 DME miles from LFK. A pilot would have no way, other than timing, without ADF or DME to establish this intersection.

Further examples of chart elements are represented by the Houston area enroute low altitude chart (Fig. 6-14) and the Conroe, Texas (CXO), LOC RWY 14 approach chart (Fig. 6-15). The present interest in the CXO RWY 14 approach is three transition fixes (Navasota, Hobby, and Daisetta) to the initial approach fix (IAF), the Alibi LOM. Alibi, in this case is also the final approach fix (FAF). The FAF is indicated by the maltese cross in the profile view of the approach.

During IFR flight planning we want to ensure that the route is planned to allow, in case of radio communication failure, a published route to our destination's FAF. Let's say we're departing Lake Charles (LCH) for Conroe. We note that V306 goes between LCH and the Daisetta (DAS) VORTAC. Daisetta provides a transition to the CXO approach (Fig. 6-15). (The route would be filed: LCH V306 DAS CXO.) Note that V552 coincides with V306 as far as the Ofers intersection. Recall that this route segment is not V306-552, but V306 and V552. If thunderstorm activity was reported and forecast for the Daisetta area, a pilot might file LCH V20 HUB CXO. Victor 20 goes to Beaumont, then to Hobby, where we have a published transition to the approach. An approach from the west would require filing a route to the Navasota VORTAC for its transition to the approach. An ADF receiver is unnecessary to accomplish the transitions, substituted by the outer marker or DME to establish Alibi. ATC would expect the flight to progress as filed, on the published route, in case of failed radio communications.

The MEA between LCH and DAS on V306 (Fig. 6-14) varies from 2,000 feet at each end (LCH-Ofers and Silbe-DAS) to 2,200 feet in the middle (Ofers to Silbe); note the "T" on the airway indicating MEA changes. Although in controlled airspace a pilot can file for any altitude at or above the MEA, 4,000 feet might be a good choice, above the MEA and an even altitude for a westbound flight. The distance between LCH and the Strut intersection is 20 nautical miles. Strut can be established either with DME (20 nautical miles from LCH) or the Sabine Pass (SBI) 038 radial. Ofers, the next intersection along the route, is eight nautical miles from Strut, and made up of the 28 DME from LCH or the Beaumont 049 radial. The total distance between LCH and DAS is 80 nautical miles. Stop navigating

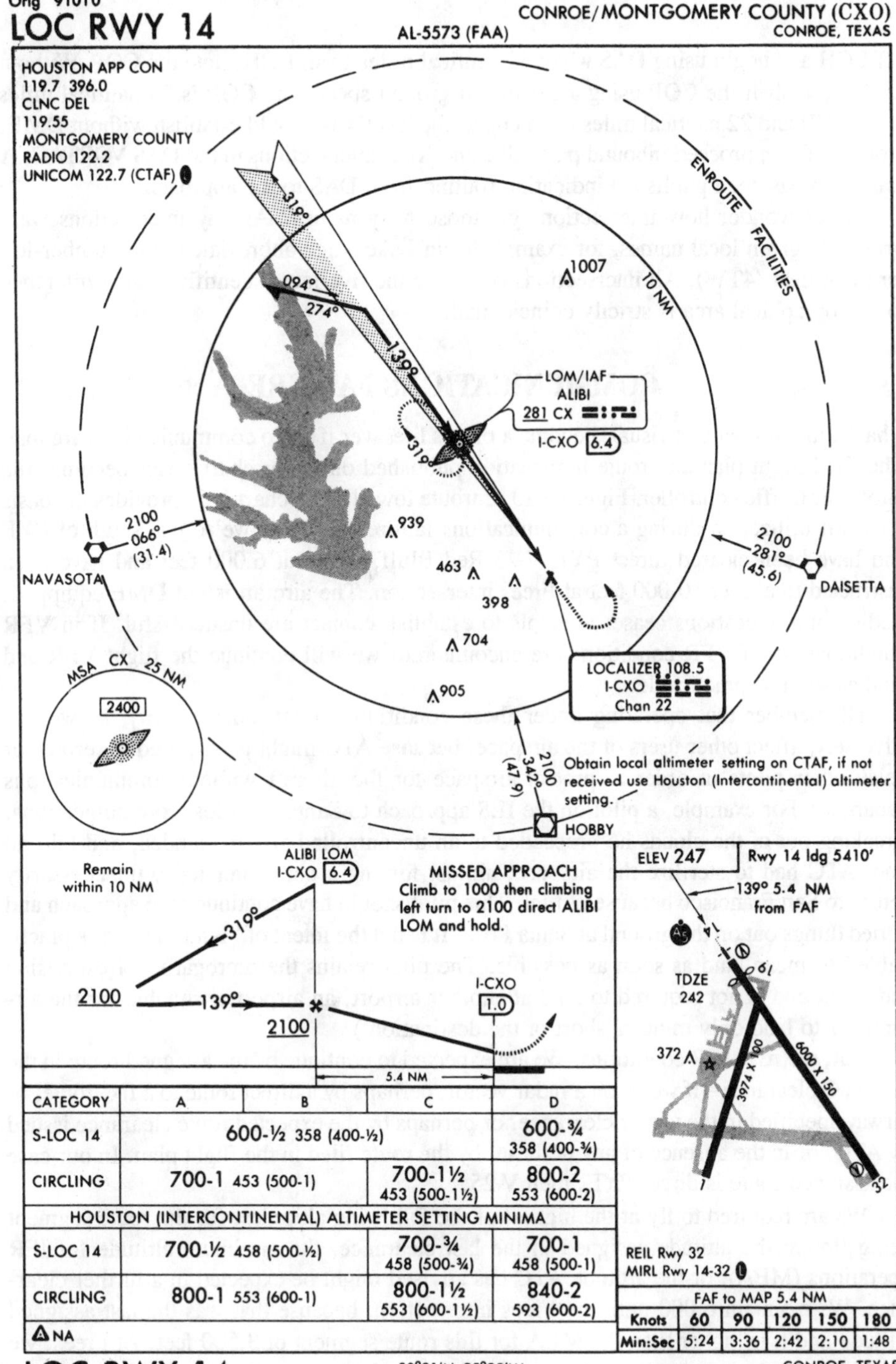

CATEGORY	A	B	C	D
S-LOC 14	600-½ 358 (400-½)			600-¾ 358 (400-¾)
CIRCLING	700-1 453 (500-1)		700-1½ 453 (500-1½)	800-2 553 (600-2)
HOUSTON (INTERCONTINENTAL) ALTIMETER SETTING MINIMA				
S-LOC 14	700-½ 458 (500-½)		700-¾ 458 (500-¾)	700-1 458 (500-1)
CIRCLING	800-1 553 (600-1)		800-1½ 553 (600-1½)	840-2 593 (600-2)

FAF to MAP 5.4 NM					
Knots	60	90	120	150	180
Min:Sec	5:24	3:36	2:42	2:10	1:48

Fig. 6-15. The Conroe LOC RWY 14 approach contains three transitions from the enroute structure to the approach procedure.

off LCH and begin using DAS when 50 nautical miles from LCH, note the COP. Without DME, establish the COP using a calculated ground speed: the COP is 50 nautical miles from LCH and 22 nautical miles from Ofers, the last fix we could establish without DME; from the COP, proceed inbound past Silbe and Kiell intersections to the DAS VORTAC. A transition has been published indicating routing from DAS to the approach.

(Ever wonder how intersections get those nifty names? Airway intersections were originally given local names, for example Twin Lakes, and abbreviated with number-letter identifiers (4TW). All intersections now have the five-letter identifiers. Any reference to a geographical area is strictly coincidental.)

COMMUNICATIONS FAILURE

Charts, instrument and visual, become a pilot's lifesaver if radio communications are lost. The filed flight plan and route information published on either chart series becomes the pilot's air traffic controller. Figure 6-13, enroute low altitude chart L-2, provides the basis for chart utilization during a communications failure. Let's say we're just south of PYE and have been cleared direct PYE V-25 Red Bluff, we're at 6,000 feet and have been instructed to expect 10,000 feet at Frees intersection. The aircraft is not DME equipped. Radio communications cease; attempts to establish contact are unsuccessful. If in VFR conditions, or if VFR conditions are encountered, we will continue the flight VFR and land as soon as practicable.

(Remember that operating under these conditions might unnecessarily, as well as adversely, affect other users of the airspace, because ATC might be required to reroute or delay other pilots in order to protect airspace for the aircraft without communications capability. For example, a pilot on the ILS approach to Santa Rosa lost communications. Breaking out of the clouds he proceeded to an uncontrolled airport, landed, and told no one. ATC had to sterilize the airspace, almost forcing a commuter jet to unnecessarily return to San Francisco because of fuel. This pilot should have continued the approach and sorted things out on the ground at Santa Rosa. It is not the intent of "land as soon as practicable" to mean land as soon as possible. The pilot retains the prerogative of exercising judgment and is not required to land at another airport, an airport unsuitable for the aircraft, or to land only minutes short of the destination.)

But, we're in IFR conditions. We are expected to continue by the assigned route in the last ATC clearance—if we're on a radar vector, perhaps by a direct route to a fix, route, or airway specified in the vector clearance, or perhaps by the expected route clearance issued by ATC, or in the absence of any of these, by the route filed in the flight plan. In our case the assigned route is direct PYE, then V-25.

We are required to fly at the highest of the following altitudes, for the route segment being flown: the altitude assigned in the last clearance, the minimum altitude for IFR operations (MEA), or the altitude ATC has advised might be expected in a further clearance. We maintain 6,000 until the Frees intersection, because that was the last assigned altitude and it is higher than the MEA for this route segment of 3,500 feet. At Frees, we were instructed to expect 10,000, so we begin a climb to that altitude at that intersection. Now, what about at Geter? The MCA northbound on V-25 is 12,000 feet. Upon reaching Geter we are expected to enter a standard holding pattern and climb to cross Geter at 12,000 feet, and maintain that altitude, the MEA for that route segment. At the Laped intersection, the MEA drops to 9,000 feet. Crossing Laped, ATC expects us to descend to the last assigned altitude, or the MEA, whichever is higher, in this case 10,000 feet.

Let's say we're southbound on V-25 north of Laped, assigned 9,000 feet. From the MCA and MRA flags at Laped we see the MRA is 9,000 feet, no factor, but the MCA is 11,000 southbound. In this case we would be expected to enter a standard holding pattern at Laped, climb and cross the intersection at 11,000, then continue the climb to the MEA of 12,000 feet.

Approach procedures during periods of radio communications failure are discussed in chapter 8.

Figure 6-16 contains an example of an enroute high altitude chart that covers the same general area as the enroute low altitude chart in Fig. 6-14. Most symbols are common to both charts; the major difference is the depiction of flight service station and NAVAID identification boxes, which were discussed earlier and illustrated in Fig. 6-2.

North of the Daisetta VOR, J180 is a single direction route during the hours between 1200 and 0400Z (zulu; coordinated universal time (UTC)). East of the Hobby VOR is Sabine Pass VOR with a class designator (L), indicating a low altitude standard service volume, following the identifier SBI. Several L/MF high altitude airways are depicted using the Galveston NDB.

The same procedures apply when transitioning from the low or high altitude structure to the approach procedure. We would plan direct or jet routes to one of the transition fixes for the Conroe approach procedure (Navasota, Daisetta, or Hobby), as shown in Fig. 6-15. Because the complexity of transitioning from the high altitude structure is often greater than the low altitude structure, standard terminal arrival routes (STARs) are commonly used. STARs are specifically addressed in chapter 7. Subsequent chapters also relate these low and high enroute charts to standard instrument departures and instrument approach procedure charts.

SUPPLEMENTAL IFR ENROUTE CHARTS

DMA produces low and high altitude chart series that cover most of the world, consisting of enroute charts and supplements, FLIPs. Charts and FLIPs are available for sale from the DMA on a one-time basis or annual subscription. DMA does not support, for public sale, charts covering the Peoples Republic of China or the former Soviet Union.

Jeppesen Sanderson produces chart series that cover the world. Planning, low and high altitude enroute, area navigation enroute, and area charts are available for sale on a one-time basis or annual subscription. Jeppesen publishes a product catalog; *see* addresses and telephone numbers in chapter 10.

Chart series that cover Canada, and selected other areas, are produced by the Canada Map Office, Department of Energy, Mines and Resources. Planning, low and high altitude enroute, and area charts are available for sale on a one-time basis or annual subscription. The Canadian Map Office also publishes a catalog; *see* addresses and telephone numbers contained in chapter 10.

DMA world chart series

Caribbean and South American low altitude charts consist of 19 charts (10 sheets) printed back-to-back, and folded to the standard 5 × 10 inches. Coverages of Mexico and

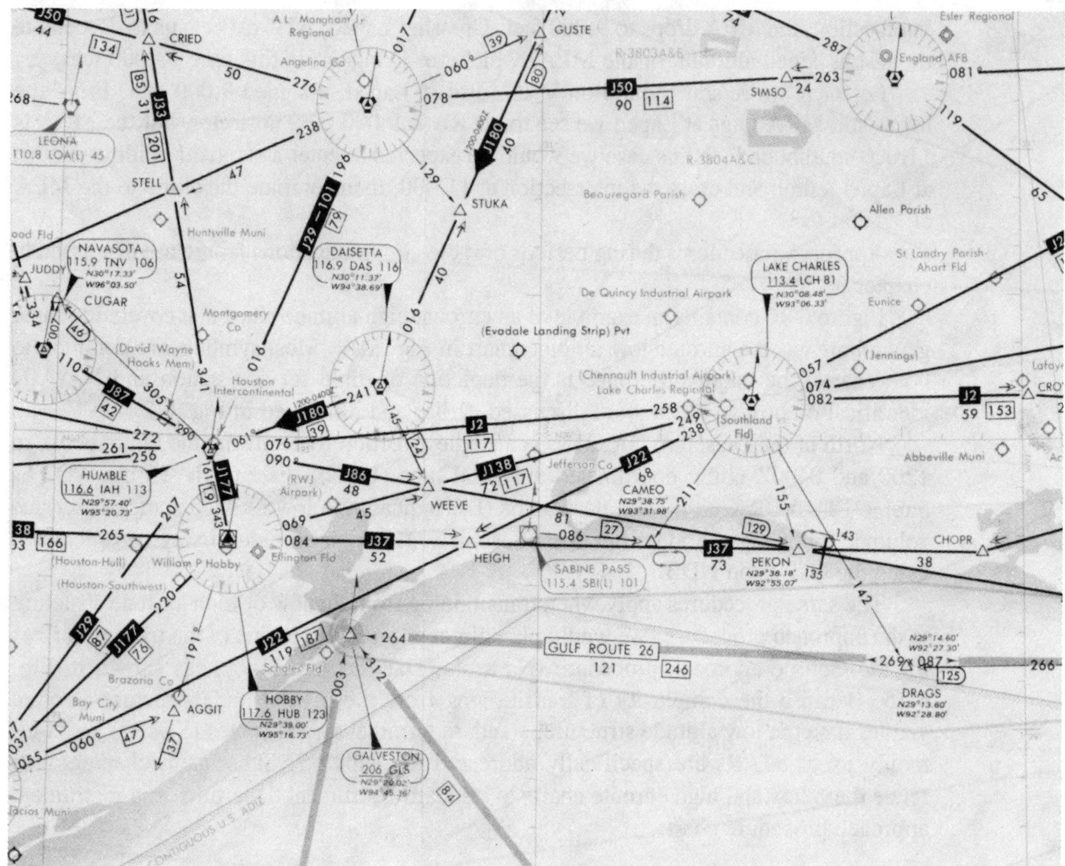

Fig. 6-16. Enroute high altitude charts provide navigational information for flight at and above 18,000 feet.

Central America are included in this series and are available as a separate subscription. They portray data required for IFR operations in this area. Two area charts are included for operations at selected terminals. The supplement is a bound book, approximately 5 x 8 inches, containing an alphabetical IFR/VFR airport/heliport facility directory, airport sketches, and data required to support the enroute and area charts.

The Caribbean and South American high altitude series covers the same geographical area as the low charts. This series consists of six charts. An inset to cover the Mexican area, not covered by the U.S. high altitude charts, is included on chart 1. An inset for Galapagos Island is included on chart 5. The charts may be joined to form a wall planning chart.

Europe, North Africa, and Middle East low altitude charts consist of 24 charts on 12 sheets that provide information required for flying the airways system in this area. Four area charts are included that contain arrival and departure routes for 29 selected terminal areas. Chart 1 includes insets for the Azores, Jan Mayen, and an enlargement for the Iceland area. The Berlin area is shown in an enlargement on chart 5. The associated high altitude series consists of 14 charts, with insets for the Azores, Iceland, and Jan Mayen. A bound supplement is also available that supports low and high enroute charts.

Africa consists of four charts that portray data required for flight in this area. Island insets include Ascension and Cape Verde. An enlargement of Johannesburg is included. There is not a separate low and high series for this area. A supplement is available with an enroute change notice because this series is not updated as often as other coverages.

Pacific, Australasia, and Antarctica series consist of 20 charts, on 10 sheets. Enlargements for Diego Garcia, Guam, Honolulu, Hulule Island, Mauritius-Reunion, McMurdo, Nandi, and Pago Pago areas are included. Chart 19 is a special Deep Freeze chart of Antarctica. Chart 1 serves as a planning chart for the area between the west coast of the U.S. and Asia. Two area charts are included. There is not a separate low and high series for this area. This series is supported by a bound supplement.

DMA also produces a series of area arrival charts, which are multicolored, single sheets, approximately 15 × 20 inches, that provide pilots with a 50 nautical miles radius depiction, at a scale of 1:500,000 (sectional), of terrain data and low altitude airways for navigation enroute to terminal transition to or from selected airports. Data required for navigation under IFR are included. Selected airports, special use airspace, tint contours, and the highest spot or obstruction elevations are also shown.

Areas of coverage for DMA publications are in Fig. 10-2, FLIP supplements, in chapter 10; however, the FLIP does not distinguish between North Africa and Africa. For enroute and terminal publications this division is along a Dakar, Khartoum, Aden line.

Jeppesen enroute charts

Jeppesen Sanderson truly provides worldwide chart coverage. As well as coverages for the U.S. and Canada, Jeppesen's Denver facility produces charts for Central and South America, the Pacific Ocean, Far East, and Australasia. Flight information compiled and published for the Western Hemisphere in Denver is available for the rest of the world through Jeppesen's Frankfurt facility. Coverages include the Atlantic, Europe, Africa, and Asia, as well as many other geographic coverages. Coverages for China are also available as metric editions, distances in kilometers and heights in meters.

Jeppesen charts provide the same aeronautical information as NOS charts, but in a slightly different format with some differences in symbology. Because these products are for civil use, references to strictly military use information, such as TACANs and meteorological services, are omitted. Any pilot who is schooled in the use of NOS charts should have no trouble transitioning to Jeppesen products.

Jeppesen publishes low and high altitude planning charts. The reverse side of the low altitude planning chart contains the following information:

- Key to aviation weather reports
- Standard and emergency transponder codes
- Light gun signals
- Mileage table between major airports
- Enroute flight advisory service outlets
- Wind and temperatures aloft forecast locations

(Continued on page 182.)

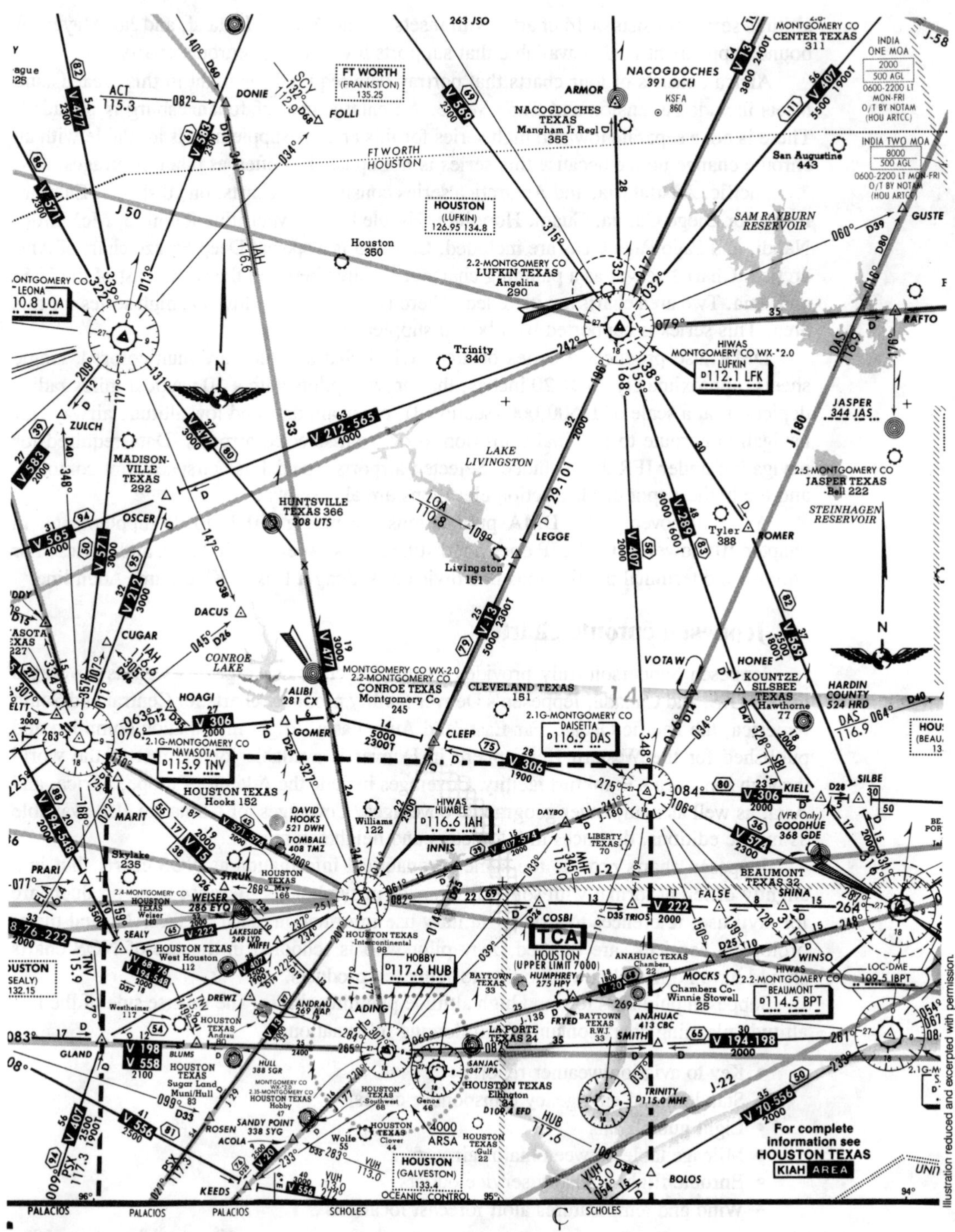

Fig. 6-17. Symbols used on Jeppesen charts are similar to symbols on NOS charts; pilots should have no trouble with the transition to Jeppesen charts.

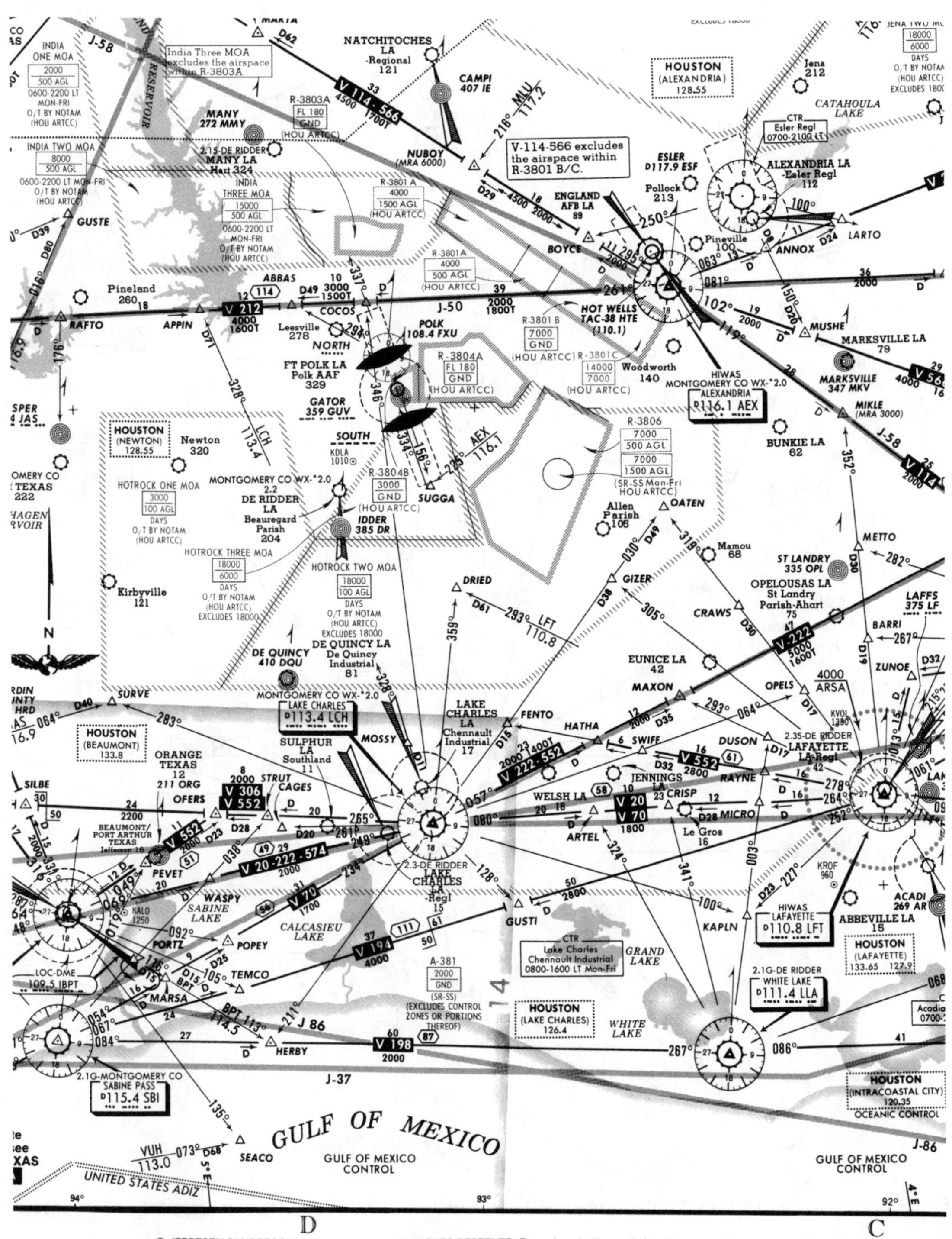

Fig. 6-17. Continued.

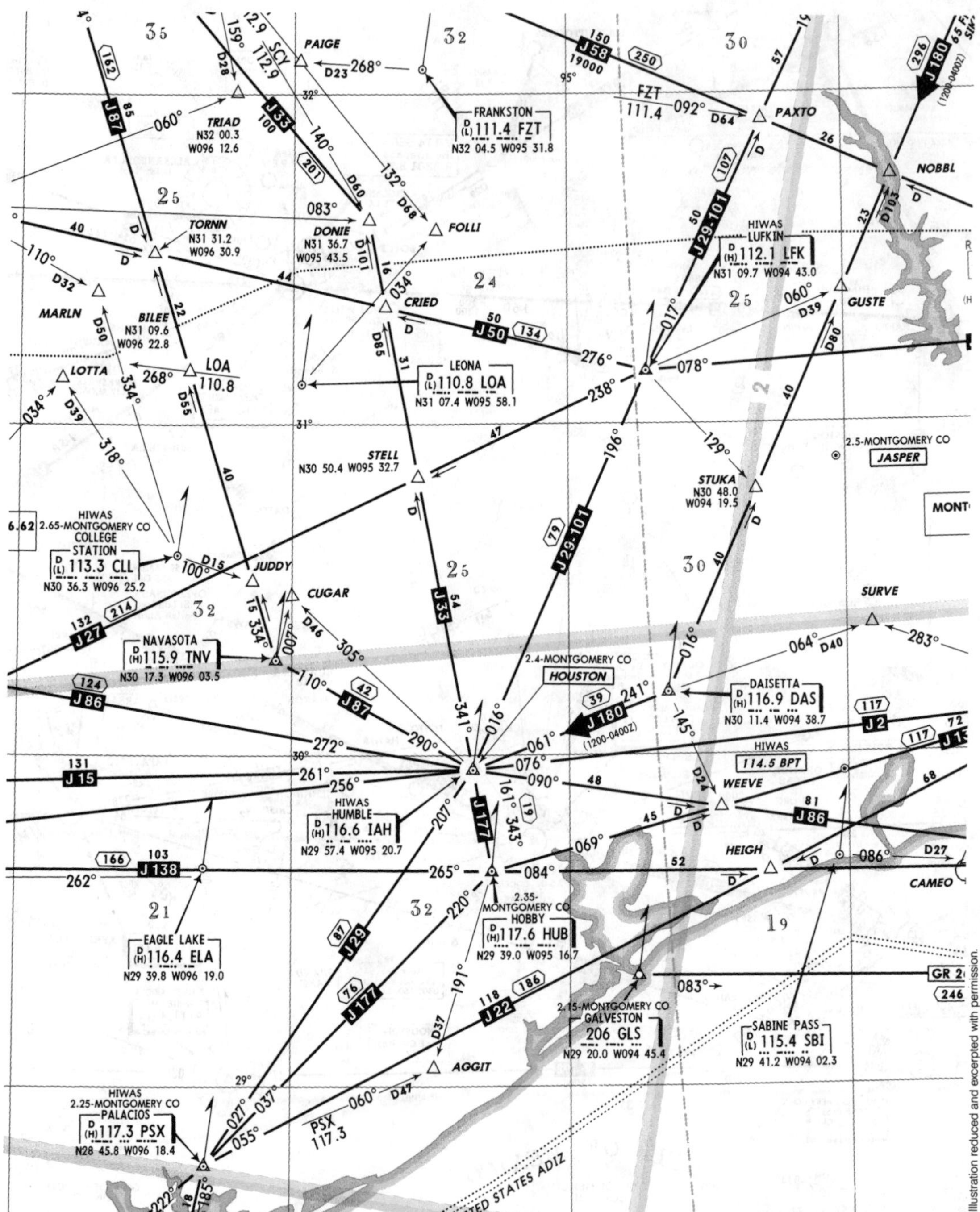

Fig. 6-18. References to military information can be omitted, reducing chart clutter, because Jeppesen charts are for civilian use.

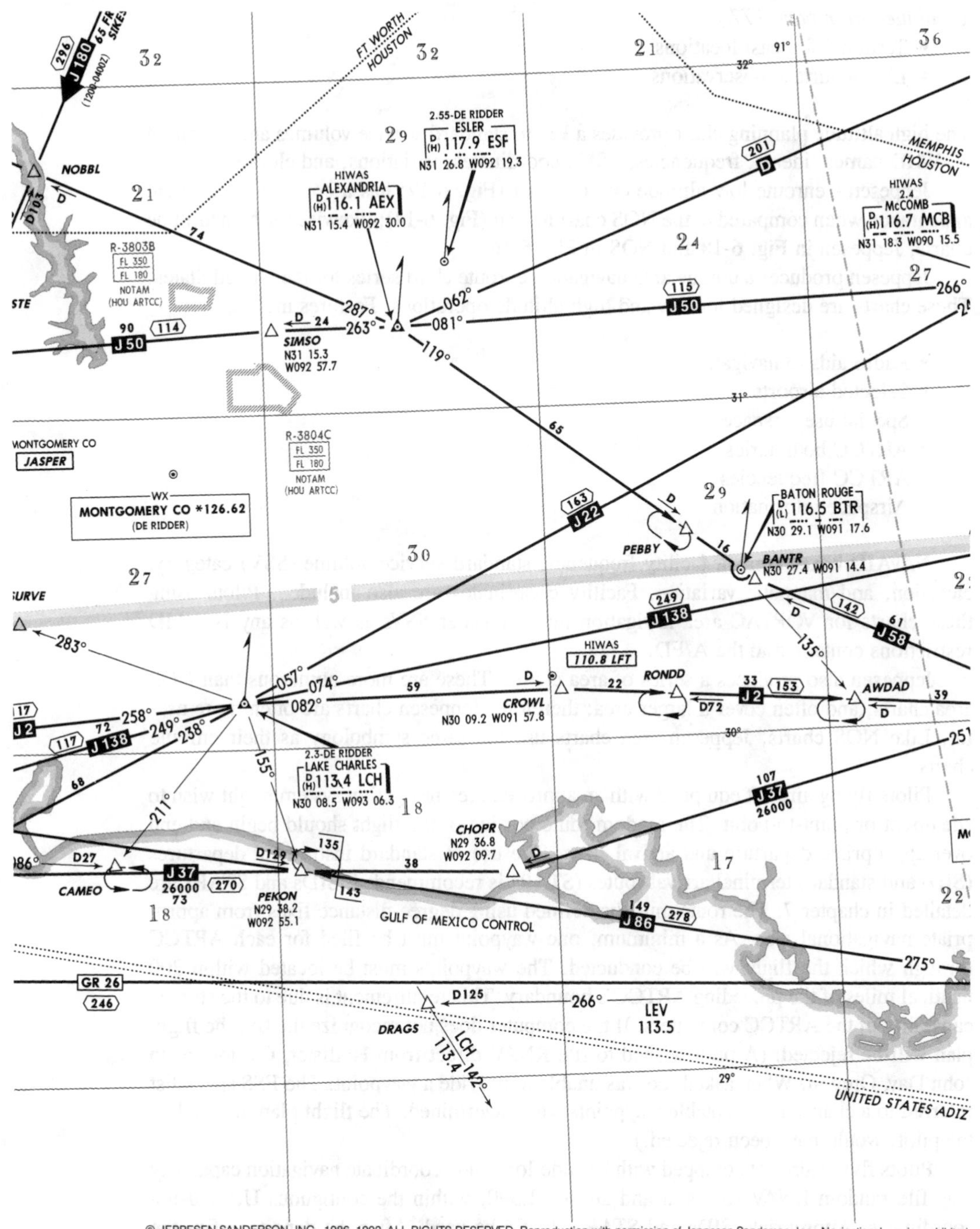

Fig. 6-18. Continued.

(Continued from page 177.)

- Terminal forecast locations
- List of surface observations

The high altitude planning chart provides a key to standard service volumes and a table of NAVAID names, idents, frequencies, SSVs, coordinates, variations, and elevations.

Jeppesen's enroute low altitude chart format (Fig. 6-17) has a completely different appearance when compared to the NOS chart format (Fig. 6-14); likewise for high altitude charts, Jeppesen in Fig. 6-18 and NOS in Fig. 6-16.

Jeppesen produces a unique area navigation enroute chart series for the United States. These charts are designed for low and high altitude operations. Features include:

- Radio aids to navigation
- Selected airports
- Special use airspace
- ARTCC boundaries
- ARTCC frequencies
- Airspace information

NAVAID boxes contain facility frequency, standard service volume (SSV) category, elevation, and magnetic variation. Facility coordinates are also included. Pilots using these charts for VORTAC area navigation must consider SSVs as well as any NAVAID restrictions contained in the A/FD.

Jeppesen also produces a series of area charts. These are more numerous than NOS area charts, and often cover a larger area; therefore, Jeppesen charts are often more useful. Like NOS charts, Jeppesen area charts use the same symbology as their enroute charts.

Pilots flying aircraft equipped with an approved area navigation system might wish to file direct or point-to-point. The random route portion of the flight should begin and end over appropriate departure and arrival fixes. The use of standard instrument departures (SID) and standard terminal arrival routes (STAR) is recommended. SIDs and STARs are detailed in chapter 7. The route must be defined using degree-distance fixes from appropriate navigational aids. As a minimum, one waypoint must be filed for each ARTCC through which the flight will be conducted. The waypoints must be located within 200 nautical miles of the preceding ARTCC's boundary. This requirement is due to the storage capability of the ARTCC computers. If the computer does not recognize the fix, the flight plan will be rejected. (A pilot wished to file RNAV direct from Hollister, California, to John Day, Oregon. When asked, he was unable to provide a waypoint. The FSS specialist referred to a chart and acceptable waypoints were determined. The flight plan, as filed by the pilot, would have been rejected.)

Pilots flying aircraft equipped with latitude/longitude coordinate navigation capability may file random RNAV routes at and above FL390, within the contiguous U.S., using coordinates. Appropriate SIDs and STARs should be used. After the departure fix, the pilot must include each turn point and the arrival fix for the destination. The arrival fix must be identified by coordinate and fix identifier. For example, OAK OAK3 OAK LIN 3910/10542 SHREW DEN. This route specifies the Oakland three departure to the Linden (LIN) VOR, direct to 3910/10542, which is the coordinate for the Shrew intersection. Shrew is the entry fix for the Denver profile descent arrival.

DUAT does not discriminate against pilots filing at any altitude for any direct route. Flight plans are accepted from any point to any point defined within the computer database. When necessary, DUAT inserts the latitude/longitude coordinates for the destination. Even when accepted by the computer, these flight plans don't work; someone, somewhere has to fix them, which results in delays to everyone.

Canadian enroute charts

Canada publishes and distributes several enroute chart series. The enroute low and high altitude charts provide aeronautical information for instrument navigation in the low and high airway structures. Charts cover Canada, Greenland, and portions of the contiguous U.S. and Alaska. Charts are also available for the Azores, Bermuda, and Iceland. These charts are revised every 56 days. Charts are printed back to back on eight sheets. Terminology and symbols are similar to those used on U.S. instrument charts.

Enroute low altitude charts, labeled "LE," depict aeronautical radio navigation information, special use airspace, and selected airports. Stations with communications are tabulated in alphabetical order. Like U.S. enroute low altitude charts, these charts provide coverage from the surface up to, but not including, 18,000 feet MSL. These charts are supplemented by a series of terminal area charts that depict aeronautical radio navigation information in congested areas at a larger scale. Enroute high altitude charts, labeled "HE," depict aeronautical radio navigation information, selected airports, and special use airspace, for flights at and above 18,000 feet MSL.

7
SIDs and STARs

CLEARANCE DELIVERY PROCEDURES HAVE BEEN SIMPLIFIED AT CERTAIN airports with establishment of standard instrument departures (SID) and standard terminal arrival routes (STAR). SIDs facilitate transition between takeoff and the enroute phase of flight; STARs aid the transition between the enroute phase and the approach segment. They provide graphic and textual descriptions of departure and arrival clearances. This reduces controller and pilot workload by allowing the controller to assign an often complex route including altitude restrictions in a single phrase, the name of the SID or STAR. The pilot then has a printed copy of the instruction. Filing any available SIDs and STARs is recommended. If, for whatever reason, a pilot does not have SIDs or STARs the contraction SSNO (SIDs STARs no) should be entered in the remarks section of the flight plan.

SIDs and STARs are published along with instrument approach procedures (IAP) charts in one of 16 top-bound NOS terminal procedure publication (TPP) volumes, $5^3/_8 \times 8^1/_4$ inches, covering the contiguous U.S., Puerto Rico, and the Virgin Islands. Alaskan SIDs and STARs are available in the terminal Alaska publication, which is similar to the TPP. For the Pacific area, SIDs and STARs are in the Pacific chart supplement, discussed in chapter 9. Charts are generally not to scale. These publications are designed for civil and military use. SIDs and STARs, along with profile descent charts and charted visual flight procedures (CVFP), are listed in the index of the TPP by city and airport name. Figure 7-1 contains U.S. terminal procedures publication coverage. Features include:

- Inoperative components table
- Explanation of terms/landing minima format
- Index of terminal charts and minimums
- IFR takeoff and departure procedures, including SIDs

- Rate of climb table
- IFR alternate minimums
- General information and abbreviations
- Chart legends
- Frequency pairing
- Radar minimums
- STARs
- Terminal charts
- Loran TD correction table
- Rate of descent table

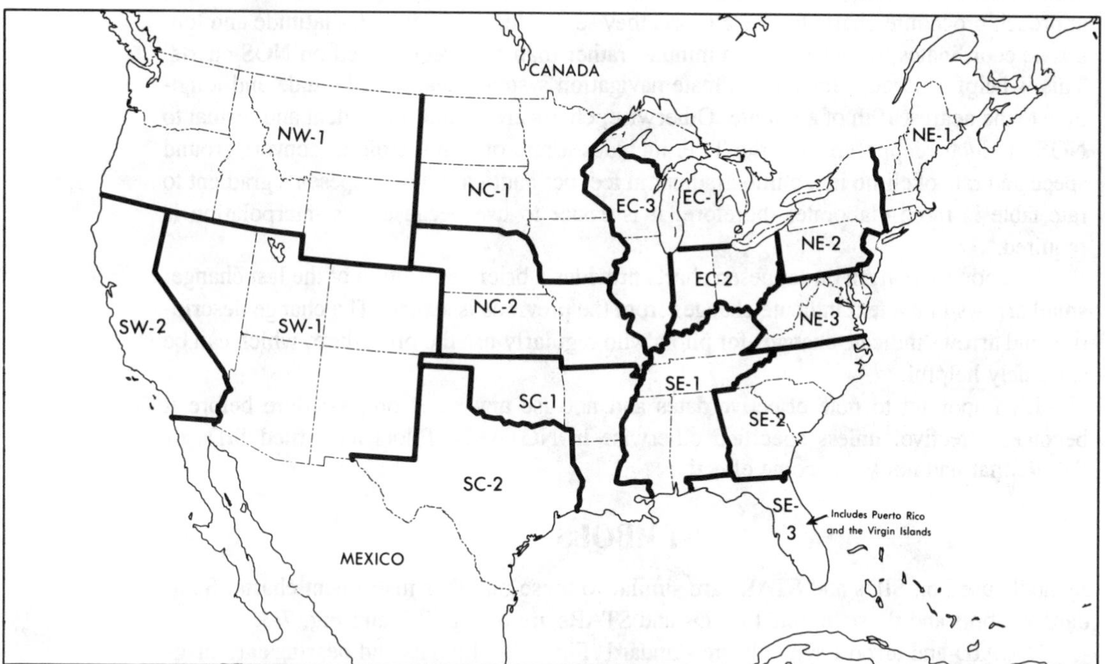

Fig. 7-1. The terminal procedures publication contains SIDs and STARs.

Volumes are published every 56 days, with a 28-day midcycle change notice volume. Changes that occur between the 28-day cycles are published in the form of FDC NOTAMs and incorporated in NOTAMs.

This chapter discusses the following charts and sections in the terminal procedures publication:

- Standard instrument departure charts
- Rate of climb table
- Standard terminal arrival charts
- Profile descent charts
- Charted visual flight procedures

The profile descent allows a pilot an uninterrupted descent, except for level flight required for speed adjustments, from the enroute structure to the point of glide slope intercept, or a specific minimum altitude. Charted visual flight procedures have been created to move air traffic safely and expeditiously during periods of relatively good weather. Charted visual flight procedures supplement conventional visual approach procedures by providing specific routes and altitudes, at times for noise abatement purposes.

Jeppesen SIDs and STARs use a different chart indexing system than the NOS terminal procedures publication. Charts are indexed alphabetically by city name within each state. Each airport listed under the city name is given an index number, enclosed in an oval, centered in the chart top margin. Unlike the TPP, which places STARs in a separate section of the volume, Jeppesen SID and STAR charts are filed along with the instrument approach procedure charts for the airports they serve. Jeppesen provides latitude and longitude coordinates, to the 10th of a minute, rather than hundredths used on NOS charts. This is helpful because most coordinate navigation systems deal with latitude and longitude to the nearest 10th of a minute. Otherwise, charts are similar in content and format to NOS products. Jeppesen, like the TPP, includes a rate of climb table to convert ground speed and rate of climb into climb gradient in feet per nautical mile. Jeppesen's gradient to rate table is more elaborate; therefore, it is easier to use because less interpolation is required.

The bottom margin on Jeppesen charts provides a brief description of the last change; small arrows in the text indicate changes from the previous issuance. The change description and arrows indicate changes for pilots who regularly use the procedure, which can be extremely helpful.

It's important to note effective dates and not use any chart or procedure before it becomes effective, unless specified otherwise in NOTAMs. Pilots have filed SIDs or STARs that had not yet become effective.

SYMBOLS

Symbols used on SIDs and STARs are similar to those on other instrument charts. Standard symbols and those unique to SIDs and STARs are in Fig. 7-2 and Fig. 7-3.

NAVAID and airport symbols are standard (Fig. 7-2). Radials and bearings are magnetic, mileages nautical, and altitude and elevations are feet above mean sea level. Mileage between fixes is shown in parenthesis adjacent to the route segment. A (T) or (L) after the frequency indicates frequency protection range, the facility is restricted by standard service volumes (SSV) described in Table 6-1. A (Y) following the TACAN channel indicates that the TACAN receiver must be placed in "Y" mode to receive distance information; this would only apply to aircraft equipped with TACAN receivers, normally only the military.

NAVAID boxes do not indicate FSS frequencies, but primary NAVAIDs indicate name, frequency, frequency protection range (SSV), identification, and RNAV coordinate information. RNAV coordinate information is omitted from secondary NAVAIDs. Departure routes are depicted by a heavy black line with arrow, transition routes a medium black

(Continued on page 191.)

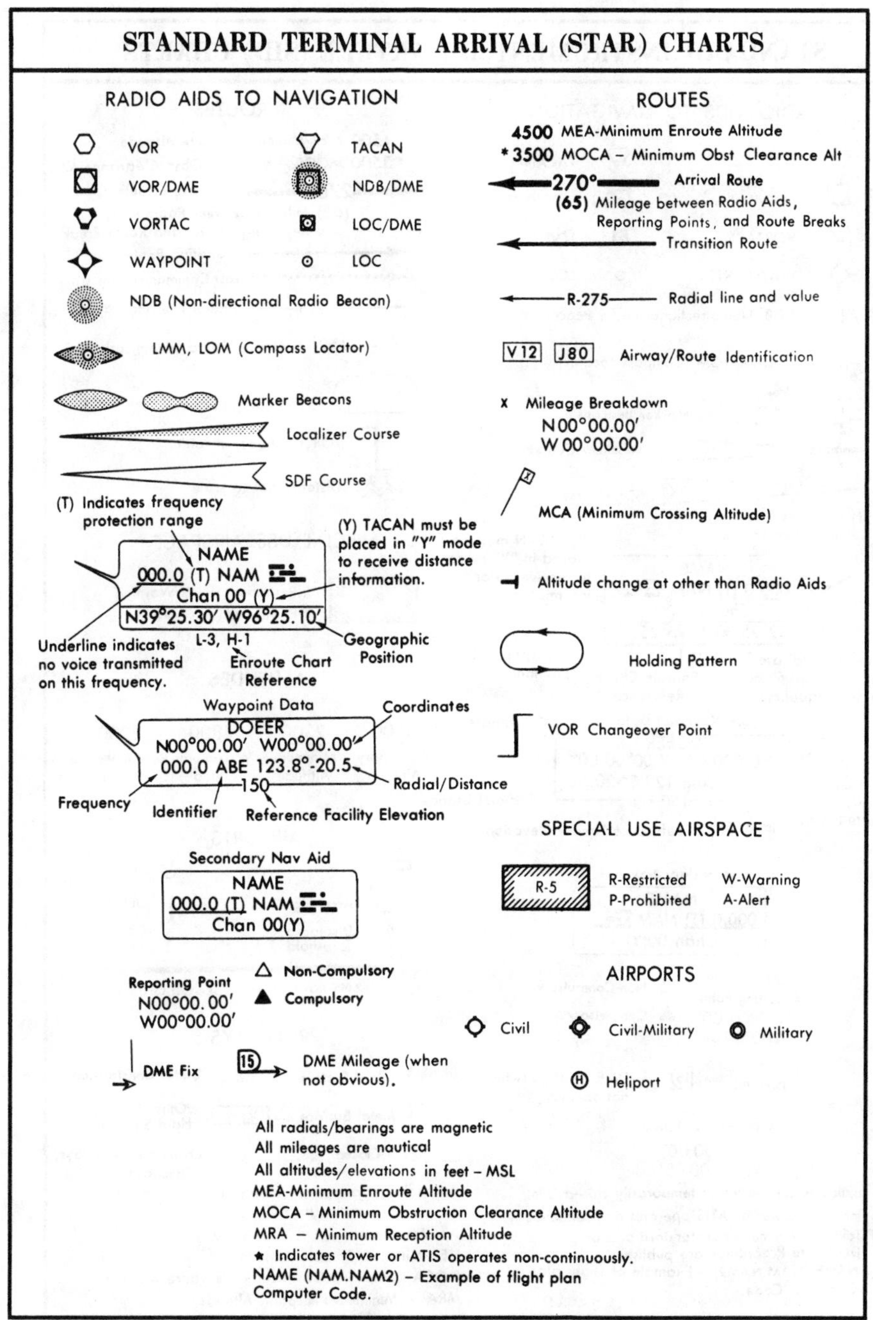
STANDARD TERMINAL ARRIVAL (STAR) CHARTS
RADIO AIDS TO NAVIGATION
VOR
TACAN
VOR/DME
NDB/DME
VORTAC
LOC/DME
WAYPOINT
LOC
NDB (Non-directional Radio Beacon)
LMM, LOM (Compass Locator)
Marker Beacons
Localizer Course
SDF Course
(T) Indicates frequency protection range
NAME
000.0 (T) NAM
Chan 00 (Y)
N39°25.30′ W96°25.10′
(Y) TACAN must be placed in "Y" mode to receive distance information.
Underline indicates no voice transmitted on this frequency.
L-3, H-1
Enroute Chart Reference
Geographic Position
Waypoint Data
Coordinates
DOEER
N00°00.00′ W00°00.00′
000.0 ABE 123.8°-20.5
150
Radial/Distance
Frequency
Identifier
Reference Facility Elevation
Secondary Nav Aid
NAME
000.0 (T) NAM
Chan 00(Y)
Reporting Point
N00°00.00′
W00°00.00′
Non-Compulsory
Compulsory
DME Fix
15
DME Mileage (when not obvious).
ROUTES
4500 MEA-Minimum Enroute Altitude
*3500 MOCA – Minimum Obst Clearance Alt
270°
Arrival Route
(65) Mileage between Radio Aids, Reporting Points, and Route Breaks
Transition Route
R-275
Radial line and value
V12
J80
Airway/Route Identification
X Mileage Breakdown
N00°00.00′
W00°00.00′
MCA (Minimum Crossing Altitude)
Altitude change at other than Radio Aids
Holding Pattern
VOR Changeover Point
SPECIAL USE AIRSPACE
R-5
R-Restricted
W-Warning
P-Prohibited
A-Alert
AIRPORTS
Civil
Civil-Military
Military
Heliport
All radials/bearings are magnetic
All mileages are nautical
All altitudes/elevations in feet – MSL
MEA-Minimum Enroute Altitude
MOCA – Minimum Obstruction Clearance Altitude
MRA – Minimum Reception Altitude
* Indicates tower or ATIS operates non-continuously.
NAME (NAM.NAM2) – Example of flight plan Computer Code.

Fig. 7-2. SID and STAR symbols are similar to those used on other instrument charts.

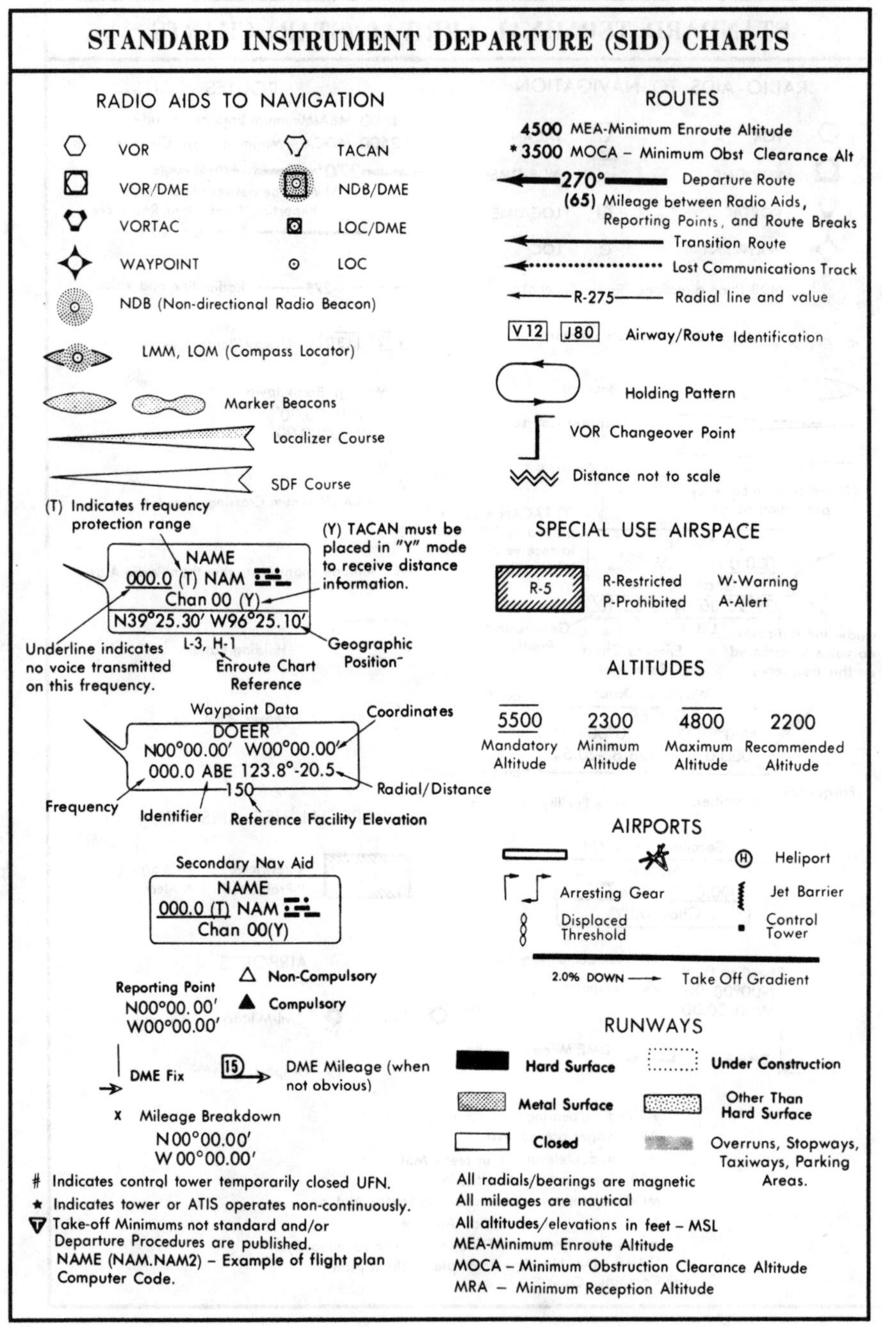

Fig. 7-2. Continued.

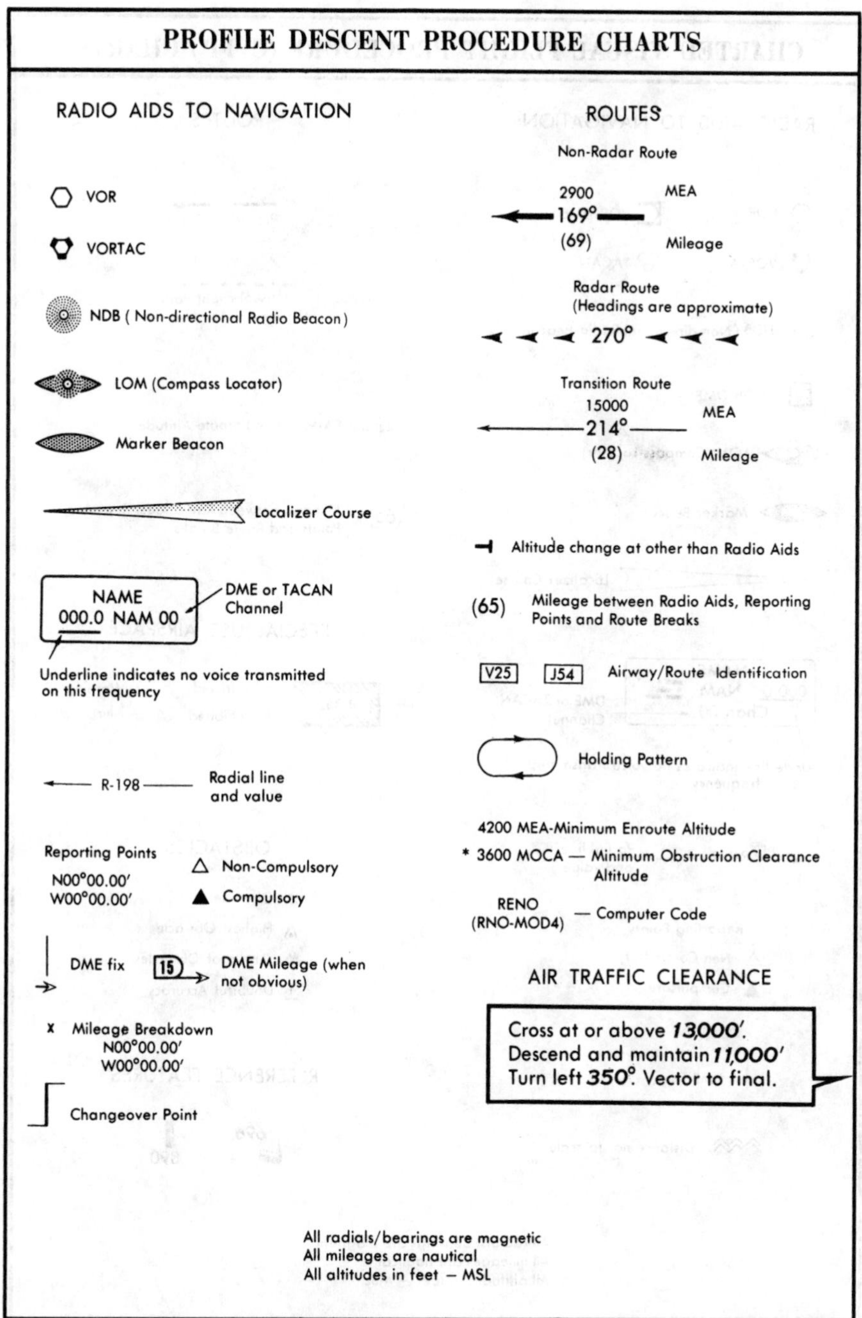

Fig. 7-3. Profile descents allow an uninterrupted descent; charted visual flight procedures are published visual approaches.

CHARTED VISUAL FLIGHT PROCEDURE (CVFP) CHARTS

RADIO AIDS TO NAVIGATION

VOR
VOR/DME
VORTAC
TACAN
NDB (Non-directional Radio Beacon)
NDB/DME
LOM (Compass Locator)
Marker Beacon
Localizer Course

NAME
000.0 NAM
Chan 00

DME or TACAN Channel

Underline indicates no voice transmitted on this frequency

R-117 Radial line and value

Reporting Points
Non-Compulsory
Compulsory

DME fix

Distance not to scale

ROUTES

Procedure Track

Visual Flight Path

4200 MEA-Minimum Enroute Altitude

(65) Mileage between Radio Aids, Reporting Points and Route Breaks

SPECIAL USE AIRSPACE

R-352
R-Restricted
P-Prohibited
W-Warning
A-Alert

OBSTACLES

Obstacle
Highest Obstacle
Group of Obstacles
± Doubtful Accuracy

REFERENCE FEATURES

696
390

All radials/bearings are magnetic
All mileages are nautical
All altitudes in feet – MSL

Fig. 7-3. Continued.

(Continued from page 186.)
line, and radials with a thin black line. Lost communication tracks are shown as dotted lines. Mandatory altitudes are indicated by lines above and below the altitude, minimum altitude by a line below the altitude, maximum altitude by a line above the altitude; the omission of lines above and below indicate recommended altitudes.

Figure 7-3 shows symbols unique to profile descents and charted visual flight procedures. Profile descents are merely a special case STAR, symbols are similar. Nonradar routes are indicated by a solid heavy black line with arrow; radar vector routes, with approximate heading, are designated by a series of arrowheads. Air traffic clearance instructions appear on these charts.

CVFP charts depict prominent landmarks, courses, recommended altitudes to specific runways, and NAVAID information for supplemental navigational guidance. If landmarks used for navigation will not be visible at night, the approach will be annotated: "PROCEDURE NOT AUTHORIZED AT NIGHT." Procedure track is indicated by a solid black line, visual flight path by a dashed black line. Reference features and obstacles are designated, sometimes along with their height. CVFPs usually begin within 15 miles of the airport, with published weather minimums based on minimum vectoring altitudes; they are not instrument approaches, and do not contain missed approach procedures.

STANDARD INSTRUMENT DEPARTURES

The use of a SID requires the pilot to have at least the textual description of the procedure. If the SID contains the departure control frequency, it might be omitted from the clearance. Features include:

- Radio aids to navigation
- Communication frequencies
- Reporting points
- Airways and mileages
- Holding patterns
- Special use airspace
- Airports
- IFR takeoff minimums and departure procedures table
- Airport sketch
- Departure route description
- Geographic positions of NAVAIDs and reporting points
- Mileage breakdown points
- Changeover points
- Computer codes for filing flight plans
- Data for coordinate navigation systems

SIDs assist pilots in avoiding obstacles during climbout. Obstacle clearance is based on the aircraft climbing at the rate of at least 200 feet per nautical mile (fpnm), crossing the end of the runway at least 35 feet agl, and climbing to 400 feet agl before turning, unless otherwise specified in the procedure. Climb gradients are specified when required for obstacle clearance. Crossing restrictions might be established for traffic separation or obstacle clearance. When no gradient is specified, the pilot is expected to climb at a rate of at least 200 feet per nautical mile to the MEA. To assist pilots in determining required climb rate the TPP contains a rate of climb table on page D1 (Fig. 7-4). The table converts rate of climb in feet per minute to rate of climb in feet per nautical mile, based on known or approximate ground speed.

For example, at Bakersfield, California, the Wring Three Departure requires a minimum climb gradient of 345 fpnm to 5,400 feet MSL. The 1967 Cessna 172 manual advertises a sea level climb rate of 645 fpm: at 5,000 feet, 435 fpm, at an indicated airspeed of 70 knots. Assume a ground speed of 70 knots—which cannot necessarily be presumed in the real world—for interpolating a rate of climb (Fig. 7-4) of approximately 410 fpm that would be required. Based on standard conditions, the 1967 Cessna 172 should be able to comply with this SID's climb gradient. It's extremely important for pilots to carefully consider ground speeds and aircraft performance before accepting a SID with an increased climb gradient.

Pilot navigation SIDs

The pilot assumes the primary responsibility for navigation with pilot navigation SIDs. Some pilot nav SIDs might contain vector instructions, with which the pilot is expected to comply, until instructions are received to resume normal navigation on the filed or assigned route, or SID procedure.

Flight plans with incorrect SID or transition codes are rejected. Pilots filing SIDs must use the correct departure codes and exit fix or transition; the filed airway structure must begin at the exit or transition fix.

Figure 7-5 shows the Houston Intercontinental Scholes Two Departure (Pilot NAV). Note the computer code (VUH2 METZY) above the SID name. Scholes Two is the departure route and METZY is the exit fix. The departure route is indicated by the heavy black line. The Scholes Two Departure has three transitions: Lafayette, Leeville, and White Lake. The transitions are indicated by the medium black lines. Thin lines identify intersection radials, such as the Lake Charles 204 and 142 radials that make up the Metzy and Drags intersections. The Lake Charles 204 radial should not be confused with a transition route.

Appropriate communication frequencies are provided in the upper left portion of the plan view. Two notes appear. All notes should be studied, they always contain pertinent information. In the example, the notes state: "This procedure requires over water flight" and "Chart not to scale." Standard low altitude NAVAID boxes are used, with primary NAVAIDs containing latitude and longitude coordinates for pilots using approved RNAV equipment. L-17 and H-5 appear below the coordinates for the Lake Charles (LCH) VORTAC. This indicates that LCH appears on the L-17 enroute low altitude chart and the H-5 enroute high altitude chart (Verified by reviewing Fig. 6-14 and Fig. 6-16.). Leeville

(Continued on page 196.)

RATE OF CLIMB TABLE

A rate of climb table is provided for use in planning and executing takeoff procedures under known or approximate ground speed conditions.

(ft. per min.)

REQUIRED CLIMB RATE (ft. per NM)	GROUND SPEED (KNOTS)						
	30	60	80	90	100	120	140
200	100	200	267	300	333	400	467
250	125	250	333	375	417	500	583
300	150	300	400	450	500	600	700
350	175	350	467	525	583	700	816
400	200	400	533	600	667	800	933
450	225	450	600	675	750	900	1050
500	250	500	667	750	833	1000	1167
550	275	550	733	825	917	1100	1283
600	300	600	800	900	1000	1200	1400
650	325	650	867	975	1083	1300	1516
700	350	700	933	1050	1167	1400	1633

REQUIRED CLIMB RATE (ft. per NM)	GROUND SPEED (KNOTS)					
	150	180	210	240	270	300
200	500	600	700	800	900	1000
250	625	750	875	1000	1125	1250
300	750	900	1050	1200	1350	1500
350	875	1050	1225	1400	1575	1750
400	1000	1200	1400	1600	1700	2000
450	1125	1350	1575	1800	2025	2250
500	1250	1500	1750	2000	2250	2500
550	1375	1650	1925	2200	2475	2750
600	1500	1800	2100	2400	2700	3000
650	1625	1950	2275	2600	2925	3250
700	1750	2100	2450	2800	3150	3500

CLIMB TABLE

Fig. 7-4. This table converts climb rate in fpm to rate of climb in feet per nautical miles, based on ground speed.

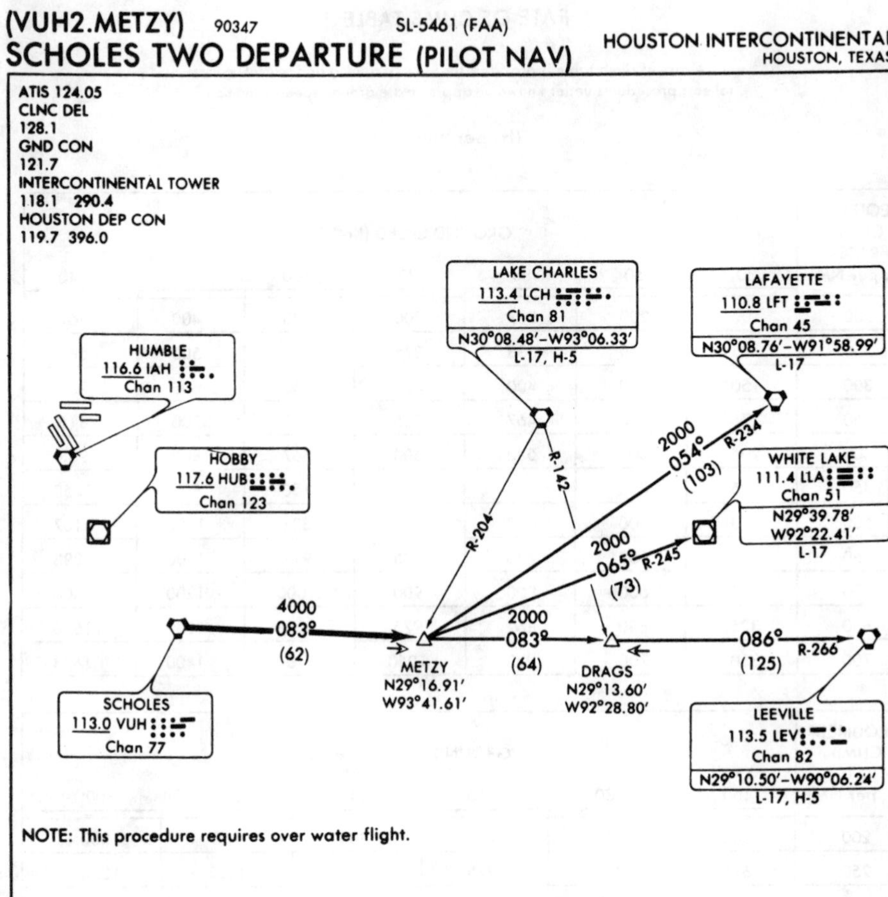

DEPARTURE ROUTE DESCRIPTION

All aircraft cleared as filed. From over VUH VORTAC then via VUH R-083 to METZY INT. Then via (transition) or (assigned route). Expect further clearance to requested altitude 10 minutes after departure.

LAFAYETTE TRANSITION (VUH2.LFT): From METZY INT via LFT R-234 to LFT VORTAC.

LEEVILLE TRANSITION (VUH2.LEV): From METZY INT via VUH R-083 and LEV R-266 to LEV VORTAC.

WHITE LAKE TRANSITION (VUH2.LLA): From METZY INT via LLA R-245 to LLA VOR/DME.

SCHOLES TWO DEPARTURE (PILOT NAV) HOUSTON, TEXAS
(VUH2.METZY) HOUSTON INTERCONTINENTAL

Fig. 7-5. Pilot NAV SIDs: The pilot assumes primary responsibility for navigation; vector SIDs: ATC assumes primary responsibility for navigation.

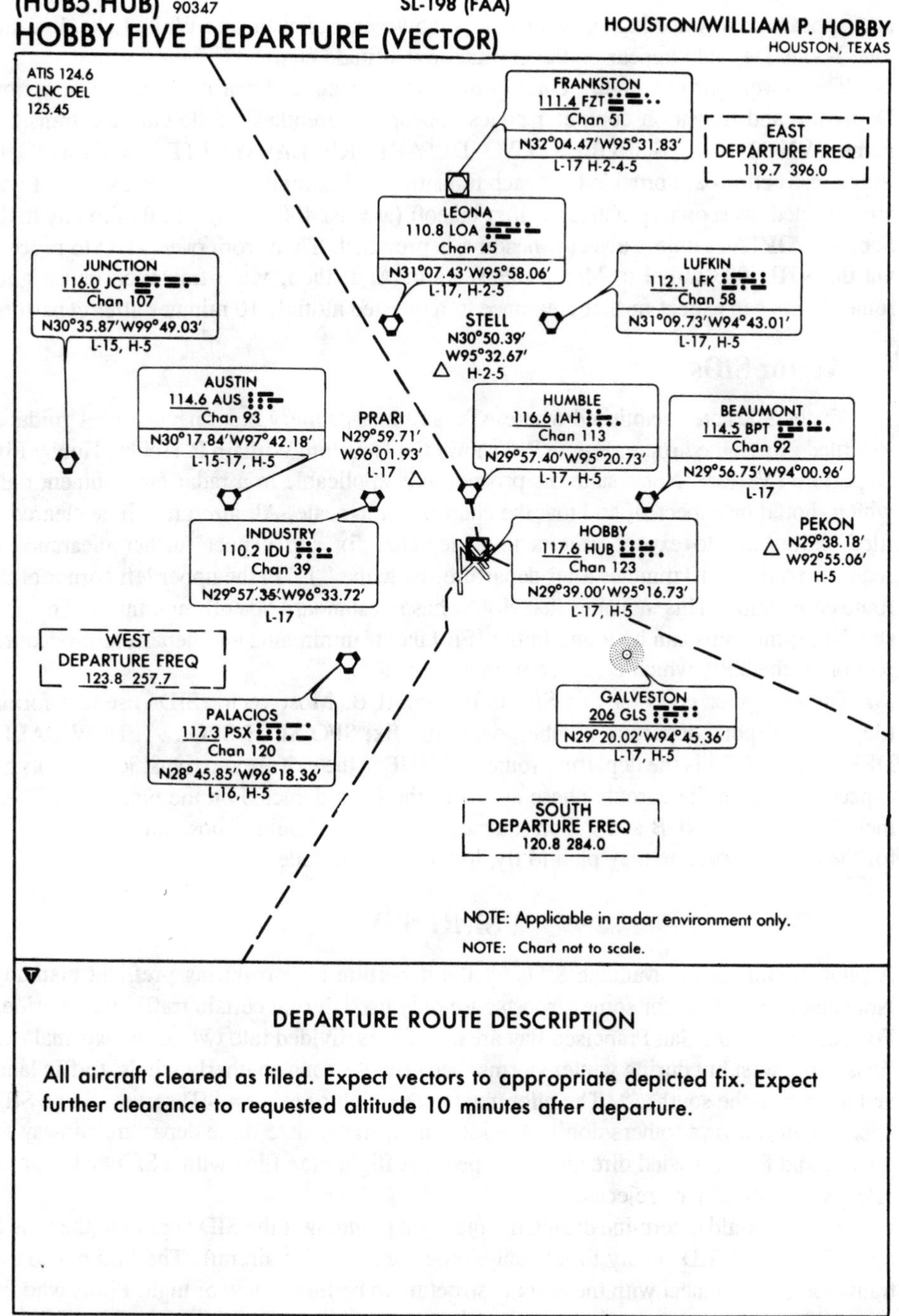
(HUB5.HUB) 90347
SL-198 (FAA)
HOBBY FIVE DEPARTURE (VECTOR)
HOUSTON/WILLIAM P. HOBBY
HOUSTON, TEXAS
ATIS 124.6
CLNC DEL 125.45
FRANKSTON 111.4 FZT Chan 51 N32°04.47′W95°31.83′ L-17 H-2-4-5
EAST DEPARTURE FREQ 119.7 396.0
LEONA 110.8 LOA Chan 45 N31°07.43′W95°58.06′ L-17, H-2-5
JUNCTION 116.0 JCT Chan 107 N30°35.87′W99°49.03′ L-15, H-5
LUFKIN 112.1 LFK Chan 58 N31°09.73′W94°43.01′ L-17, H-5
STELL N30°50.39′ W95°32.67′ H-2-5
AUSTIN 114.6 AUS Chan 93 N30°17.84′W97°42.18′ L-15-17, H-5
PRARI N29°59.71′ W96°01.93′ L-17
HUMBLE 116.6 IAH Chan 113 N29°57.40′W95°20.73′ L-17, H-5
BEAUMONT 114.5 BPT Chan 92 N29°56.75′W94°00.96′ L-17
INDUSTRY 110.2 IDU Chan 39 N29°57.35′W96°33.72′ L-17
HOBBY 117.6 HUB Chan 123 N29°39.00′W95°16.73′ L-17, H-5
PEKON N29°38.18′ W92°55.06′ H-5
WEST DEPARTURE FREQ 123.8 257.7
GALVESTON 206 GLS N29°20.02′W94°45.36′ L-17, H-5
PALACIOS 117.3 PSX Chan 120 N28°45.85′W96°18.36′ L-16, H-5
SOUTH DEPARTURE FREQ 120.8 284.0
NOTE: Applicable in radar environment only.
NOTE: Chart not to scale.
DEPARTURE ROUTE DESCRIPTION
All aircraft cleared as filed. Expect vectors to appropriate depicted fix. Expect further clearance to requested altitude 10 minutes after departure.
HOBBY FIVE DEPARTURE (VECTOR)
(HUB5.HUB)
HOUSTON, TEXAS
HOUSTON/WILLIAM P. HOBBY

Fig. 7-5. Continued.

(Continued from page 192.)
also appears on low and high altitude charts; however, Lafayette and White Lake, low altitude NAVAIDs, only appear on the enroute low altitude chart.

The lower portion of the chart provides a narrative departure route description. Departure and transition computer codes appear in parenthesis following the transition name (VHU2 LFT, SCHOLES TWO DEPARTURE LAFAYETTE TRANSITION). Route instructions are provided for each transition. All aircraft are cleared as filed. Pilots are expected, as soon as practicable after takeoff (at least 400 feet agl) to fly directly to the Scholes VORTAC, unless a vector heading is provided. Then from over VHU to proceed via the VHU 083 radial to Metzy intersection. And, then, via a transition or assigned route. Pilots can expect further clearance to requested altitude 10 minutes after departure.

Vector SIDs

Vector SIDs are established where ATC assumes primary radar navigational guidance to a filed or assigned route. Figure 7-5 shows the Houston/William P. Hobby, Hobby Five Departure (Vector). Notes state the procedure is applicable in a radar environment only, which should be expected, and that the chart is not to scale. All aircraft will be cleared as filed. Pilots are to expect vectors to a depicted fix and expect further clearance to requested altitude 10 minutes after departure. Note the "T" in the upper left corner of the route description. This indicates that Hobby has nonstandard takeoff minimums. The nonstandard minimums can be found in the IFR takeoff minimums and departure procedures section of the TPP, which is reviewed in chapter 8.

The computer code for this SID is HUB5.HUB. Most vector SIDs use this format (departure airport, departure number, departure fix; SFO SFO3 SFO . . ., DFW DALL7 DFW . . .). HUB5 is the departure route and HUB is the exit fix; no transitions. Pilots are expected to begin the enroute phase at one of the fixes depicted on the plan view. These include Galveston NDB and Pekon intersection. Pilots should choose an appropriate fix for the airway structure they plan to fly, low or high altitude.

USING SIDs

A pilot should review available SIDs for the departure airport during preflight planning. Note that some SIDs, for some airports, are only used during certain traffic flow periods. For example, in the San Francisco Bay area, traffic is divided into two plans. Normal traffic is to the west but during winter storms, because of strong southerly winds, traffic lands and departs to the southeast. The pilot must note which runways a SID serves. Some SIDs apply to all runways, others don't. A pilot can normally determine departure runway for present and forecast wind direction and speed. A flight plan filed with a SID that is not in effect will normally be rejected.

A pilot should determine during the preflight planning if the SID applies to the type of operation; some SIDs apply to jets only, others exclude jet aircraft. The SID exit fix or transition must connect with the enroute structure to be flown, low or high. Pilots who fail to consider these factors might be assigned departure procedures that they did not file, or encounter delays while ATC attempts to find and assign a suitable route. When in an unfamiliar area, local pilots or the FSS can recommend appropriate departure routes.

The pilot must also ensure that the aircraft is equipped to comply with the procedure.

Some SIDs require the use of special facilities, such as radar, ADF, or DME. Pilots should not accept SIDs with which their aircraft performance or equipment cannot comply. For example, departing San Francisco International I was assigned, by clearance delivery, a SID that required the use of DME to establish a turning point on the departure. The aircraft was not equipped with DME, and I informed the controller; however, the controller was able to provide me with a VOR radial, not on the SID chart, to establish the fix, and I was able to accept the departure.

Selecting a suitable SID allows the pilot to study the procedure, becoming familiar with NAVAIDs, route, and altitudes. This will assure that once assigned by ATC, the procedure can be flown, especially if the SID has an increased climb gradient or special NAVAID requirement.

References to Fig. 6-14 and Fig. 6-16 might be helpful while reading this section. Pilots planning to use the Scholes Two Departure and transition to the high altitude structure would normally only file the exit fix Metzy (IAH VHU2 METZY) or the Leeville Transition (IAH VHU2 LEV). The other transitions terminate at low altitude NAVAIDs. For example, a pilot could file IAH VHU2 METZY LCH J2 . . .. Here the pilot plans to fly the Scholes Two Departure to the Metzy intersection, then direct LCH, and proceed via J2. Notice in Fig. 6-16 that LCH is on J2. A flight plan filed IAH VHU2 LCH J2 . . . would be rejected. LCH is not part of the SID; the SID must have an exit fix or transition. A pilot planning to pickup J86 at LEV would file IAH VHU2 LEV J86 . . .; however, IAH VHU2 J86 . . . would be rejected. Again, the filed SID must have an exit fix, and the exit fix must connect to the enroute structure.

Altitude would be as assigned, with the explicit instruction to "expect further clearance to requested altitude 10 minutes after departure." This would be important in the case of radio communications failure. Let's say we were assigned 2,000 feet. We would fly that altitude until reaching Scholes, then climb to 4,000 feet, the MEA, and 10 minutes after departure climb to our requested altitude.

A pilot planning to fly Gulf Route 26 (Fig. 6-16) could file HOU HUB5 HUB GLS GR26 . . .. If the pilot plans to fly J86, file HOU HUB5 HUB PEKON J86 . . .. Please note how the route structure connects, from the departure airport HOU via the HUB5 HUB SID route to the Pekon intersection, which is on J86. The HOU HUB5 HUB J86 route would not be accepted because the computer knows that HUB is not on J86. Although, a pilot could file HOU HUB5 HUB IAH J86. This is acceptable because IAH is on J86. For example, a pilot planning a flight eastbound below 18,000 feet might file to Beaumont (HOU HUB5 HUB BPT—connecting airway structure) to pickup the Victor airways eastbound, such as V552, V20, V222, or V574 (Fig. 6-14); however, HOU HUB5 BPT J20 . . . would be rejected because the exit fix HUB is missing.

The important point for the pilot and ATC is that routes connect. Departure airport, appropriate SID, exit fix or transition (to the desired low or high airway structure), direct to other fixes (within the desired airway structure), or airways that connect to the exit fix or transition.

One final thought about departures. Extensive use is made of IFR to VFR on top. These clearances might include a SID. In affect they are clearances to nowhere. Let's take, for example, a clearance to the XYZ VOR via the Headsup Two departure XYZ transition, climb and maintain 3,000, if not on top at 3,000, maintain 3,000 and advise. After departure we lose radio communications, arrive at XYZ at 3,000 and we're still in the clouds.

FARs don't cover this possibility. A prudent pilot would be very cautious accepting such a clearance unless tops were known. If not, it might be wise to file to a destination, then cancel when on top.

STANDARD TERMINAL ARRIVAL ROUTES

STARs provide graphic and textual descriptions of preplanned IFR air traffic control arrival procedures. They reduce pilot and controller workload and communications plus they minimize potential errors in delivery and receipt of clearances. The use of STARs facilitates aircraft transition from the enroute environment to instrument approaches. Like SIDs, pilots are requested to use associated STAR codes when filing flight plans. Features include:

- Geographic positions for NAVAIDs and reporting points
- Radio aids to navigation
- Communication frequencies
- Routes
- Assigned altitudes
- Special use airspace
- Mileages
- Mileage breakdown points
- Airports
- Reporting points/fixes
- Holding patterns
- Minimum enroute altitudes
- Minimum obstruction clearance altitudes
- Computer codes for filing flight plans
- Changeover points
- Data for inertial navigation systems

Standard STARs

Figure 7-6 shows the Houston Daisetta Nine Arrival. The chart contains procedure name followed by computer code (DAS DAS9), appropriate communication frequencies, and a graphic presentation of the procedure in the plan view. The Daisetta (DAS) VORTAC is the entry fix and Daisetta Nine the arrival route. Appropriate secondary and primary NAVAIDs, with latitude and longitude coordinates, and reference to low and high altitude charts (L-17, H-4-5, and the like) are provided. This STAR contains some narrative information in the plan view. Note under the DAS NAVAID box: "Turbojets cross at 250K IAS. Turbojet vertical navigation planning information, expect clearance to cross at 10,000′." Pilots of turbojet aircraft can expect to cross DAS at 10,000 feet at 250 knots indicated airspeed.

The narrative portion of the chart describes transitions, shown as medium black lines, from enroute NAVAIDs and fixes to the entry fix Daisetta, and the arrival, shown as a heavy black line. Alexandria, Lake Charles, McComb, and Stuka transitions are available. Computer codes follow the transition name (AEX DAS9).

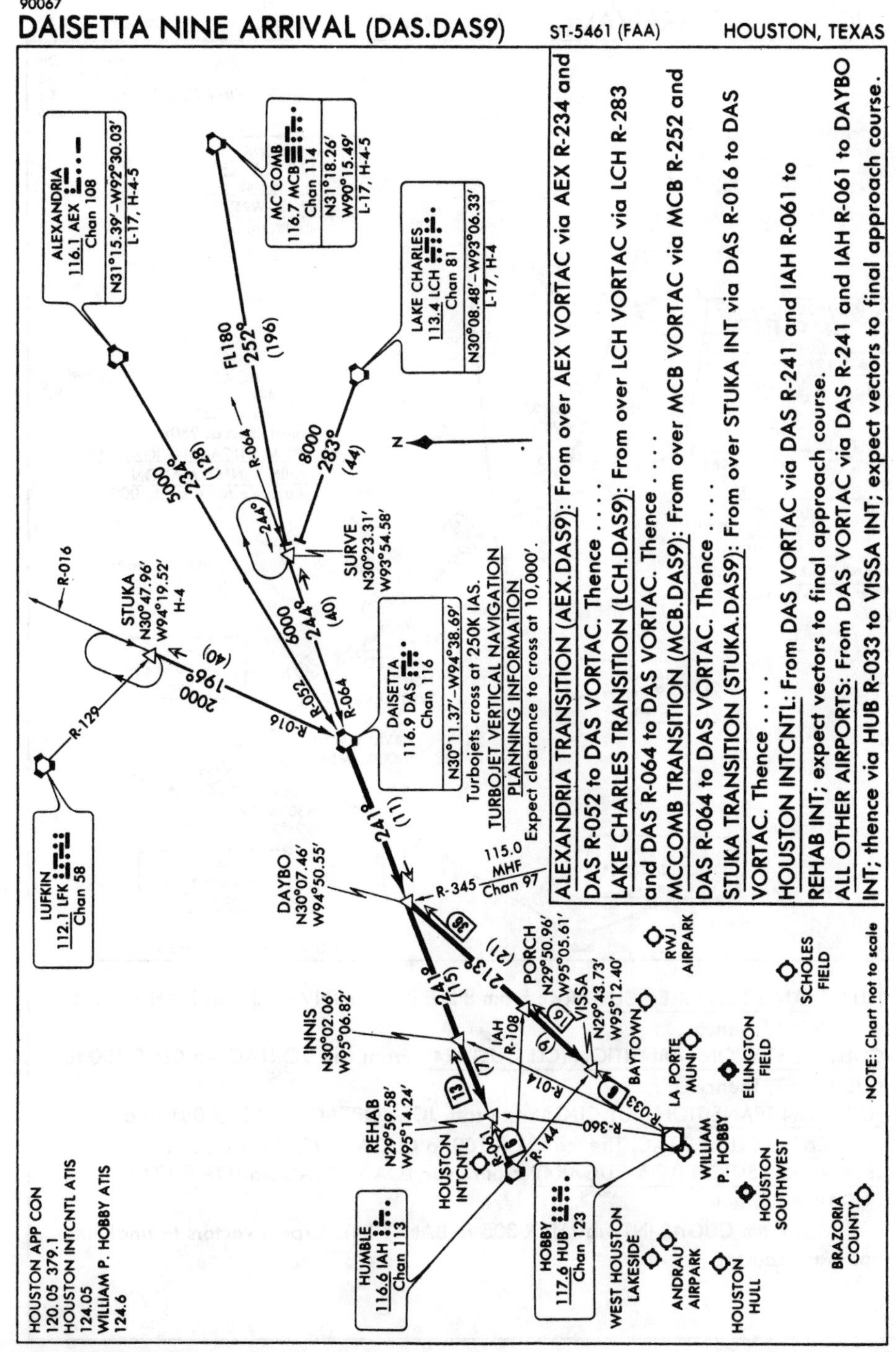

Fig. 7-6. STARs facilitate aircraft transition from the enroute structure to the instrument approach.

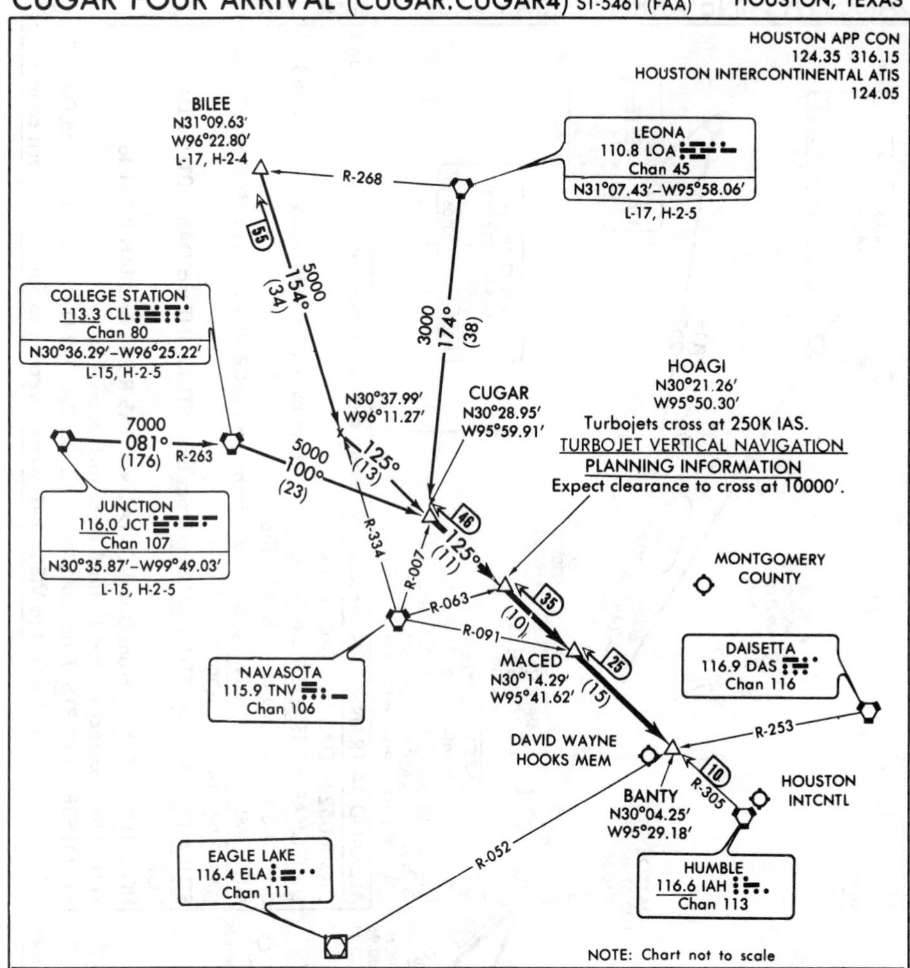

BILEE TRANSITION (BILEE.CUGAR4): From BILEE INT via TNV R-334 and IAH R-305 to CUGAR INT. Thence . . . .

COLLEGE STATION TRANSITION (CLL.CUGAR4): From CLL VORTAC via CLL R-100 to CUGAR INT. Thence . . . .

JUNCTION TRANSITION (JCT.CUGAR4): From JCT VORTAC via JCT R-081 and CLL R-263 to CLL VORTAC. Thence CLL R-100 to CUGAR INT. Thence . . . .

LEONA TRANSITION (LOA.CUGAR4): From over LOA VORTAC via LOA R-174 to CUGAR INT. Thence . . . .

. . . . From CUGAR INT via IAH R-305 to BANTY INT. Expect vectors to final approach course.

CUGAR FOUR ARRIVAL (CUGAR.CUGAR4) HOUSTON, TEXAS

Fig. 7-6. Continued.

Pilots destined for Houston Intercontinental (IAH) are instructed to proceed to the Rehab intersection and expect vectors to the final approach course. Pilots planning to land at any of the other airports in the plan view are instructed to fly from the DAS VORTAC via the DAS 241 and IAH 061 radials to Daybo intersection, then via the HUB 033 radial to the Vissa intersection, and expect vectors to their destinations final approach course.

Figure 7-6 also shows the Cugar Four Arrival. This STAR provides an arrival procedure to the Houston Intercontinental airport, from the northwest. Note that the Bilee intersection is depicted on high and low altitude charts (L-17, H-2-4); therefore, unlike the Stuka intersection and transition on the Daisetta Nine, which only appears on the high altitude chart, the Bilee transition of the Cugar Four can be utilized by high and low altitude flights. The Cugar Four Arrival has Junction (JCT) and College Station (CLL) transitions. These transitions overlap between CLL and Cugar. A pilot wishing to transition from the enroute structure at JCT should file the Junction transition (. . . JCT CUGAR4 IAH). Transitioning from CLL, the College Station transition should be filed (. . . CLL CUGAR4 IAH).

PROFILE DESCENT STARs

Profile descent clearances are subject to traffic and may be changed by ATC when necessary. Issuance, and acceptance by the pilot of a profile descent clearance (". . . cleared for the Runway 28 profile descent. . .") requires that the pilot adhere to all charted procedures. Clearance for a profile descent does not constitute clearance to fly an instrument approach procedure (IAP). The pilot is expected to comply with the altitudes specified in the procedure, and not descend below specified altitudes until the issuance of an approach clearance. Any subsequent ATC revision of altitude or route cancels the remaining portion of the charted profile descent procedure. ATC will then assign altitudes, routes, and speeds as required. If any subsequent revision of depicted speed restrictions is issued, all charted speed restrictions are void. Route and altitudes are not affected. A pilot that cannot comply with charted restrictions because of a revised speed restriction must advise ATC.

Figure 7-7 shows the San Francisco profile descent Runway 28. (This procedure has subsequently been cancelled and is used for illustration purposes only.) The entry fix is the Big Sur (BSR) VORTAC, and the computer code for this arrival is . . . BSR BSR5 SFO. A pilot cleared for the profile descent Runway 28 from BSR would descend to cross Anjee intersection at 16,000 feet. The pilot would then descend to cross the Skunk intersection between 12,000 and 10,000 feet; subsequently, continuing descent to cross Boldr at 10,000 feet at an indicated airspeed of 250 knots. After Boldr the descent continues to and maintains 6,000 feet to the Menlo intersection. At Menlo the pilot is expected to maintain 6,000 feet and fly heading 330° for radar vectors to the final approach course. The pilot would not descend below 6,000 or begin an approach until cleared by ATC.

CHARTED VISUAL FLIGHT PROCEDURES

The Dallas-Fort Worth Stadium Visual RWY 31R procedure (Fig. 7-7) notes that radar is required, the chart is not to scale, weather minimums are 2,500 foot ceiling, visibility 5 miles, and that the procedure is not authorized at night. The procedure starts over Texas

(Continued on page 204.)

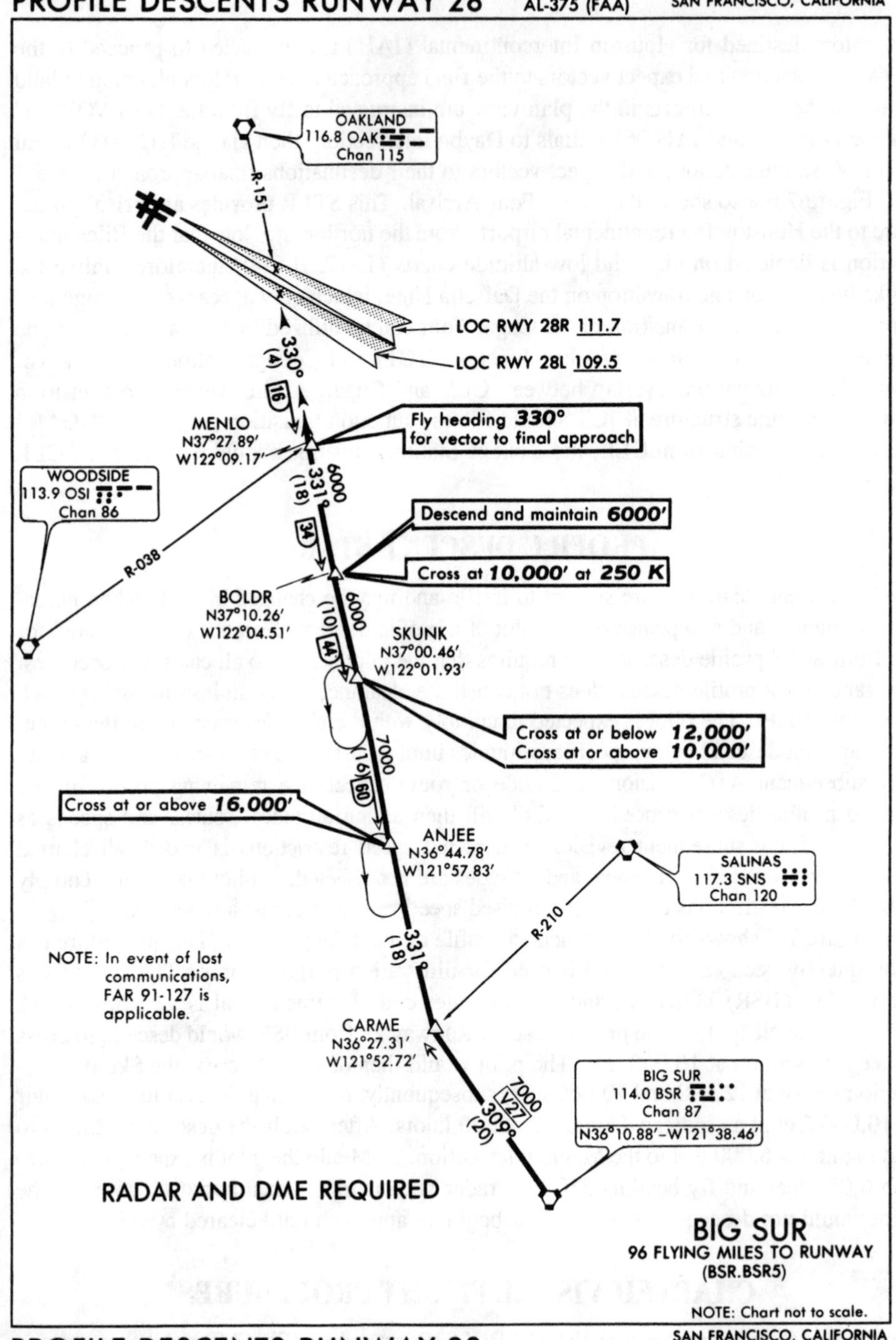

Fig. 7-7. Profile descents and charted visual flight procedures are special-case arrival procedures.

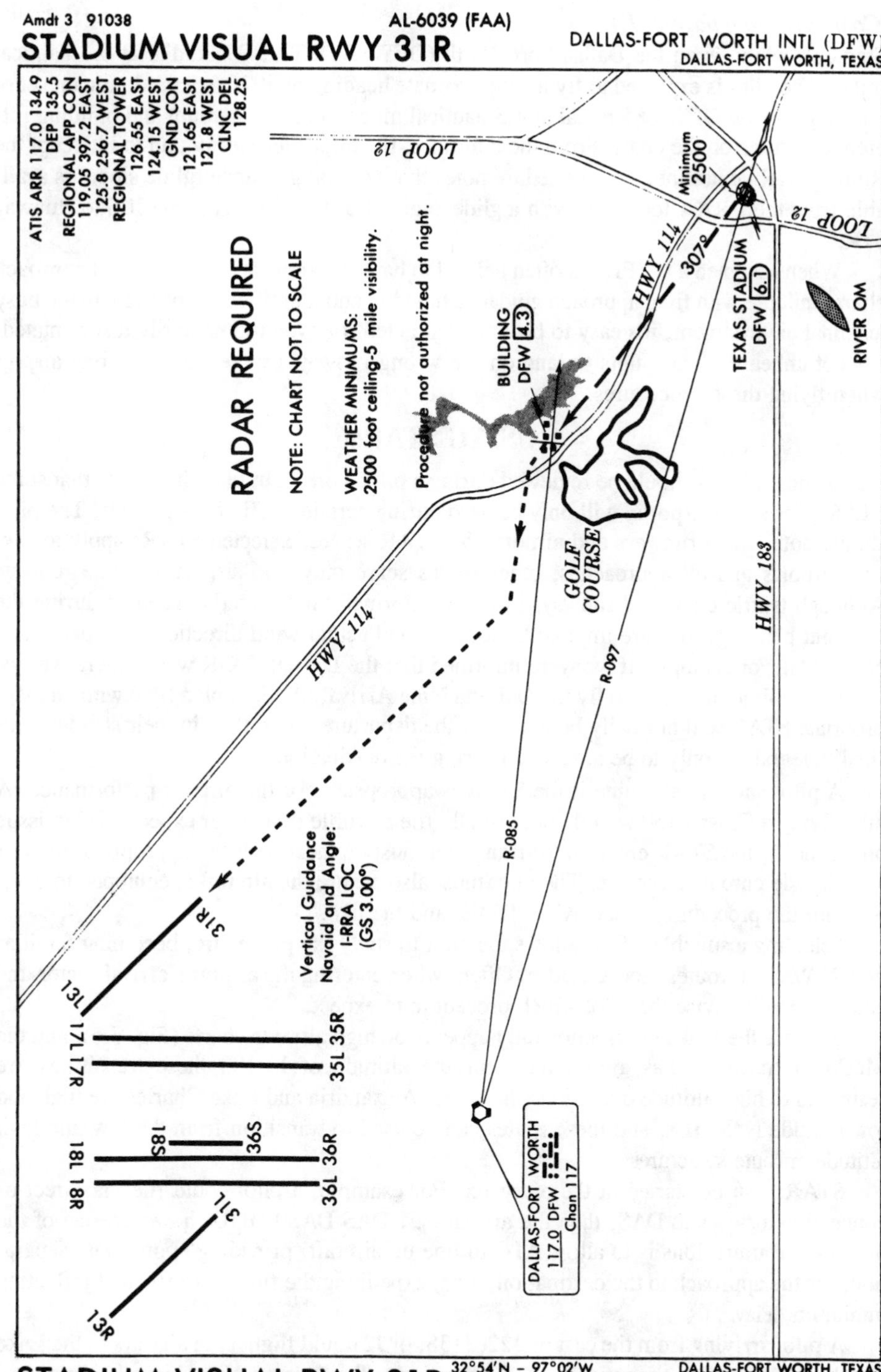
Amdt 3 91038
AL-6039 (FAA)
STADIUM VISUAL RWY 31R
DALLAS-FORT WORTH INTL (DFW)
DALLAS-FORT WORTH, TEXAS
ATIS ARR 117.0 134.9
DEP 135.5
REGIONAL APP CON
119.05 397.2 EAST
125.8 256.7 WEST
REGIONAL TOWER
126.55 EAST
124.15 WEST
GND CON
121.65 EAST
121.8 WEST
CLNC DEL
128.25
RADAR REQUIRED
NOTE: CHART NOT TO SCALE
WEATHER MINIMUMS:
2500 foot ceiling-5 mile visibility.
Procedure not authorized at night.
LOOP 12
Minimum
2500
HWY 114
307°
BUILDING
DFW 4.3
TEXAS STADIUM
DFW 6.1
RIVER OM
GOLF COURSE
HWY 183
R-097
R-085
Vertical Guidance
Navaid and Angle:
I-RRA LOC
(GS 3.00°)
31R
13L
17L 17R
35L 35R
18S
36S
18L 18R
36L 36R
31L
13R
DALLAS-FORT WORTH
117.0 DFW
Chan 117
STADIUM VISUAL RWY 31R
32°54'N - 97°02'W
DALLAS-FORT WORTH, TEXAS
DALLAS-FORT WORTH INTL (DFW)
Fig. 7-7. Continued.

(Continued from page 201.)
Stadium, which is on the Dallas-Fort Worth (DFW) VORTAC 097 radial at 6.1 nautical miles. The pilot is expected to fly an approximate heading of 307° to a predominant set of buildings on the DFW 085 radial at 4.3 nautical miles; note the stadium and building reference features on the chart. From the buildings, the pilot flies westbound to intercept the Runway 31R centerline. The procedure notes that vertical guidance (glide slope) is available from the I-RAA localizer, with a glide angle of 3.00°. I-RAA is the ILS to Runway 31R.

When assigned a CVFP, it's often helpful to have the appropriate instrument approach chart available with final approach guidance tuned in and identified. Especially in the busy terminal environment, it's easy to become distracted, or even momentarily disorientated. It's not unheard of for pilots to land on the wrong runway, or even at the wrong airport when flying these procedures.

USING STARs

Destination STARs should be reviewed during a pilot's preflight planning. Note that some STARs, for some airports, will only be used during certain traffic flow periods. The pilot should note which runways and airports the STAR serves. Selected STARs apply to several airports and all approaches, other routes serve only one airport and one runway. Although traffic plans and runways in use are normally not directly available during the preflight briefing, they are implied by forecast and actual wind direction and speed, and NOTAMs. For example, if we were informed that the Daisetta VOR was inoperative, we would not file, nor expect to fly the Daisetta Nine Arrival. A flight plan filed with an inappropriate STAR will normally be issued by the departure controllers in their standard "as filed" clearance, only to be amended nearing the destination.

A pilot should determine if the STAR is appropriate for the aircraft performance. A pilot flying a Cessna 150 would not normally file a profile descent or expect ATC to issue one. Finally, the STAR entry fix or transition must connect with the appropriate high or low altitude enroute structure. The pilot must also ensure the aircraft is equipped to comply with the procedure: radar, ADF, DME, and the like.

Selecting a suitable STAR allows the pilot to study the procedure, becoming familiar with NAVAIDs, routes, and altitudes. Often, when entering the airport's arrival sector, the controller will advise the pilot which procedure to expect.

Because the Stuka intersection only appears on high altitude charts (Fig. 7-6), and the McComb transition has a minimum enroute altitude of FL180 these transitions are restricted to high altitude operations; however, Alexandria and Lake Charles are high and low altitude NAVAIDs, and these routes can be used to transition from the low and high altitude enroute structures.

STARs can be started at the entry fix. For example, a pilot could file via direct or connecting airways to DAS, then the arrival . . . DAS DAS9 HOU; however, one of the purposes of transitions is to allow ATC to line up aircraft, providing appropriate separation, for the approach to the destination; thus, expediting the flow of traffic and inflecting minimum delay.

A pilot arriving from the east on J22, J138, or J2 could flight plan airways to the Lake Charles (LCH) VOR, then via the LCH transition to the Daisetta Nine Arrival to Intercontinental . . . J2 LCH DAS9 IAH (Fig. 6-16). Because LCH is a low and high altitude

NAVAID, it is equally suitable to provide a transition from Victor airways to the destination. For example, a pilot inbound on V70 from the east, destined for Hobby, could file . . . V70 LCH DAS9 HOU (Fig. 6-14). Route and transitions must connect; . . . V70 DAS DAS9 HOU or . . . V70 DAS9 HOU would be rejected; in the first case, V70 does not connect to DAS, in the second, an entry fix is missing.

A profile descent is filed as if it were a STAR. The enroute structure should be filed to a transition or entry fix, then the profile descent name to the destination: . . . J501 BSR BSR5 SFO. CVFPs are not instrument procedures as such, and do not have computer codes; their purpose is to expedite the flow of traffic. When destination weather is forecast to be at or above CVFP minimums, pilots can expect to be issued the procedure; however, it is the pilot's decision to accept a CVFP, or request an instrument approach procedure.

Certain STARs are designed to allow the pilot to navigate to an approach segment where the instrument approach will begin, others, as in the previous examples, end in radar vectors to the instrument approach procedure. In the case of radio communications failure, in the first instance, when the STAR ends where the approach begins pilots would apply assigned route and altitudes as required by FARs. The route would have been assigned by ATC either as the original clearance if communications were lost immediately after departure, or subsequently as an expected specific approach clearance for the destination if communications were lost during the arrival segment. Last assigned or minimum altitudes would apply.

If the filed or assigned STAR is a vector procedure, the pilot would be expected to comply with FAR radar vector procedures. The pilot would fly a direct route from the point of radio failure to the fix, route, or airway specified in the vector clearance. If communications are lost immediately after departure, the pilot would fly the filed STAR to the point where radar vectors were to begin. From this point the pilot would fly, as direct as possible, to the final approach course. For example, the Cugar Four Arrival also serves the Conroe Montgomery County Airport (Fig. 6-15).

Let's say we've been assigned Cugar Four and have lost radio communications. We would fly the STAR as published to Banty intersection, where we were supposed to expect vectors to the final approach course. Two options would be available to comply with FARs: proceed direct to Alibi LOM if the aircraft is equipped with an ADF; without an ADF, pick up a heading to intercept the Hobby 342 radial. The enroute low altitude chart in Fig. 6-14 reveals that the Hobby 342 radial is only a few miles east of Banty; last assigned or minimum altitudes would apply.

JEPPESEN SIDs AND STARs

Unlike the terminal procedures publication, which places STARs in a separate section of the volume, Jeppesen SID and STAR charts are filed along with the instrument approach procedure (IAP) charts for the respective airports. Jeppesen, like the TPP, includes a rate of climb table to convert ground speed and rate of climb into climb gradient in feet per nautical mile. Jeppesen's gradient to rate table is more detailed; hence, easier to use because fewer interpolations are required.

Figure 7-8 is the Houston Scholes Two Departure and Hobby Five Departure published by Jeppesen (NOS versions, Fig. 7-5). These charts use month, day, year issuance

(Continued on page 211.)

SID

JEPPESEN

JUL 26-91 (70-3A)

HOUSTON, TEXAS

HOUSTON INTERCONTINENTAL

(ALSO SERVES ❶)

Fig. 7-8.

HOUSTON Departure (R) North 119.7 East 133.6 South 134.45 West 123.8

SCHOLES TWO DEPARTURE (VUH2.METZY) (PILOT NAV)

(This procedure requires overwater flight.)

SCHOLES TWO DEPARTURE (VUH2.METZY) (PILOT NAV)
(This procedure requires overwater flight.)
All aircraft cleared as filed. Scholes VORTAC, then via Scholes R-083 to Metzy Int. Then via (transition) or (assigned route). EXPECT further clearance to requested altitude 10 minutes after departure.

TRANSITIONS
Lafayette (VUH2.LFT): From Metzy Int to Lafayette VORTAC (103 nm): Via Lafayette R-234.
Leeville (VUH2.LEV): From Metzy Int to Leeville VORTAC (189 nm): Via Scholes R-083 and Leeville R-266.
White Lake (VUH2.LLA): From Metzy Int to White Lake VORDME (73 nm): Via White Lake R-245.

Direct distance from Houston Intercontinental to:
Scholes VORTAC **49 nm**
Direct distance from AIRPORTS SERVED to Scholes VORTAC:
❶ AIRPORTS SERVED
Hooks Meml Apt **60 nm**
Montgomery Co Apt **71 nm**

CONROE TEX
Montgomery Co
247
HOUSTON TEX
Hooks Meml
150
Houston
Intercontinental
98
HOUSTON TEX
West Houston
-Lakeside
112
HOUSTON TEX
Andrau
80
HUMBLE
D (H) 116.6 IAH
N29 57.4 W095 20.7
HOUSTON TEX
Hobby
47
LA PORTE TEX
La Porte Mun
29
HOBBY
D (H) 117.6 HUB
N29 39.0 W095 16.7
HOUSTON TEX
Ellington
34
HOUSTON TEX
-Hull
83
HOUSTON TEX
-Southwest
66
ANGLETON/
LAKE JACKSON
TEX
Brazoria Co
25
083°
GALVESTON TEX
Scholes
7
SCHOLES
D (L) 113.0 VUH
N29 16.1 W094 52.1
62
D
METZY
N29 16.9
W093 41.6
LAKE CHARLES
D (H) 113.4 LCH
N30 08.5 W093 06.3
204°
054°
065°
083°
2000
64
LAFAYETTE
(VUH2.LFT)
2000
103
R234°
LAFAYETTE
D (L) 110.8 LFT
N30 08.8 W091 59.0
WHITE LAKE
(VUH2.LLA)
2000
73
R245°
WHITE LAKE
D (L) 111.4 LLA
N29 39.8 W092 22.4
086°
D
LEEVILLE
(VUH2.LEV)
2000
125
R266°
DRAGS
N29 13.6
W092 28.8
LCH
142°
113.4
LEEVILLE
D (H) 113.5 LEV
N29 10.5 W090 06.2
N
NOT TO SCALE

CHANGES: Departure frequencies.

Fig. 7-8.
Continued.

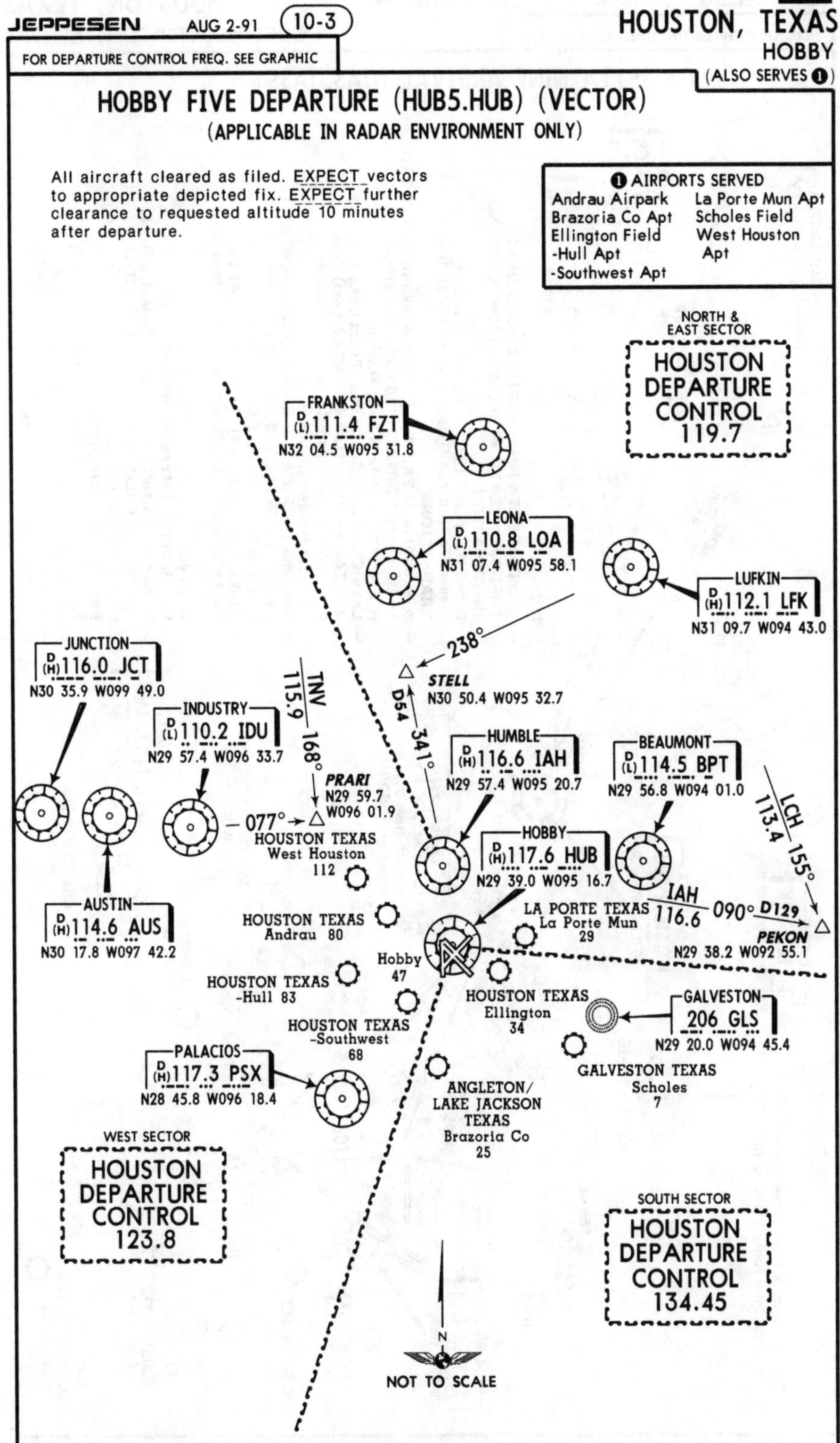

SID
JEPPESEN
AUG 2-91
10-3
HOUSTON, TEXAS
HOBBY
(ALSO SERVES ❶)
FOR DEPARTURE CONTROL FREQ. SEE GRAPHIC
HOBBY FIVE DEPARTURE (HUB5.HUB) (VECTOR)
(APPLICABLE IN RADAR ENVIRONMENT ONLY)
All aircraft cleared as filed. EXPECT vectors to appropriate depicted fix. EXPECT further clearance to requested altitude 10 minutes after departure.
❶ AIRPORTS SERVED
Andrau Airpark
Brazoria Co Apt
Ellington Field
-Hull Apt
-Southwest Apt
La Porte Mun Apt
Scholes Field
West Houston Apt
NORTH & EAST SECTOR
HOUSTON DEPARTURE CONTROL 119.7
FRANKSTON
111.4 FZT
N32 04.5 W095 31.8
LEONA
110.8 LOA
N31 07.4 W095 58.1
LUFKIN
112.1 LFK
N31 09.7 W094 43.0
238°
STELL
N30 50.4 W095 32.7
D54
341°
JUNCTION
116.0 JCT
N30 35.9 W099 49.0
TNV
115.9
168°
INDUSTRY
110.2 IDU
N29 57.4 W096 33.7
PRARI
N29 59.7
W096 01.9
077°
HUMBLE
116.6 IAH
N29 57.4 W095 20.7
BEAUMONT
114.5 BPT
N29 56.8 W094 01.0
LCH
113.4
155°
HOUSTON TEXAS
West Houston
112
HOBBY
117.6 HUB
N29 39.0 W095 16.7
AUSTIN
114.6 AUS
N30 17.8 W097 42.2
HOUSTON TEXAS
Andrau 80
LA PORTE TEXAS
La Porte Mun
29
IAH
116.6
090°
D129
PEKON
N29 38.2 W092 55.1
Hobby
47
HOUSTON TEXAS
-Hull 83
HOUSTON TEXAS
Ellington
34
GALVESTON
206 GLS
N29 20.0 W094 45.4
HOUSTON TEXAS
-Southwest
68
PALACIOS
117.3 PSX
N28 45.8 W096 18.4
ANGLETON/
LAKE JACKSON
TEXAS
Brazoria Co
25
GALVESTON TEXAS
Scholes
7
WEST SECTOR
HOUSTON DEPARTURE CONTROL 123.8
SOUTH SECTOR
HOUSTON DEPARTURE CONTROL 134.45
N
NOT TO SCALE
CHANGES: South sector departure control frequency.
© JEPPESEN SANDERSON, INC., 1985, 1991. ALL RIGHTS RESERVED.

JEPPESEN SEP 15-89 (70-2B) Eff Sep 21

STAR

HOUSTON, TEXAS

HOUSTON INTERCONTINENTAL ATIS 124.05

HOUSTON INTERCONTINENTAL

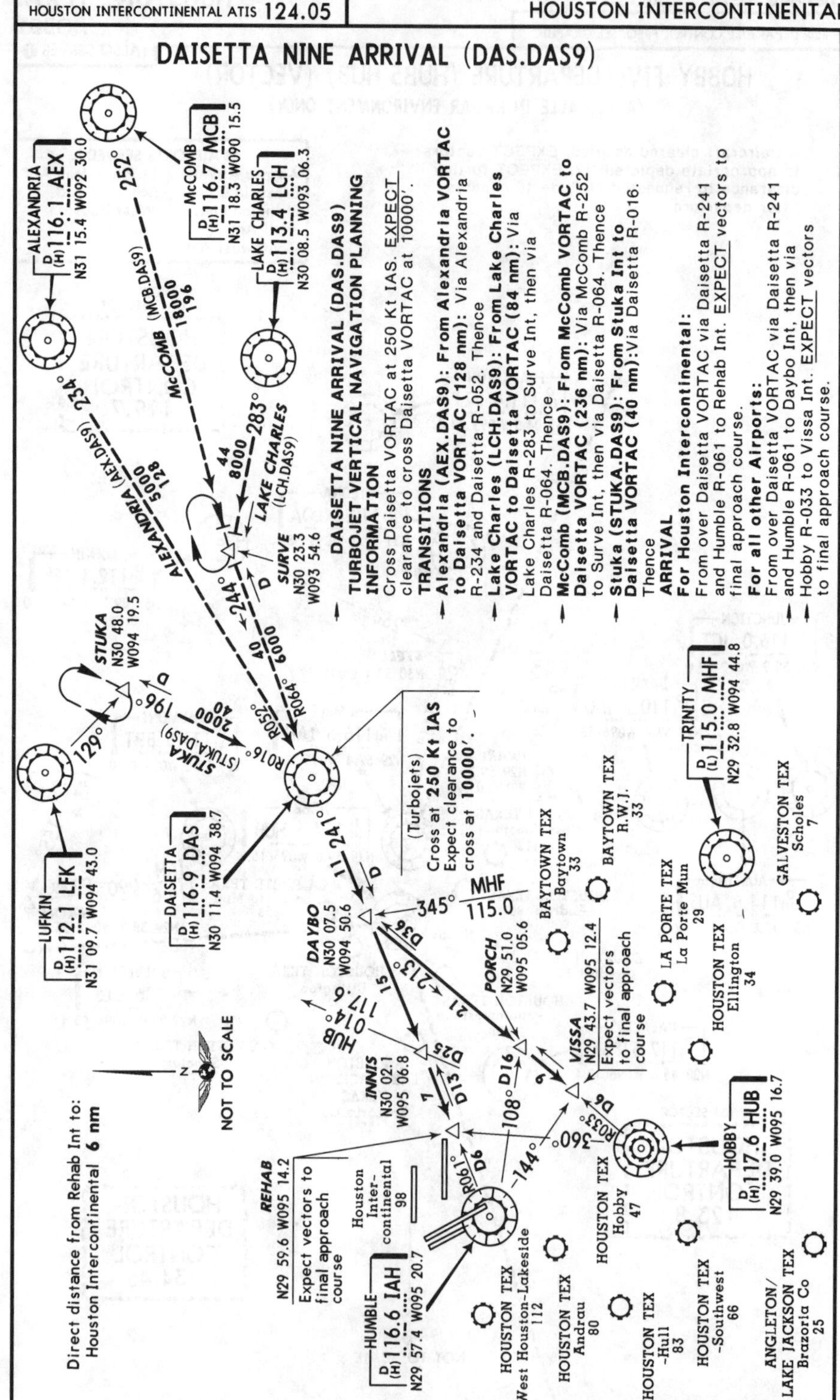

Fig. 7-9.

CHANGES: Arrival revised.

Fig. 7-9.
Continued.

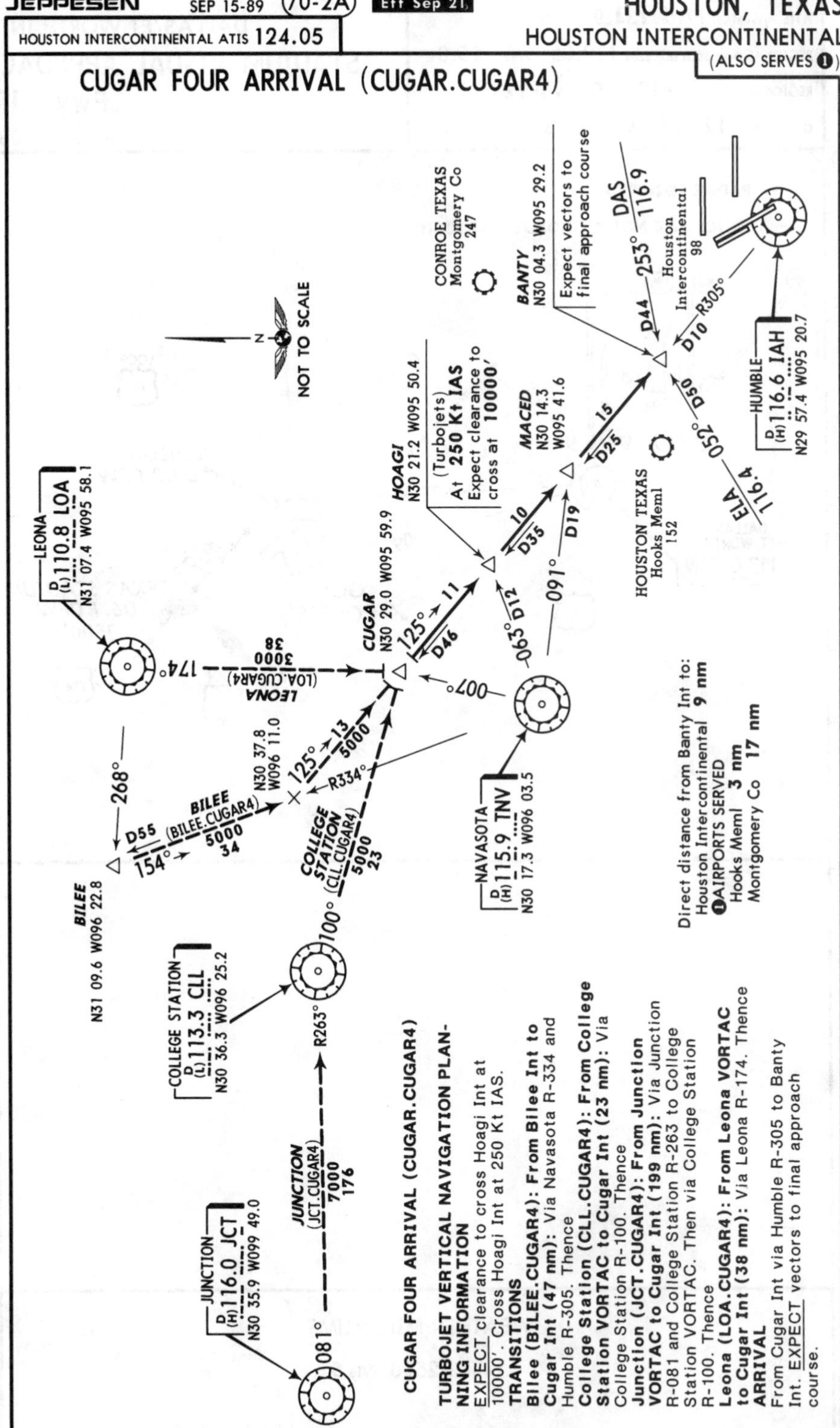

STAR
JEPPESEN
SEP 15-89
70-2A
Eff Sep 21
HOUSTON, TEXAS
HOUSTON INTERCONTINENTAL ATIS 124.05
HOUSTON INTERCONTINENTAL
(ALSO SERVES ❶)
CUGAR FOUR ARRIVAL (CUGAR.CUGAR4)
NOT TO SCALE
CONROE TEXAS
Montgomery Co
247
BANTY
N30 04.3 W095 29.2
Expect vectors to final approach course
DAS 116.9
253°
D44
Houston Intercontinental
98
R305°
D10
D50
HUMBLE
D (H) 116.6 IAH
N29 57.4 W095 20.7
052°
ELA 116.4
HOUSTON TEXAS
Hooks Meml
152
MACED
N30 14.3 W095 41.6
15
D25
HOAGI
N30 21.2 W095 50.4
(Turbojets)
At 250 Kt IAS
Expect clearance to cross at 10000'
10
D35
D19
091°
D12
063°
CUGAR
N30 29.0 W095 59.9
125° → 11
D46
007°
LEONA
D (L) 110.8 LOA
N31 07.4 W095 58.1
174°
LEONA
(LOA.CUGAR4)
3000
38
268°
D55
154°
BILEE
(BILEE.CUGAR4)
5000
34
BILEE
N31 09.6 W096 22.8
N30 37.8
W096 11.0
125° → 13
5000
R334°
COLLEGE STATION
(CLL.CUGAR4)
5000
23
NAVASOTA
D (H) 115.9 TNV
N30 17.3 W096 03.5
COLLEGE STATION
D (L) 113.3 CLL
N30 36.3 W096 25.2
100°
R263°
JUNCTION
(JCT.CUGAR4)
7000
176
JUNCTION
D (H) 116.0 JCT
N30 35.9 W099 49.0
081°
Direct distance from Banty Int to:
Houston Intercontinental 9 nm
❶AIRPORTS SERVED
Hooks Meml 3 nm
Montgomery Co 17 nm
CUGAR FOUR ARRIVAL (CUGAR.CUGAR4)
TURBOJET VERTICAL NAVIGATION PLANNING INFORMATION
EXPECT clearance to cross Hoagi Int at 10000'. Cross Hoagi Int at 250 Kt IAS.
TRANSITIONS
Bilee (BILEE.CUGAR4): From Bilee Int to Cugar Int (47 nm): Via Navasota R-334 and Humble R-305. Thence
College Station (CLL.CUGAR4): From College Station VORTAC to Cugar Int (23 nm): Via College Station R-100. Thence
Junction (JCT.CUGAR4): From Junction VORTAC to Cugar Int (199 nm): Via Junction R-081 and College Station R-263 to College Station VORTAC. Then via College Station R-100. Thence
Leona (LOA.CUGAR4): From Leona VORTAC to Cugar Int (38 nm): Via Leona R-174. Thence
ARRIVAL
From Cugar Int via Humble R-305 to Banty Int. EXPECT vectors to final approach course.
CHANGES: See other side.
© JEPPESEN SANDERSON, INC., 1987, 1989. ALL RIGHTS RESERVED

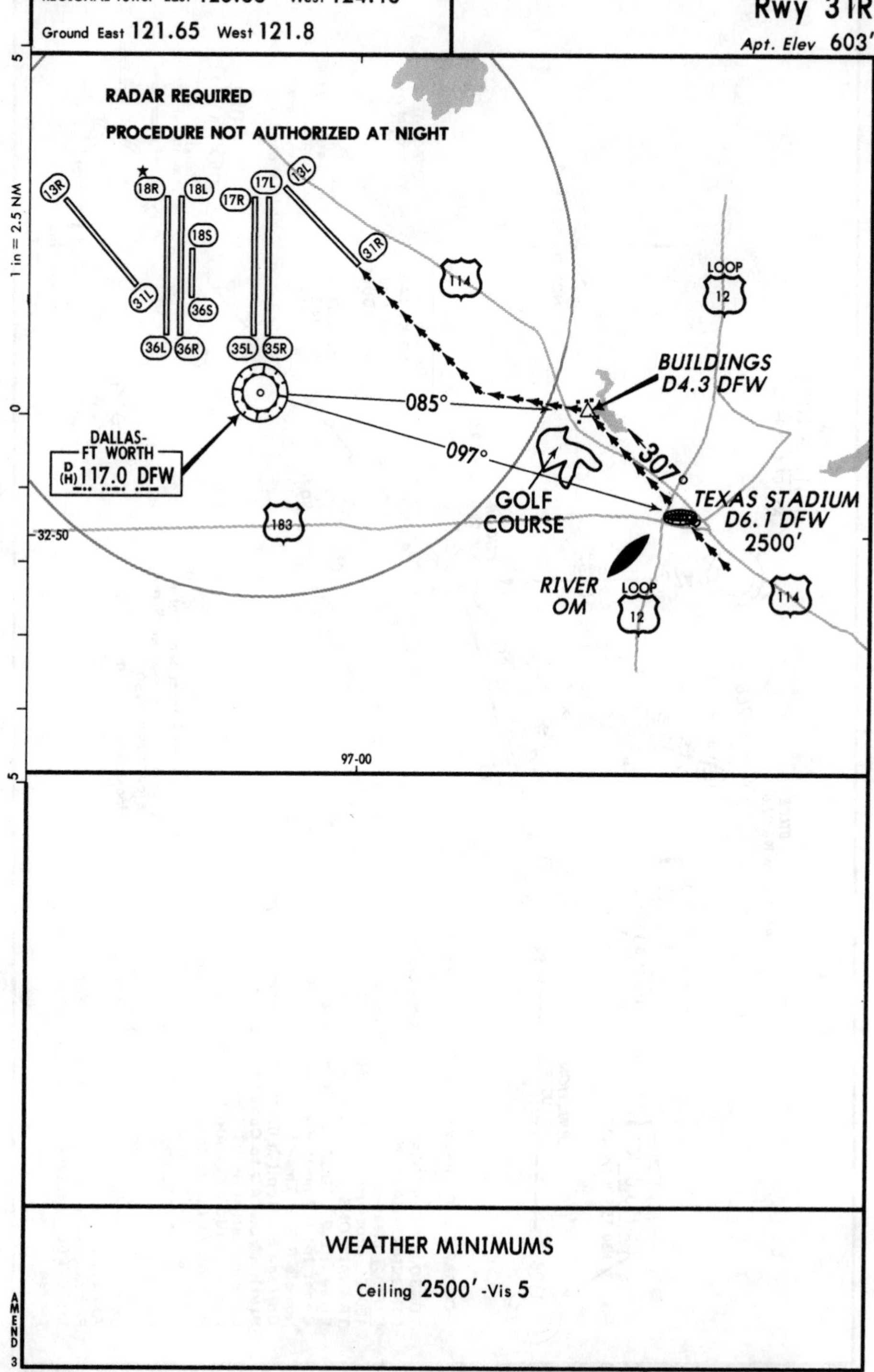

JEPPESEN
SEP 23-88 69-1
DALLAS-FT WORTH, TEXAS
DALLAS/FT WORTH INTL
STADIUM VISUAL APPROACH
Rwy 31R
Apt. Elev 603'
ATIS Arrival 117.0 134.9
REGIONAL Approach (R) East 119.05 West 125.8
REGIONAL Tower East 126.55 West 124.15
Ground East 121.65 West 121.8
RADAR REQUIRED
PROCEDURE NOT AUTHORIZED AT NIGHT
1 in = 2.5 NM
13R
18R
18L
17R
17L
13L
18S
31R
31L
36S
36L
36R
35L
35R
114
LOOP
12
BUILDINGS
D4.3 DFW
085°
097°
307°
DALLAS-
FT WORTH
D(H) 117.0 DFW
GOLF
COURSE
TEXAS STADIUM
D6.1 DFW
2500'
183
32-50
RIVER
OM
97-00
WEATHER MINIMUMS
Ceiling 2500'-Vis 5
AMEND 3
CHANGES: Chart reindexed.
© JEPPESEN SANDERSON, INC., 1988.
ALL RIGHTS RESERVED

Fig. 7-10.

(Continued from page 205.)
and effective dates, rather than Julian dates. It's important to note effective dates, and not use a chart before it becomes effective; pilots have filed for procedures that are not effective.

The margin contains departure frequencies, except in the case of the Hobby Five Departure, where frequencies are prominently displayed in respective sectors in the plan view. In this instance, ATC might not issue the departure frequency. Notes are similar to those on TPP charts; however, an additional note informs the pilot of other airports served by this SID. For example, the Scholes Two Departure also serves the Hooks Memorial and Montgomery County Airports.

The plan view is similar to TPP charts. Coordinates, to the 10th of a minute, appear below the NAVAID box. This is helpful because most coordinate navigation systems deal with latitude and longitude to the nearest 10th of a minute. The departure is depicted with a heavy black line, with transitions as heavy dashed lines. Departure and transition instructions, names, and computer codes are contained on the plan view, as well as in the text of the procedure.

The bottom margin advises that the last change to these charts involved a revised departure procedure. Small arrows in the text indicate changes from the previous issuance. The arrows flag changes for pilots that regularly use the procedure, which can be extremely helpful.

Figure 7-9 is the Houston Daisetta Nine Arrival and Cougar Four Arrival published by Jeppesen (NOS versions, Fig. 7-6). The margin contains only ATIS frequencies; approach, tower, and ground frequencies are on appropriate IAP charts. Note the bottom margin on the Cugar Four Arrival. The Cugar Four has not been amended, but the pilot is referred to the chart on the other side, which has been changed.

Figure 7-10 is the Dallas-Fort Worth Stadium Visual Approach RWY 31R chart from Jeppesen (NOS version, Fig. 7-7). CVFPs are located with and following IAPs for the respective airports. Appropriate notes are prominently displayed and ATIS, approach, tower, and ground frequencies are provided.

8 Approach procedure charts

DATA REQUIRED TO EXECUTE A DESCENT FROM THE IFR ENROUTE environment to a point where a safe landing can be made are presented on instrument approach procedure (IAP) charts. National Ocean Service IAP charts are contained in the terminal procedures publication (TPP), as discussed in chapter 7, and shown in Fig. 7-1. Changes that occur between revision cycles are advertised as FDC NOTAMs and incorporated in the NOTAM publication described in chapter 2. The Defense Mapping Agency (DMA), Canada, and private vendors also produce instrument approach procedure charts.

DMA provides instrument approach procedure charts for those areas supported with enroute charts. Caribbean and South America low and high altitude IAPs are contained in one volume. SIDs, airport diagrams, and radar minimums are included for certain civil and military airports. Europe, North Africa, and the Middle East low altitude IAPs are published in two volumes, with high altitude IAPs in a separate book. SIDs for this area are contained in one low and one high altitude volume. Africa low and high altitude IAPs are published in one volume; SIDs, airport diagrams, and radar minimums are included for certain civil and military airports. Pacific, Australasia, and Antarctica low and high altitude IAPs are combined in three volumes; STARs, SIDs, radar minimums, and airport diagrams for certain civil and military airports are included. Two additional volumes provide coverage for Canada and the North Atlantic. Similar to the TPP, these are bound volumes, approximately 5 × 8 inches.

IAPs are established by the FAA after careful analysis of obstructions, terrain, and navigational facilities. Procedures authorized by the FAA are published in the *Federal Register* as rule-making action under FAR Part 97, which covers standard instrument approach procedures. Based on this information, NOS and other charting agencies publish IAPs. FAR 91.175 (a), regarding instrument approaches to civil airports states: "unless

INSTRUMENT APPROACH PROCEDURE (IAP) CHARTS

88154 Julian Date of Last Revision   GENERAL INFORMATION & ABBREVIATIONS

★ Indicates control tower or ATIS operates non-continuously, or non-standard Pilot Controlled Lighting.
Distances in nautical miles (except visibility in statute miles and Runway Visual Range in hundreds of feet).
Runway Dimensions in feet. Elevations in feet Mean Sea Level (MSL). Ceilings in feet above airport elevation.
Radials/bearings/headings/courses are magnetic

# Indicates control tower temporarily closed UFN.

ADF . . . . Automatic Direction Finder
ALS . . . . Approach Light System
ALSF . . . . Approach Light System with Sequenced Flashing Lights
APP CON . . . . Approach Control
ARR . . . . Arrival
ASLAR . . . . Airport Surge Launch and Recovery
ASR/PAR . . . . Published Radar Minimums at this Airport
ATIS . . . . Automatic Terminal Information Service
AWOS . . . . Automated Weather Observing System
AZ . . . . Azimuth
BC . . . . Back Course
C . . . . Circling
CAT . . . . Category
CCW . . . . Counter Clockwise
Chan . . . . Channel
CLNC DEL . . . . Clearance Delivery
CTAF . . . . Common Traffic Advisory Frequency
CW . . . . Clockwise
DH . . . . Decision Heights
DME . . . . Distance Measuring Equipment
DR . . . . Dead Reckoning
ELEV . . . . Elevation
FAF . . . . Final Approach Fix
FM . . . . Fan Marker
GPI . . . . Ground Point of Interception
GS . . . . Glide Slope
HAA . . . . Height Above Airport
HAL . . . . Height Above Landing
HAT . . . . Height Above Touchdown
HIRL . . . . High Intensity Runway Lights
IAF . . . . Initial Approach Fix
ICAO . . . . International Civil Aviation Organization
IM . . . . Inner Marker
Intcp . . . . Intercept
INT . . . . Intersection
LDA . . . . Localizer Type Directional Aid
Ldg . . . . Landing
LDIN . . . . Lead in Light System
LIRL . . . . Low Intensity Runway Lights
LOC . . . . Localizer
LR . . . . Lead Radial. Provides at least 2 NM (Copter 1 NM) of lead to assist in turning onto the intermediate/final course
MALS . . . . Medium Intensity Approach Light System
MALSR . . . . Medium Intensity Approach Light Systems with RAIL
MAP . . . . Missed Approach Point
MDA . . . . Minimum Descent Altitude
MIRL . . . . Medium Intensity Runway Lights
MLS . . . . Microwave Landing System
MM . . . . Middle Marker
NA . . . . Not Authorized
NDB . . . . Non-directional Radio Beacon
NM . . . . Noutical Miles
NoPT . . . . No Procedure Turn Required (Procedure Turn shall not be executed without ATC clearance)
ODALS . . . . Omnidirectional Approach Light System
OM . . . . Outer Marker
R . . . . Radial
RA . . . . Radio Altimeter setting height
Radar Required . . . . Radar vectoring required for this approach
RAIL . . . . Runway Alignment Indicator Lights
RBn . . . . Radio Beacon
RCLS . . . . Runway Centerline Light System
REIL . . . . Runway End Identifier Lights
RNAV . . . . Area Navigation
RPI . . . . Runway Point of Intercept(ion)
RRL . . . . Runway Remaining Lights
Runway Touchdown Zone..... First 3000′ of Runway
Rwy . . . . Runway
RVR . . . . Runway Visual Range
S . . . . Straight-in
SALS . . . . Short Approach Light System
SSALR . . . . Simplified Short Approach Light System with RAIL
SDF . . . . Simplified Directional Facility
TA . . . . Transition Altitude
TAC . . . . TACAN
TCH . . . . Threshold Crossing Height (height in feet Above Ground Level)
TDZ . . . . Touchdown Zone
TDZE . . . . Touchdown Zone Elevation
TDZ/CL . . . . Touchdown Zone and Runway Centerline Lighting
TDZL . . . . Touchdown Zone Lights
TLv . . . . Transition Level
VASI . . . . Visual Approach Slope Indicator
VDP . . . . Visual Descent Point
WPT . . . . Waypoint (RNAV)
X . . . . Radar Only Frequency

Fig. 8-1. Instrument approach procedure charts are divided into five sections: margin identification, plan view, profile diagram, landing minima data, and airport sketch.

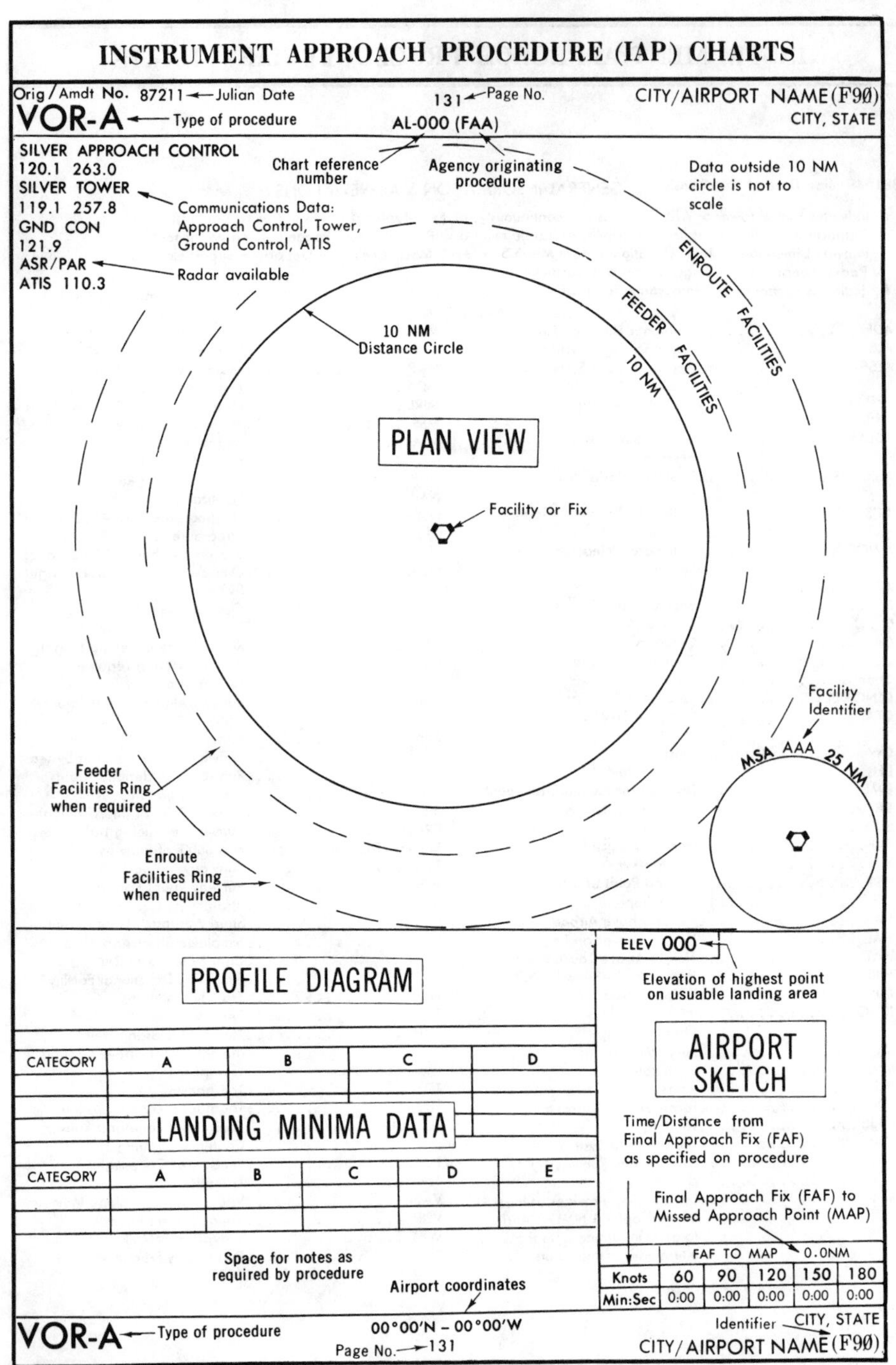

Fig. 8-1. Continued.

otherwise authorized by the administrator, when an instrument letdown to a civil airport is necessary, each person operating an aircraft . . . shall use a standard instrument approach procedure prescribed for that airport in Part 97 of this chapter." Appropriate maneuvers, which include altitude, courses, and other limitations, are prescribed in these procedures. Based on many years of accumulated experience, IAPs provide a safe letdown during instrument flight conditions. It is important that all pilots thoroughly understand these procedures and their use.

This chapter discusses the following sections and charts included in the terminal Alaska and terminal procedures publications:

- Inoperative components table
- Explanation of terms/landing minima format
- Index of terminal charts and minimums
- IFR takeoff and departure procedures (except SIDs)
- IFR alternate minimums
- General information and abbreviations
- Chart legends
- Frequency pairing
- Radar minimums
- Terminal charts
- Loran TD correction table
- Rate of descent table

IAP charts depict all related navigational data, communications information, and airport layout. The scale on IAPs varies, usually 1:500,000 or 1:750,000; individual charts are to scale, except where concentric rings appear. Features include:

- Related navigational data
- Communications frequencies
- Reporting points/fixes
- Transitional data
- Obstacles
- Drainage
- Minimum safe altitude
- Holding patterns
- Missed approach procedures
- Approach minima data
- Airport sketches
- Airport lighting information

APPROACH PROCEDURE CHART

The IAP chart is divided into five sections: margin identification, plan view, profile diagram, landing minima data, and airport sketch. Figure 8-1 contains standard abbreviations and definitions used on IAP charts, and shows their location. The margin identification provides procedure identification, type, and number, along with other information. The plan view contains a bird's eye look at the procedure. The profile diagram shows a side view of the approach. The landing minima data contains approach minimums information and procedure notes. The airport sketch provides an airport diagram, along with runway and airport information.

Bearings and courses are expressed in degrees magnetic. Radials are identified by the letter "R" (R-087) and expressed as magnetic bearing from the facility. Heights are expressed in feet above mean sea level (MSL) below 18,000 feet, and as flight levels at and above 18,000 (FL190). Distances are nautical miles and tenths (10.4), except visibilities, which are expressed in statute miles and fractions (2½). Runway visual range (RVR) is expressed in feet (2,400). Aircraft speeds are in knots.

An IAP might have up to four separate segments: initial, intermediate, final, and missed approach. Approach segments begin and end at designated fixes; however, under some circumstances certain segments might begin at specified points where no fixes are available. The fixes are named to coincide with the associated segment. For example, the intermediate segment begins at the intermediate fix and ends at the final approach fix, where the final approach segment begins.

An instrument approach begins at the initial approach fix (IAF). Feeder routes provide a transition from the enroute structure to the IAF; however, when the IAF is part of the enroute structure there might be no need to designate feeder routes. This is the point where the aircraft departs the enroute phase and maneuvers to enter an intermediate segment. An initial approach may be made along an arc, radial, course, heading, radar vector, or any combination. Procedure turns, holding pattern descents, and high altitude penetrations are initial segments. Positive course guidance is required, except when dead reckoning courses can be established over limited distances. Altitudes are established in 100-foot increments. Normally a minimum of 1,000 feet obstacle clearance is provided.

The intermediate approach segment blends the initial approach into the final approach segment. In this segment aircraft configuration, speed, and positioning adjustments are made for transition to the final approach. The intermediate segment begins at the intermediate fix (IF), or point, and ends at the final approach fix (FAF). There are two types of intermediate segments: the radial or course intermediate segment, and the arc intermediate segment. Altitudes are established in 100-foot increments. The optimum descent gradient is 250 feet per nautical mile, but where a higher descent gradient is necessary, 500 feet per nautical mile is allowed. Normally a minimum of 500 feet obstacle clearance is provided.

Alignment and descent for landing are accomplished in the final approach segment, which begins at the final approach fix. The FAF is designated on IAP charts with a Maltese cross for a nonprecision approach and a lightning bold symbol for a precision approach. When a NAVAID, such as a VOR or NDB, is located on the airport, an FAF is not designated. The final approach point (FAP) on these procedures designates the point where the aircraft is established inbound on the final approach course from the procedure turn and the final approach descent commences. The FAP serves as the FAF and identifies the beginning of the final approach segment for a nonprecision approach. Descent gradients generally vary from 300 to 400 feet per nautical mile for nonprecision approaches. Obstacle clearance begins at the FAF or FAP and ends at the runway or missed approach point. Obstacle clearance for nonprecision approaches normally varies between 250 and 350 feet depending upon the type of procedure (LOC, VOR, NDB, and the like). An aircraft should never descend below the published altitude on the final approach segment, except for final descent to landing. Glide slope on a precision approach is normally 3°, but can vary slightly on individual procedures; the glide path is normally 1° thick, which represents a vertical distance of approximately 920 feet when the aircraft is 10 miles from touchdown, narrowing to a few feet at touchdown. Obstacle clearance for precision approaches is based on a complicated formula; basically, the closer to the runway, the lower the obstacle clearance. The pilot should never allow a full-scale deflection below the glide slope while on the final approach segment.

Some nonprecision approaches have designated visual descent points (VDP). When an approach incorporates a VDP, it is identified by a navigational fix. Where a visual approach slope indicator (VASI) is installed, the VDP is located at the point where the lowest VASI glide slope intersects the lowest MDA; where a VASI is not installed, the VDP is located at the point on the final approach course at the MDA, normally, where a descent gradient to the threshold of 300 to 400 feet per nautical mile begins.

Missed approach procedures are established for all instrument approaches. The missed approach procedure begins at the decision height for precision approaches, or at a specified point for nonprecision procedures. This segment is designed to be simple, specify an altitude, and whenever practical, provide a clearance limit. Whenever possible, the missed approach track will be a continuation of the final approach course. A turn of 15° or fewer is considered straight. Missed approach obstacle clearance begins from the missed approach point (MAP) and ultimately provides 1,000 feet clearance. It is important for the pilot to immediately establish the aircraft in a positive rate of climb at the MAP. Obstacle clearance protected areas are predicated on the assumption that the missed approach is executed at the prescribed minimum descent altitude (MDA) or decision height (DH). If visual reference is lost at any point during a circling approach, the missed approach procedure must be immediately executed. To become established on the missed approach course, the pilot should make an initial climbing turn toward the landing runway and continue to turn until established on the missed approach course. Because the circling maneuver may be accomplished in more than one direction, different flight paths will be required to establish the aircraft on the missed approach course.

The MAP might be the intersection of an electronic glide path with a DH, a NAVAID, a fix, or a specified distance from the FAF. The specified distance will not be greater than the distance from the FAF to the runway, or prior to the VDP. The MAP for a full ILS, microwave landing system (MLS), and precision approach radar (PAR) is at the DH, the localizer-only MAP is usually over the runway threshold. In some nonprecision procedures, the MAP might be prior to reaching the runway threshold in order to clear obstructions in the missed approach climb. The pilot determines the MAP by timing from the FAF for some nonprecision approaches. The distance from FAF to MAP, and time and speed table, are provided below the airport sketch.

When the missed approach procedure specifies holding at a facility or fix, holding is accomplished in accordance with the depicted pattern on the plan view, and at the altitude in the missed approach instructions, unless otherwise specified by ATC. ATC might specify an alternate missed approach procedure.

Various terms are used in the missed approach procedure that have specific meanings with respect to climbs and turns, usually for obstruction avoidance:

- "Climb to . . . ." means a normal climb along the prescribed course
- "Climbing right turn . . . ." means climbing right turn as soon as safety permits, normally to avoid obstructions straight ahead
- Climb to 2,400 turn right . . . ." means climb to 2,400 feet prior to making the right turn, normally to clear obstructions

A Category I ILS approach procedure provides an approach to a decision height of not fewer than 200 feet. The Category I ILS consists of the following components: localizer (LOC), glide slope (GS), outer marker (OM), and middle marker (MM). A complete Category I ILS consists of these components. When the localizer fails an ILS approach cannot be used. When the glide slope becomes inoperative, the ILS becomes a nonprecision approach. When other components become inoperative the ILS may continue to be used, normally with increased landing minimums. Nondirectional radio beacons, called compass locators, might be installed at the outer and middle marker sites, but are not considered as basic components of the ILS; however, when installed, they may be substituted for the outer or middle marker. DME might also be associated with an ILS. When installed with the ILS, DME may be substituted for the outer marker. When a unique operational requirement exists, DME information from a separate facility might also be used to provide DME arc initial approaches, a FAF for back course approaches, or substituted for the outer marker. At one time, DME information from a separate facility was used to establish fixes inside the FAF. On several occasions fatal accidents resulted from pilots tuning the receiver to the wrong facility or misinterpreting the information. Pilots should use extra caution when obtaining DME information from a facility not being used for course guidance; positive aural identification of a navigational facility must be incorporated into every such frequency change; identifying enroute NAVAIDs was presented in chapter 4.

An ILS is identified by the Morse code letter "I" preceding the airport location identifier; the ILS at the Stockton, California (SCK), is identified ISCK. Compass locators normally have a two-letter Morse code identifier. The outer compass locater (LOM) uses the first two letters of the airport identifier (SC). The middle compass locator (LMM), when installed, uses the last two letters of the airport identifier (CK). Most LOMs are given five-letter names, similar to intersections. Because large airports have more than one ILS procedure, random three-letter identifiers, preceded by the letter "I," are used. Microwave landing systems are identified by the use of the letter "M."

Margin identification

IAP identification is designed to be meaningful for the pilot, and permit ready identification to air traffic controllers. The procedure identification is derived from the type facility providing final approach course guidance (ILS, VOR, NDB, and the like). A solidus (/) indicates that more than one type of navigational equipment is required to execute the procedure (VOR/DME: The aircraft must be equipped with operational VOR and

DME receivers.). When DME arcs and fixes are used in the procedure, but the approach identification does not include DME, the procedure may be used by aircraft not equipped with DME. In such cases arc transitions could not be used, and minimums might be higher. Where high altitude—jet penetration— procedures are available the procedure identification is prefixed with the letters "HI" (HI-VOR RWY 5).

The MLS provides precision navigational guidance—azimuth, elevation, and distance—for exact alignment and descent of aircraft on approach to a runway. The MLS initially supplements and is supposed to eventually replace ILS as the standard landing system in the United States; however, the system requires special radio equipment aboard the aircraft that most owners have not installed because ILS facilities are more prevalent, among other reasons.

When any approach procedure meets criteria for straight-in landing (approach course within 30° of runway centerline), the runway number identifies the procedure, ILS RWY 14 for Runway 14. Only a circling approach will be authorized when the approach course is more than 30° from runway centerline. Circling approaches are identified by the type of navigational aid and an alphabetical suffix (VOR-A, NDB-C, and the like).

In cases where different approaches use the same final approach guidance to the same runway, a number differentiates the procedures (LDA/DME-1 RWY 18 and LDA/DME-2 RWY 18). In the preceding example there are two LDA/DME approaches to Runway 18 at the South Lake Tahoe Airport in California. The LDA/DME-1 has lower minimums than the LDA/DME-2 because the missed approach utilizes the Squaw Valley (SWR) VORTAC. Should SWR be inoperative, the procedure is not authorized; however, the LDA/DME-2, with higher minimums not utilizing SWR, could be executed with the SWR VORTAC out of service.

At airports where a program has been specifically approved, ATC might conduct simultaneous converging instrument approaches to runways having an included angle from 15° to 100°. The procedures require separate instrument approach procedures for each converging runway. MAPs must be at least three miles apart, and missed approach procedures must ensure that missed approach protected airspace does not overlap. Other requirements include, radar, nonintersecting final approach courses, precision approach systems, and if the runways intersect, the controller must be able to apply visual separation and intersecting runway separation. Intersecting runways also require minimums of at least 700 feet and 2 miles. Straight-in approaches and landings are required. Whenever simultaneous converging approaches are in progress, pilots will be informed by the controller or automatic terminal information service (ATIS).

IAPs are labeled as original (Orig) or with an amendment number (Amdt 7), which helps determine chart currency. FDC NOTAMs reference the original or amended chart: FDC . . . ILS RWY 7 AMDT 9. STRAIGHT-IN MINIMUMS NA; this FDC NOTAM changes the procedures for the ILS RWY 7 amendment 9 approach, straight-in minimums for the approach are not authorized. If the approach chart is not ILS RWY 7, AMDT 9, it is obsolete. If this is a permanent change, the ILS RWY 7, AMDT 10 should contain the amended procedure.

In addition to procedure identification, margins contain chart amendment number and Julian date, TPP page number, reference number and approving authority, airport name and location with three-letter identifier, and airport coordinates, as shown in Fig. 8-1.

Plan view

The plan view page provides a horizontal picture of the procedure, plus communication frequencies, minimum safe altitudes within 25 nautical miles of a central facility or fix, and denotes transition routes from enroute and feeder facilities. Figure 8-2 contains plan view symbols, which are similar to those used on other instrument charts.

The plan view is shown to scale within the 10-nautical-mile circle. Dashed concentric circles are used when information will not fit to scale within the limits of the plan view area, labeled enroute facilities and feeder facilities; enroute facilities are NAVAIDs, fixes, and intersections that are part of the enroute low altitude structure; feeder facilities are NAVAIDs, fixes, and intersections used by ATC to direct aircraft to intermediate facilities, or fixes between the enroute structure and the initial approach segment.

When it is necessary to reverse direction to establish the aircraft inbound on an intermediate or final approach course, a procedure turn is prescribed, symbolized in Fig. 8-2. A procedure turn is required, except when one of four conditions applies:

- The symbol "NoPT" is shown
- Radar vectoring is being provided
- A one-minute holding pattern is published in lieu of a procedure turn
- The procedure turn is not authorized

The altitude prescribed for the procedure turn is the minimum altitude until the aircraft is established on the inbound course. The maneuver must be completed within the distance specified in the profile diagram. A barb indicates the direction or side of the outbound course on which the course reversal is to be made. Headings are provided for the 45° procedure turn; however, the point at which the turn may be commenced, and type and rate of turn is left to the discretion of the pilot.

For example, the 45° procedure turn, racetrack pattern, teardrop procedure, or 80°–260° course reversal are acceptable. When a teardrop procedure or holding pattern replaces the procedure turn, the pilot must accomplish the published procedure, unless otherwise authorized by ATC. The absence of the procedure turn barb in the plan view indicates the procedure turn is not authorized.

When an approach course is published on an ILS procedure that does not require a procedure turn (NoPT), the following applies. In the case of a dogleg track where no fix is depicted at the point of interception on the localizer course, the total distance is shown from the facility or fix to the LOM. The minimum altitude applies until glide slope intercept, at which point the aircraft begins descent. When the glide slope is not utilized, the minimum altitude is maintained to the LOM.

Minimum safe altitudes provide at least 1,000 feet obstacle clearance for emergency use within 25 miles of the procedure navigation facility. These will normally be the VOR or NDB, or FAF for ILS and LOC approaches. Altitudes are rounded to the next higher 100-foot increments.

Emergency safe altitudes, normally only used in military procedures, are established with a 100 mile radius and feature a common altitude for the entire area, providing 2,000 feet of obstacle clearance when in designated mountainous areas.

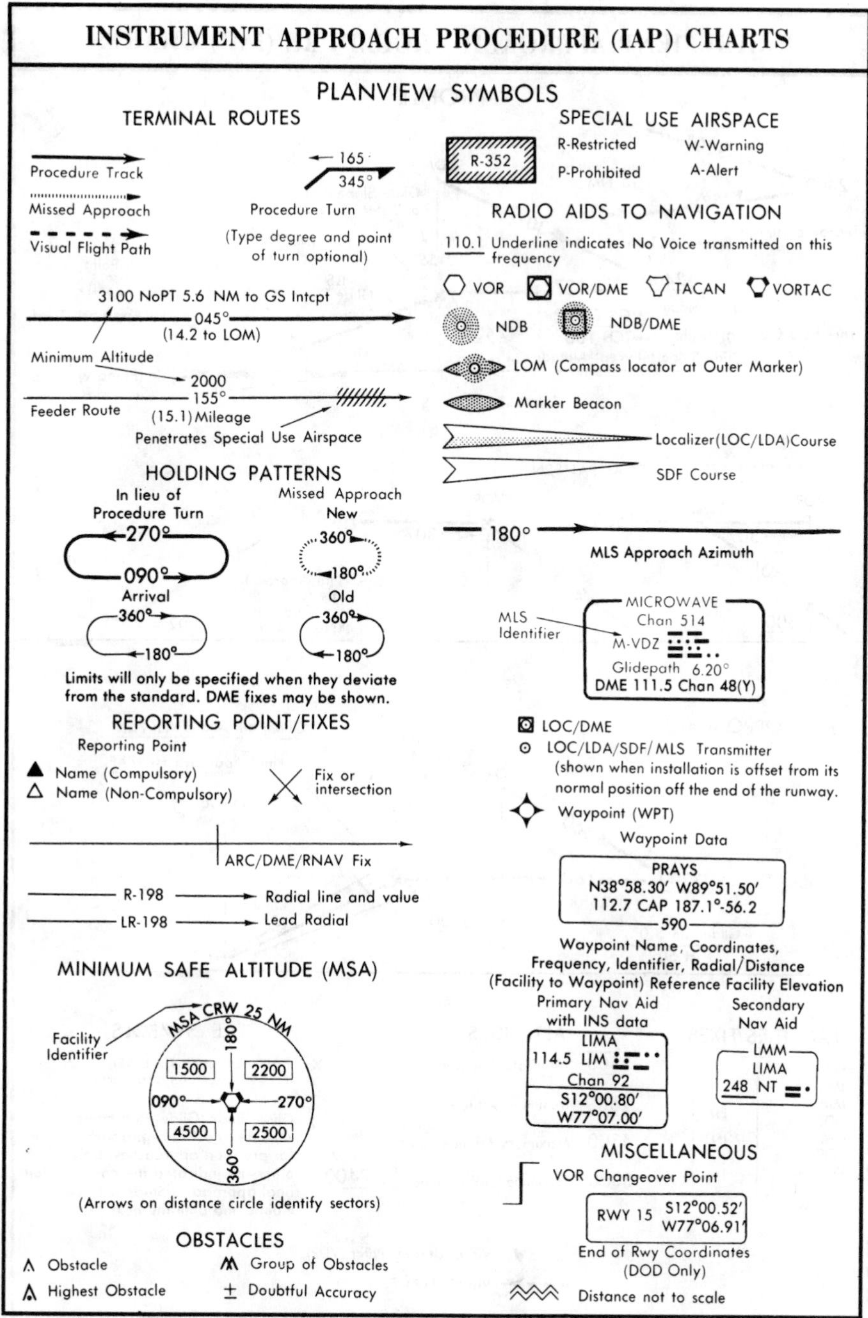

Fig. 8-2. Symbols on instrument approach procedure charts are similar to those used on other instrument charts.

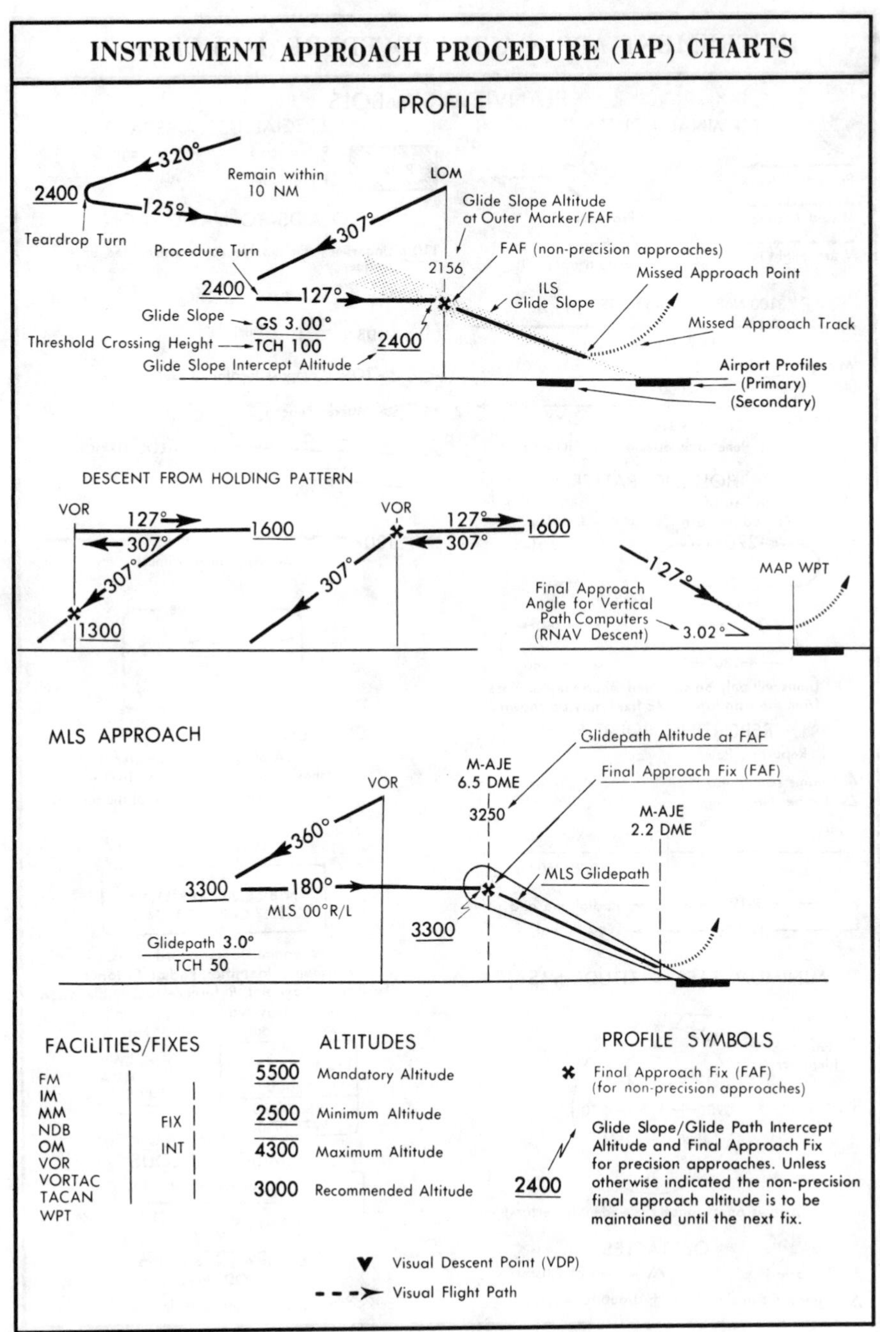
INSTRUMENT APPROACH PROCEDURE (IAP) CHARTS
PROFILE
320°
2400
125°
Remain within
10 NM
Teardrop Turn
LOM
307°
Procedure Turn
Glide Slope Altitude
at Outer Marker/FAF
FAF (non-precision approaches)
2156
Missed Approach Point
2400
127°
ILS
Glide Slope
Glide Slope
GS 3.00°
Missed Approach Track
Threshold Crossing Height
TCH 100
2400
Glide Slope Intercept Altitude
Airport Profiles
(Primary)
(Secondary)
DESCENT FROM HOLDING PATTERN
VOR
127°
1600
307°
307°
1300
VOR
127°
1600
307°
307°
127°
MAP WPT
Final Approach
Angle for Vertical
Path Computers
(RNAV Descent)
3.02°
MLS APPROACH
Glidepath Altitude at FAF
M-AJE
6.5 DME
Final Approach Fix (FAF)
VOR
3250
M-AJE
2.2 DME
360°
MLS Glidepath
3300
180°
MLS 00°R/L
3300
Glidepath 3.0°
TCH 50
FACILITIES/FIXES
FM
IM
MM
NDB
OM
VOR
VORTAC
TACAN
WPT
FIX
INT
ALTITUDES
5500 Mandatory Altitude
2500 Minimum Altitude
4300 Maximum Altitude
3000 Recommended Altitude
PROFILE SYMBOLS
Final Approach Fix (FAF)
(for non-precision approaches)
Glide Slope/Glide Path Intercept
Altitude and Final Approach Fix
for precision approaches. Unless
otherwise indicated the non-precision
final approach altitude is to be
maintained until the next fix.
2400
Visual Descent Point (VDP)
Visual Flight Path

Fig. 8-2. Continued.

Profile diagram

The profile diagram contains minimum altitudes, provides maximum distance for procedure turns, altitudes over fixes, distances between fixes, glide slope angle for precision approaches, and missed approach procedures. Figure 8-2 contains profile diagram symbols. When a navigational facility (VOR, NDB, marker beacon, and the like) establishes the fix, a solid vertical line is depicted. An intersection fix is shown as a dashed vertical line. Mandatory, minimum, maximum, and recommended altitudes are depicted in the same manner as on STAR charts. Note the Maltese cross and lightning bolt symbols used to identify the FAF or FAP.

The precision approach glide slope intercept altitude is the minimum altitude for glide slope interception after completion of the procedure turn. The intercept altitude applies to precision approaches and, except where otherwise prescribed, becomes the minimum altitude for crossing the final approach fix if the glide slope is inoperative or not used.

Stepdown fixes in nonprecision procedures might be provided between the final approach fix and the airport for the purpose of allowing a lower minimum descent altitude after passing an obstruction. Normally, there is only one step down fix between the final approach fix and the MAP. If the step down fix cannot be identified for any reason, the altitude at the step down fix becomes the MDA.

Landing minima data

Figure 8-3 shows landing minima data symbols. Values contained within parentheses do not apply to civil operations, only military operations. Landing minimums are expressed as minimum descent altitudes (MDA) or decision height (DH), and visibility in statute miles or RVR. MDA means the lowest altitude, expressed in feet above mean sea level, to which descent is authorized on final approach, where no electronic glide slope is provided, or during circle-to-land maneuvering. DH means the height at which a decision must be made during an approach with an ILS/MLS or PAR to continue the approach only if the runway environment is visible or immediately execute a missed approach. This height is expressed in feet above mean sea level and as a radar altimeter setting (RA) for Category II and III procedures, which require additional aircraft equipment and aircrew certification. Height above airport (HAA) and height above touchdown (HAT) are also published; HAA indicates the height of the MDA above the published airport elevation, published in conjunction with circling minimums for all types of approaches; HAT indicates the height of the DH or MDA above the highest runway elevation in the touchdown zone, the first 3,000 feet of runway, and is published in conjunction with straight-in minimums.

Figure 8-3 also shows a visibility conversion from RVR to statute miles. To authorize RVR minimums, the procedure must have, in addition to basic components, RVR reported for the runway, high intensity runway lights (HIRL), and all weather runway markings for precision procedures, or instrument runway markings for nonprecision approaches. If RVR minimums for takeoff or landing are published, but RVR equipment is inoperative, RVR must be converted and applied as ground visibility as shown in Fig. 8-3; symbols contained in Fig. 8-3 are included in section K of the TPP.

(Continued on page 226.)

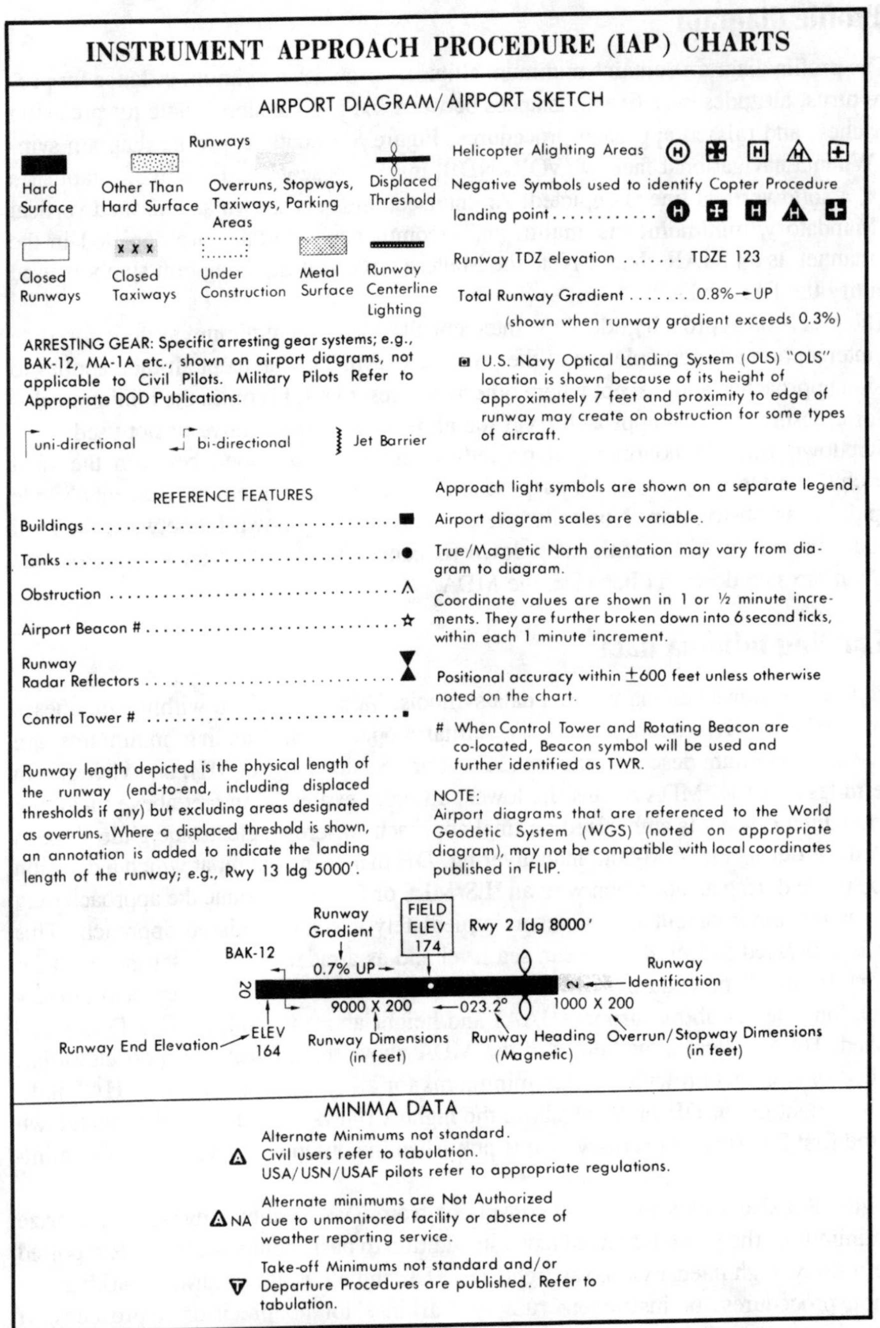

INSTRUMENT APPROACH PROCEDURE (IAP) CHARTS
AIRPORT DIAGRAM/AIRPORT SKETCH
Runways
Hard Surface
Other Than Hard Surface
Overruns, Stopways, Taxiways, Parking Areas
Displaced Threshold
Helicopter Alighting Areas
Negative Symbols used to identify Copter Procedure landing point
Closed Runways
Closed Taxiways
Under Construction
Metal Surface
Runway Centerline Lighting
Runway TDZ elevation ...... TDZE 123
Total Runway Gradient .... 0.8%→UP
(shown when runway gradient exceeds 0.3%)
ARRESTING GEAR: Specific arresting gear systems; e.g., BAK-12, MA-1A etc., shown on airport diagrams, not applicable to Civil Pilots. Military Pilots Refer to Appropriate DOD Publications.
U.S. Navy Optical Landing System (OLS) "OLS" location is shown because of its height of approximately 7 feet and proximity to edge of runway may create on obstruction for some types of aircraft.
uni-directional
bi-directional
Jet Barrier
REFERENCE FEATURES
Buildings
Tanks
Obstruction
Airport Beacon #
Runway Radar Reflectors
Control Tower #
Approach light symbols are shown on a separate legend.
Airport diagram scales are variable.
True/Magnetic North orientation may vary from diagram to diagram.
Coordinate values are shown in 1 or ½ minute increments. They are further broken down into 6 second ticks, within each 1 minute increment.
Positional accuracy within ±600 feet unless otherwise noted on the chart.
When Control Tower and Rotating Beacon are co-located, Beacon symbol will be used and further identified as TWR.
Runway length depicted is the physical length of the runway (end-to-end, including displaced thresholds if any) but excluding areas designated as overruns. Where a displaced threshold is shown, an annotation is added to indicate the landing length of the runway; e.g., Rwy 13 ldg 5000′.
NOTE:
Airport diagrams that are referenced to the World Geodetic System (WGS) (noted on appropriate diagram), may not be compatible with local coordinates published in FLIP.
Runway Gradient
FIELD ELEV 174
Rwy 2 ldg 8000′
BAK-12
0.7% UP
20
2
Runway Identification
9000 X 200
023.2°
1000 X 200
Runway End Elevation
ELEV 164
Runway Dimensions (in feet)
Runway Heading (Magnetic)
Overrun/Stopway Dimensions (in feet)
MINIMA DATA
Alternate Minimums not standard. Civil users refer to tabulation. USA/USN/USAF pilots refer to appropriate regulations.
NA
Alternate minimums are Not Authorized due to unmonitored facility or absence of weather reporting service.
Take-off Minimums not standard and/or Departure Procedures are published. Refer to tabulation.

Fig. 8-3. Airport symbols describe the runway environment; landing minima are expressed in feet MSL or a radar altimeter height, and visibility in statute miles or RVR in feet.

INSTRUMENT APPROACH PROCEDURE (IAP) CHARTS

AIRCRAFT APPROACH CATEGORIES

Speeds are based on 1.3 times the stall speed in the landing configuration at maximum gross landing weight. An aircraft shall fit in only one category. If it is necessary to maneuver at speeds in excess of the upper limit of a speed range for a category, the minimums for the next higher category should be used. For example, an aircraft which falls in Category A, but is circling to land at a speed in excess of 91 knots, should use the approach Category B minimums when circling to land. See following category limits:

MANEUVERING TABLE

Approach Category	A	B	C	D	E
Speed (Knots)	0-90	91-120	121-140	141-165	Abv 165

LANDING MINIMA FORMAT

In this example airport elevation is 1179, and runway touchdown zone elevation is 1152.

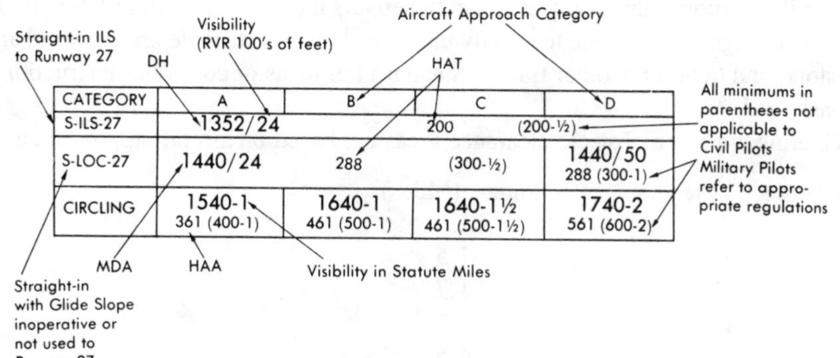

RVR/Meteorological Visibility Comparable Values

The following table shall be used for converting RVR to meteorlogical visibility when RVR is not reported for the runway of intended operation. Adjustment of landing minima may be required – see Inoperative Components Table.

RVR (feet)	Visibility (statute miles)	RVR (feet)	Visibility (statute miles)
1600	1/4	4500	7/8
2400	1/2	5000	1
3200	5/8	6000	1 1/4
4000	3/4		

Fig. 8-3. Continued.

(Continued from page 223.)

Aircraft approach categories are shown with landing minima established for six aircraft approach categories: A, B, C, D, E, and COPTER. In the absence of COPTER MINIMA, helicopters may use category A minimums. Where the airport landing surface is not adequate, or other restrictions prohibit certain categories of aircraft, a not authorized (NA) notation appears. And, approach category E is normally only published on high altitude procedures. Pilots must be aware, as the note in Fig. 8-3 explains, ". . . aircraft which falls in category A, but is circling to land at a speed in excess of 91 knots, should use the approach category B minimums when circling to land." Minimums are based on obstruction clearance because higher speeds command consideration of larger obstruction clearance areas. That's why MDA and visibility increases with higher approach speeds.

Circling minimums provide adequate obstruction clearance and the aircraft pilot should not descend below this altitude until in a position to make the final descent for landing. The pilot must determine the exact maneuver for this procedure, based on airport design, aircraft position, altitude, and airspeed. This requires good judgment and a knowledge of aircraft capabilities. In general, the following basic rules should be applied. Maneuver the shortest path to base or downwind considering weather conditions. There is no restriction from passing over the airport or other runways. Keep in mind, especially at uncontrolled airports, that many circling maneuvers may be made while VFR flying is in progress. It might be desirable to overfly uncontrolled airports to determine wind and turn indicators, and to observe other traffic. Standard left turns or controller instructions must be considered.

Circling approach obstacle clearance areas are based on aircraft approach category:

Approach Category	Radius (Miles)
A	1.3
B	1.5
C	1.7
D	2.3
E	4.5

This explains the following FDC NOTAM:

FDC 1/1521 SCK FI/T /SCK/ STOCKTON METROPOLITAN, STOCKTON, CA.
ILS RWY 29R AMDT 18..VOR RWY 29R AMDT 17..NDB RWY 29R AMDT 14: CATS B/C CIRCLING MDA 580/HAA 550. 265 FT MSL CRANE 1.4 NM WEST OF AER 11L.

The circling MDA for aircraft categories B and C have been raised because of a 265-foot MSL crane 1.4 miles west of approach end of runway (AER) 11L. Category A is not affected because the crane is outside of the obstruction clearance radius. Category D minimums are already at 580 feet because of other obstructions; therefore, this FDC NOTAM only applies to category B and C operations.

The instrument pilot must know which category applies to the aircraft flown during this trip. Most often FSS specialists are not aware of aircraft approach categories. In such cases pilots can expect to receive information that does not apply; however, as always, it's the pilot's responsibility to determine whether the information will affect the operation.

The same minimums apply to day and night operations, unless different minimums are specified at the bottom of the box in the space provided for symbols or notes. Minimums for full ILS and LOC only, straight-in, and circling appear directly under each aircraft category. When there is no division line between minimums for each category, the minimums apply to two or more categories. Normally, when a step down fix is within three miles of the airport, dual equipment will be required to establish the fix. Under these conditions two sets of minima will appear, one will be labeled DUAL VOR or DME MINIMUMS.

ATC might authorize a side-step maneuver to an adjacent parallel runway when separated by 1,200 feet or fewer. When so cleared, pilots are expected to commence the side-step maneuver as soon as possible after the runway or runway environment is sighted. Landing minimums to adjacent runways will be higher than those for the primary runway, but will normally be lower than circling minimums.

Category of operation, with respect to the operation of aircraft, means a straight-in ILS approach to a runway under special instrument approach procedures issued by the FAA or other authority. Category II and Category III operations permit lower landing minimums for specially trained crews operating specially equipped aircraft with radios that receive specially certificated ILS systems; the DH is expressed as a radar altimeter setting (RA 100 [feet]) as opposed to mean sea level.

Space is provided below the landing minima data box for notes as required by the procedure. Notes contain additional information about the procedure, such as altimeter setting source, runway lighting activation procedures, and nonstandard alternate and takeoff minimums, or published departure procedure.

FAR 91.169 (c), regarding IFR alternate airport weather minimums, specifies standard alternate criteria. "Precision approach procedure: Ceiling 600 feet and visibility 2 statute miles. Nonprecision approach procedure: Ceiling 800 feet and visibility 2 statute miles." Alternate minimums require ceiling and visibility. A triangle enclosing an "A" indicates that alternate minimums are not standard. If the airport is not authorized for use as an alternate, the letters "NA" follow the symbol.

When a pilot elects to proceed to the selected alternate airport, the alternate ceiling and visibility minimums no longer apply; published landing minimums now apply to the new destination. The alternate airport becomes the new destination, and the pilot uses the landing minimum appropriate to aircraft equipment and the type of procedure selected.

A triangle enclosing a "T" indicates nonstandard takeoff minimums or published departure procedures, or both. FAR 91.175 (f), civil airport takeoff minimums, specifies standard takeoff minimums for pilots operating under FAR Parts 121, 125, 127, 129, or 135, which are air carrier, air taxi, and large aircraft operators. Nonstandard alternate and takeoff minimums, and departure procedures are tabulated in the TPP.

Airport sketch

Airport diagrams are specifically designed to assist pilots while maneuvering on the ground by providing runway and taxiway configurations, and provide information for updating inertial navigational systems (INS). The diagram also indicates airport elevation and time and distance from the FAF to the MAP; time and distance from FAF to MAP are omitted on ILS procedures, when the NAVAID is collocated with the MAP, or when DME is required for the procedure.

Figure 8-3 shows airport sketch symbols. The scale perspective of airport diagrams and the northern orientation varies. The symbols provide runway surface information, runway and taxiway condition, and indicate the presence of displaced thresholds. Runway touchdown elevations, and gradients are shown when they exceed 0.3 percent. Runway length is the length of the runway, end to end, including displaced thresholds, but excluding overruns. When a displaced threshold is depicted, a note is added to indicate runway available for landing. Availability of arresting gear and optical landing systems are noted for military operations.

Pilots should be familiar with the lighting systems at destination airports (Fig. 8-4), especially when operating at night. Runway approach lighting system symbols indicate the type of lighting available, and where applicable, pilot controlled lighting (PCL); if the PCL is nonstandard, the pilot must refer to the *Airport/Facility Directory* (A/FD) for activation procedures. Availability of approach lighting system is indicated by a circled letter A; a number in the circle indicates the type of system installed; Fig. 8-4, which is section L of the TPP, reveals that the simplified short approach lighting system (SSALR) is designated A3.

TERMINAL PROCEDURES PUBLICATION

Pilots that use NOS chart material must be familiar with the TPP. Beyond sections of the TPP that are discussed elsewhere, landing minimums published in the landing minima data section of IAP charts are based on the full operation of all electronic components and visual aids, such as approach lighting systems, associated with the procedure. Higher minimums apply when components or visual aids are inoperative. When more than one component is inoperative, each minimum must be raised to the highest altitude required by any single inoperative component; increases are not cumulative.

A pilot flying an aircraft in approach category D was planning an ILS approach to an airport where the middle marker (MM) and approach lighting system (MALSR) were inoperative. According to the table in Fig. 8-5, inoperative MM requires an increase in DH of 50 feet and no increase in DH with the MALSRs inoperative. The table also shows an increase in visibility of one-quarter mile for the MM and MALSR; however, these values are not cumulative, and only an increase in visibility of 1/4 mile, not 1/2 mile, is required. Individual IAP charts might contain notes that increase minimums and supersede those contained in Fig. 8-5; however, ILS glide slope inoperative minimums are always published as localizer minimums on IAP charts.

General rules commonly apply when components are inoperative:

- Runway lights are required for night operations; if inoperative, the procedure is not authorized.
- When the facility providing course guidance is inoperative, the procedure cannot be used.
- If more than one component is specified in the procedure identification (VOR/DME), when either the VOR or DME is inoperative, the procedure is not authorized.
- Localizer minimums apply when the ILS glide slope is inoperative or not used.

INSTRUMENT APPROACH PROCEDURE (IAP) CHARTS

RUNWAY APPROACH LIGHTING SYSTEMS

A dot " • " portrayed with approach lighting letter identifier indicates sequenced flashing lights (F) installed with the approach lighting system e.g., (A1) . Negative symbology, e.g., A1 , V indicates Pilot Controlled Lighting (PCL).

PILOT CONTROLLED AIRPORT LIGHTING SYSTEMS

Available pilot controlled lighting (PCL) systems are indicated as follows:

1. Approach lighting systems that bear a system identification are symbolized using negative symbology, e.g., A1, V, ✪
2. Approach lighting systems that do not bear a system identification are indicated with a negative " L " beside the name.

A star (*) indicates non-standard PCL, consult Directory/Supplement, e.g., L*

To activate lights use frequency indicated in the communication section of the chart with a L or the appropriate lighting system identification e.g., UNICOM 122.8 L, A1, V

KEY MIKE	FUNCTION
7 times within 5 seconds	Highest intensity available
5 times within 5 seconds	Medium or lower intensity (Lower REIL or REIL-off)
3 times within 5 seconds	Lowest intensity available (Lower REIL or REIL-off)

(P) PRECISION APPROACH PATH INDICATOR PAPI

(V1) "T"-VISUAL APPROACH SLOPE INDICATOR "T"-VASI

"T" ON BOTH SIDES OF RWY
ALL LIGHTS VARIABLE WHITE.
CORRECT APPROACH SLOPE-ONLY CROSS BAR VISIBLE.
UPRIGHT "T" - FLY UP
INVERTED "T" - FLY DOWN
RED "T" - GROSS UNDERSHOOT.

(V2) PULSATING VISUAL APPROACH SLOPE INDICATOR PVASI

CAUTION: When viewing the pulsating visual approach slope indicators in the pulsating white or pulsating red sectors, it is possible to mistake this lighting aid for another aircraft or a ground vehicle. Pilots should exercise caution when using this type of system.

(VT) TRI-COLOR VISUAL APPROACH SLOPE INDICATOR

CAUTION: When the aircraft descends from green to red, the pilot may see a dark amber color during the transition from green to red.

Fig. 8-4. Coded runway approach lighting system symbols indicate the lighting system's type, availability, and activation procedure if applicable.

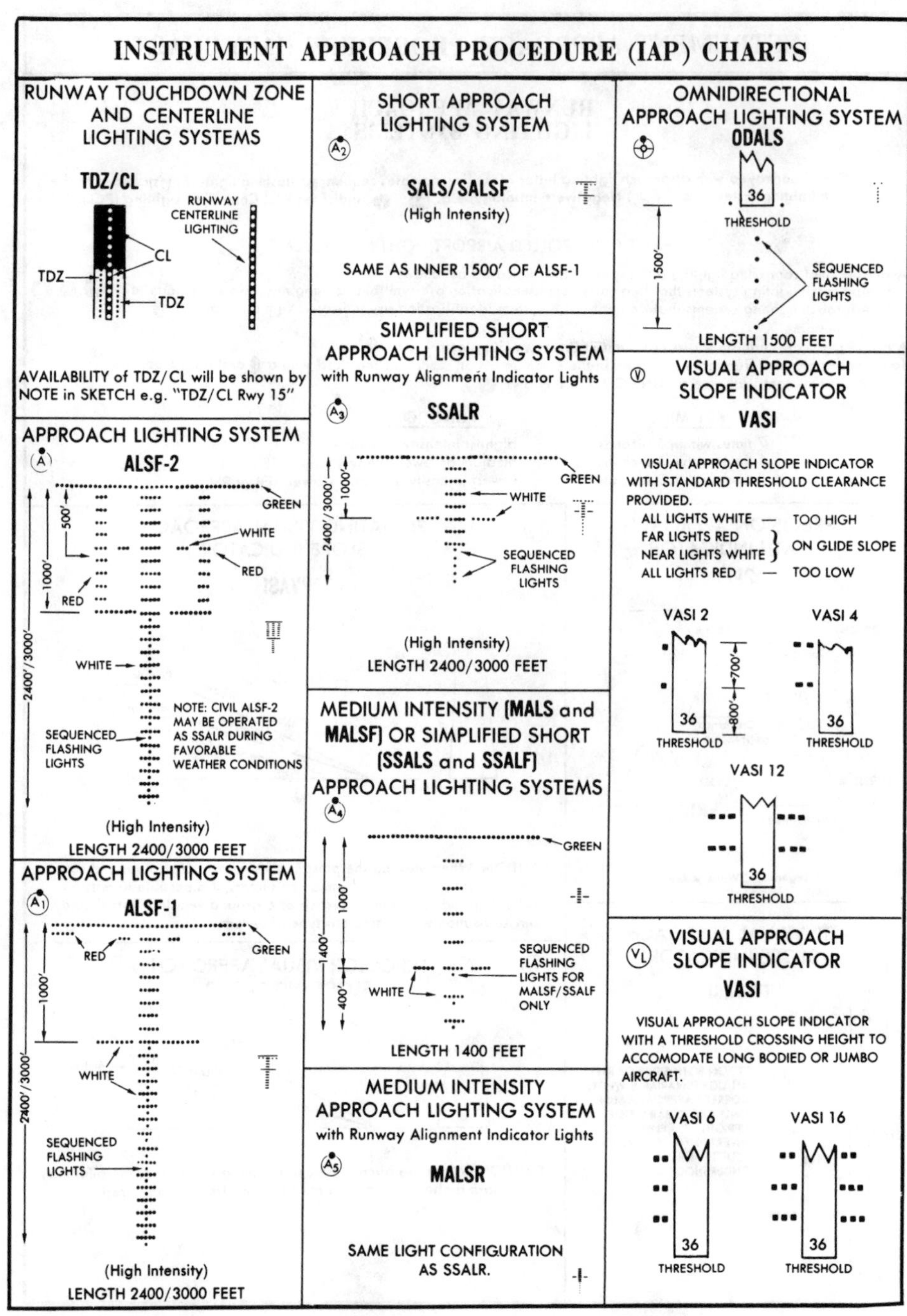

Fig. 8-4. Continued.

- Compass locator or precision radar may be substituted for the ILS outer or middle marker.
- Surveillance radar may be substituted for the ILS outer marker.
- DME, at the glide slope site, may be substituted for the outer marker when published on the ILS procedure.

Facilities that establish a step down fix (fan marker, VOR radial, and the like) are not components of the basic approach procedure. Additional methods of identifying a fix may be used when published on the procedure.

The TPP contains an index of terminal charts and minimums. Procedures are listed alphabetically by city name and airport name within the publication. For example, in Fig. 8-5, William P. Hobby is listed alphabetically with other airports associated with the Houston metropolitan area. Hobby has nonstandard takeoff minimums or a published departure procedure, which can be found in section C, IFR takeoff minimums and departure procedures, of the TPP. This airport also has nonstandard alternate minimums, published in section E, IFR alternate minimums, of the TPP. Next are listed STARs, contained in section P, and IAPs, airport diagram, and SIDs listed by page number.

IFR takeoff and departure procedures, other than SIDs, and IFR alternate minimums, are published in sections C and E, respectively, in the TPP. Houston Hobby (Fig. 8-6) has mandatory departure procedures that ensure obstruction clearance. Because no takeoff minimums appear, they are standard for Hobby. Note that Houston Andrau has nonstandard takeoff minimums, plus a specific departure procedure. Hobby also has nonstandard alternate minimums. For category E aircraft to use Hobby as a filed alternate, the forecast must be at least a ceiling of 800 feet with a visibility of $2^3/4$ miles.

Section M of the TPP contains the frequency pairing and MLS channeling table. On most civil navigation receivers, when the VHF frequency is displayed, the equipment automatically tunes the paired UHF DME frequency; this table also converts MLS or TACAN channels to the associated VHF frequency.

Radar minimums for airports with radar instrument approaches appear in Section N of the TPP. Figure 8-7 shows the radar instrument approach minimums format used in the TPP. ATC provides course, altitude, and missed approach guidance; therefore, only landing minima data, in a slightly different format from IAPs, is provided. DH is shown for precision approach radar (PAR) and MDA for airport surveillance radar (ASR), along with visibility. Minimums in parentheses are only applicable to military operations. When the pilot is cleared for an approach and instructed to circle to another runway, the circling MDA and weather minima apply. For example, in Fig. 8-7, at El Paso, Texas, the MDA for Runway 22 for category A is 4,320 feet, with a visibility of $^1/2$ mile. If the pilot is instructed to or plans to land on another runway, circling minimums of MDA 4,420, visibility 1 mile would apply.

Detailed airport diagrams are issued for large and complex airports where the airport sketch on the IAP might not be adequate. Figure 8-7 contains the Houston William P. Hobby diagram. Availability of airport diagrams is indicated in section B, index of terminal charts and minimums, of the TPP. With detailed runway, taxiway, and facility names

(Continued on page 238.)

INOP COMPONENTS

INOPERATIVE COMPONENTS OR VISUAL AIDS TABLE

Landing minimums published on instrument approach procedure charts are based upon full operation of all components and visual aids associated with the particular instrument approach chart being used. Higher minimums are required with inoperative components or visual aids as indicated below. If more than one component is inoperative, each minimum is raised to the highest minimum required by any single component that is inoperative. ILS glide slope inoperative minimums are published on instrument approach charts as localizer minimums. This table may be amended by notes on the approach chart. Such notes apply only to the particular approach category(ies) as stated. See legend page for description of components indicated below.

(1) ILS, MLS, and PAR

Inoperative Component or Aid	Approach Category	Increase DH	Increase Visibility
MM*	ABC	50 feet	None
MM*	D	50 feet	¼ mile
ALSF 1 & 2, MALSR, & SSALR	ABCD	None	¼ mile

*Not applicable to PAR, MLS, and Operators Authorized in their Operations Specifications.

(2) ILS with visibility minimum of 1,800 RVR.

MM	ABC	50 feet	To 2400 RVR
MM	D	50 feet	To 4000 RVR
ALSF 1 & 2, MALSR, & SSALR	ABCD	None	To 4000 RVR
TDZL, RCLS	ABCD	None	To 2400 RVR
RVR	ABCD	None	To ½ mile

*Not appplicable to Operators Authorized in their Operations Specifications.

(3) VOR, VOR/DME, VORTAC, VOR (TAC), VOR/DME (TAC), LOC, LOC/DME, LDA, LDA/DME, SDF, SDF/DME, RNAV, and ASR

Inoperative Visual Aid	Approach Category	Increase MDA	Increase Visibility
ALSF 1 & 2, MALSR, & SSALR	ABCD	None	½ mile
SSALS, MALS, & ODALS	ABC	None	¼ mile

(4) NDB

ALSF 1 & 2, MALSR & SSALR	C	None	½ mile
	ABD	None	¼ mile
MALS, SSALS, ODALS	ABC	None	¼ mile

CORRECTIONS, COMMENTS AND/OR PROCUREMENT

FOR CHARTING ERRORS:
Contact National Ocean Service
NOAA, N/CG31
6010 Executive Blvd.
Rockville, MD. 20852
Telephone Toll-Free 800-626-3677

FOR CHANGES, ADDITIONS, OR RECOMMENDATIONS ON PROCEDURAL ASPECTS:
Contact Federal Aviation Administration, ATO-258
800 Independence Avenue, S.W.
Washinton, D.C. 20591
Telephone (202) 267-9297

TO PURCHASE CHARTS CONTACT:
National Ocean Service
NOAA, N/CG33
Distribution Branch
Riverdale, MD. 20737
Telephone (301) 436-6993

INOP COMPONENTS

Fig. 8-5. Inoperative approach components require increased minimums. Airports are listed alphabetically by city in the terminal procedures publication.

INDEX

91094

INDEX SC-2

Fig. 8-5. Continued.

TAKE-OFF MINS

91094

NAME	TAKE-OFF MINIMUMS

GRANBURY, TX
GRANBURY MUNI
TAKE-OFF MINIMUMS: **Rwy 14,** 300-1.

HASKELL, TX
HASKELL MUNI
TAKE-OFF MINIMUMS: **Rwy 18,** 600-2 or std. with min. climb of 340′ per NM to 2300.
DEPARTURE PROCEDURE: **Rwy 18,** climb runway heading to 2300 before proceeding on course.

HENDERSON, TX
RUSK COUNTY
DEPARTURE PROCEDURE: **Rwy 34,** climb to 900 before turning eastbound.

HOUSTON, TX
ANDRAU AIRPARK
TAKE-OFF MINIMUMS: **Rwy 16,** 1000-2 or std. with min. climb of 260′ per NM to 2100.
DEPARTURE PROCEDURE: **Rwys 11, 16, 29, 34,** SE bound departures climb runway heading to 2100 before turning on course or be radar vectored.

CLOVER FIELD
TAKE-OFF MINIMUMS: **Rwys 14L, 14R,** 1500-3, **Rwy 22,** 500-2.

HOUSTON-HULL
TAKE-OFF MINIMUMS: **Rwy 35,** 400-1 or std. with min. climb of 220′ per NM to 400′.
DEPARTURE PROCEDURE: **Rwys 17, 35,** eastbound departures, climb runway heading to 2100 before proceeding on course.

HOUSTON-SOUTHWEST
DEPARTURE PROCEDURE: **Rwys 10, 28,** climb runway heading to 2100 before turning or be RADAR vectored.

WILLIAM P. HOBBY
DEPARTURE PROCEDURE: **Rwys 17, 35, 30L, 30R, 12L, 12R, 4,** climb runway heading to 600 before turning left westbound or comply with radar vectors. **Rwy 22,** climb runway heading to 2100 before turning westbound or comply with radar vectors.

HUNTSVILLE, TX
HUNTSVILLE MUNI
DEPARTURE PROCEDURE: **Rwy 18,** climb runway heading to 700 before turning.

JOHNSON CITY, TX
JOHNSON CITY
TAKE-OFF MINIMUMS: **Rwy 35,** 400-1.

NAME	TAKE-OFF MINIMUMS

JUNCTION, TX
KIMBLE COUNTY
TAKE-OFF MINIMUMS: **Rwys 8, 17, 26, 35,** 700-2.

KERRVILLE, TX
KERRVILLE MUNI (LOUIS SCHREINER FIELD)
TAKE-OFF MINIMUMS: **Rwy 2,** 400-1.

KILLEEN, TX
KILLEEN MUNI
TAKE-OFF MINIMUMS: **Rwy 19,** 1300-2 or std. with min. climb of 250′ per NM to 2400.
DEPARTURE PROCEDURE: **Rwy 19,** climb runway heading to 2400 prior to turning eastbound.

LA PORTE, TX
LA PORTE MUNI
TAKE-OFF MINIMUMS: **Rwy 30,** 600-2 or std. with min. climb gradient of 250′ per NM to 600′.
DEPARTURE PROCEDURE: **Rwy 30,** climb runway heading to 600′ before turning.

LAREDO, TX
LAREDO INTL
DEPARTURE PROCEDURE: **Rwy 27,** turn right to 350°, climb to 1500 before turning westbound. **Rwys 17L, 35R, 17R, 35L, 14, 32, 9,** climb runway heading to 1500 before turning westbound.

LEVELLAND, TX
LEVELLAND MUNI
TAKE-OFF MINIMUMS: **Rwy 35,** 600-3 or std. with min. climb of 290′ per NM to 4200.

LIBERTY, TX
LIBERTY MUNI
TAKE-OFF MINIMUMS: **Rwy 16, 34,** 2000-3 or std. with a min. climb of 270′ per NM to 2100.
DEPARTURE PROCEDURE: **Rwys 16, 34,** Eastbound departures, climb runway heading to 2100′ before proceeding on course.

McGREGOR, TX
McGREGOR MUNI
TAKE-OFF MINIMUMS: **Rwys 4, 17, 22, 35,** 2000-3 or std. with min. climb of 280′ per NM to 3000.
DEPARTURE PROCEDURE: **Rwys 17, 22,** climb via ACT R-195 to 3000 before proceeding on course. **Rwys 4, 35,** climb runway heading to 3000 before proceeding on course.

SC-2

TAKE-OFF MINS 

Fig. 8-6. Nonstandard takeoff and alternate minimums are published in separate sections of the terminal procedures publication.

▲ ALTERNATE MINS ▲

91094

NAME — ALTERNATE MINIMUMS

DALLAS LOVE FIELD RADAR-1**
VOR/DME Rwy 13R**
ILS Rwy 13R†
ILS Rwy 31R#
ILS Rwy 13L†
ILS Rwy 31L*

*Categories A, B, 1100-2, Categories C, D, 1100-3. DME equipped aircraft Categories A, B, C, standard; Category D, 800-2¼.
**Category D, 800-2¼.
†ILS, Category D, 700-2¼, LOC, Category D, 800-2¼.
#Categories A, B, 900-2, Category C, 900-2½, Category D, 900-2¾.

REDBIRD .... NDB Rwy 35†
VOR Rwy 13↓
VOR R17↓
VOR Rwy 31↓
ILS Rwy 31*

NA when control tower closed.
*ILS, Category D, 700-2¼; LOC, Category D, 800-2¼.
†Categories C, D, 800-2½
↓Category D, 800-2¼

DALLAS-FORT WORTH, TX

DALLAS/FORT WORTH
INTL ....Converging ILS-2 Rwy 31R*
Converging ILS-2 Rwy 17R↓
Converging ILS-2 Rwy 17L↓
Converging ILS-2 Rwy 18R↓
Converging ILS-2 Rwy 18L↓
Converging ILS-2 Rwy 35R↓
Converging ILS-2 Rwy 36L↓

*ILS, 900-2½, LOC, NA
↓LOC, NA

DEL RIO, TX

DEL RIO INTL .... VOR-A*
VOR/DME-B, 800-2¼

NA when Laughlin AFB control zone not in effect.
*Category C and D, 800-2½.

EL PASO, TX

EL PASO INTL .... ILS Rwy 22
RADAR-1

NA when FSS closed.

GALVESTON, TX

SCHOLES FIELD .... VOR Rwy 13*
ILS Rwy 13*†

*NA when control zone not effective except for operators with approved weather reporting service.
†ILS, Category E, 700-2¼; LOC, Category E, 800-2¼.

HARLINGEN, TX

RIO GRANDE VALLEY INTL ......NDB Rwy 17R#
ILS Rwy 17R*†
LOC BC Rwy 35L†
NDB Rwy 17L#
VOR Rwy 13†
VOR/DME Rwy 31†

*ILS Categories D, E, 700-2.
#NA when control tower closed.
†NA when control zone not in effect except for operators with approved weather reporting service.

HOUSTON, TX

ELLINGTON FIELD .... VOR Rwy 22
ILS Rwy 17R
ILS Rwy 35L
VOR/DME or TACAN Rwy 4
VOR/DME or TACAN Rwy 22
VOR/DME or TACAN Rwy 17R
VOR/DME or TACAN Rwy 35L

Category E, 800-2¼.

WILLIAM P. HOBBY ....ILS Rwy 4
ILS Rwy 12R
VOR/DME Rwy 30L

Category E, 800-2¾

HOUSTON INTERCONTINENTAL .....ILS Rwy 9*
ILS Rwy 27*
ILS Rwy 8*
ILS Rwy 32R*
ILS Rwy 26*
ILS Rwy 14L*
VOR/DME Rwy 14L†
VOR/DME Rwy 32R†

*ILS, Category E, 600-2¼; LOC, Category E, 800-2¼.
†Category E, 800-2¼.

KILLEEN, TX

KILLEEN MUNI .... VOR-A†
ILS Rwy 1*

†Category C, 800-2¼.
*LOC, NA

LAREDO, TX

LAREDO INTL ....VOR/DME or TACAN Rwy 14*
VOR or TACAN Rwy 32*
NDB Rwy 17R*†
ILS Rwy 17R*†
NDB Rwy 17L*†

*NA except for operators with approved weather reporting service.
†NA when control tower closed.

SC-2

Fig. 8-6. Continued.

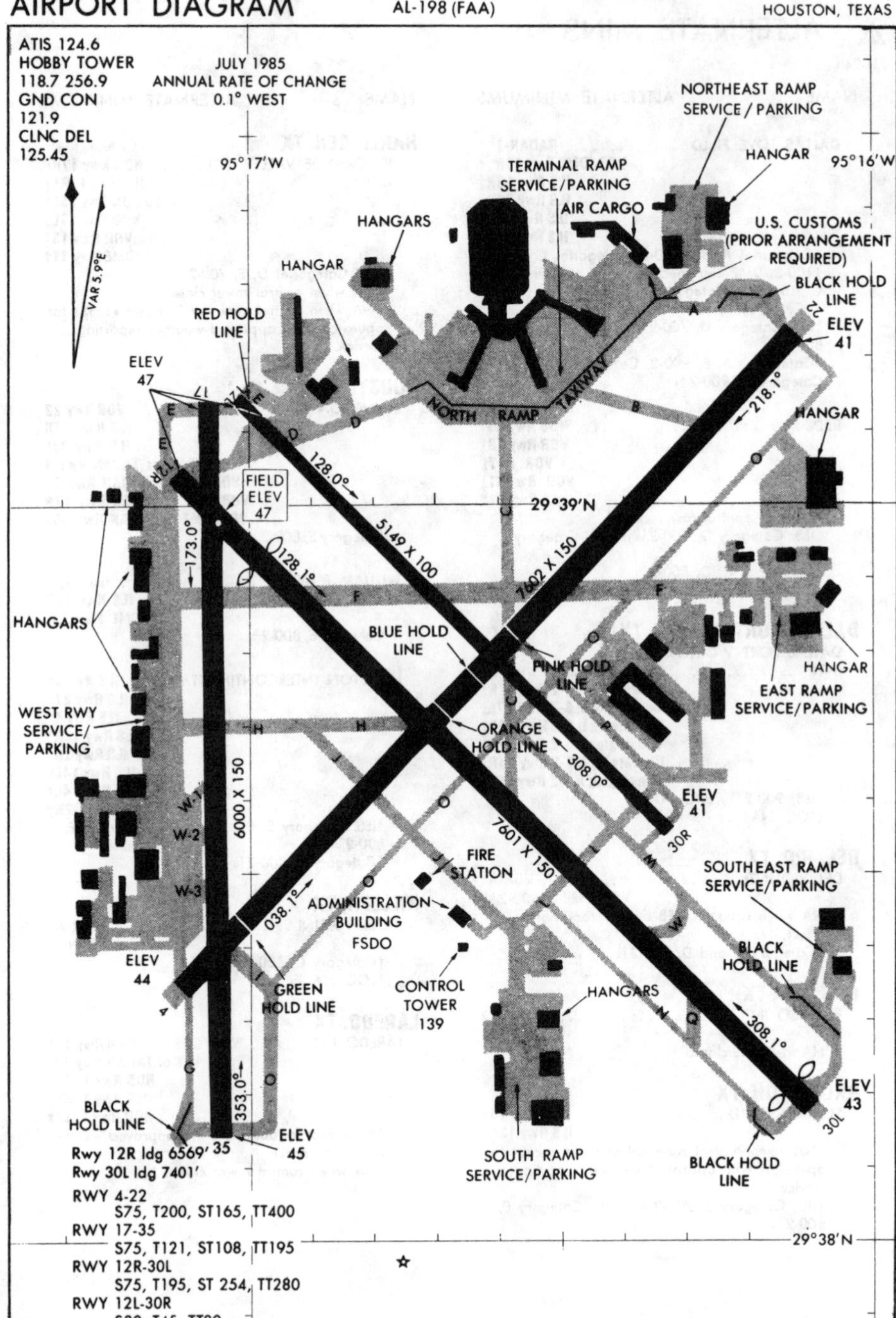

Fig. 8-7. Airport diagrams are issued for large, complex airports. Radar approach minimums appear in a separate section of the terminal procedures publication.

RADAR MINS

90347

RADAR INSTRUMENT APPROACH MINIMUMS

BEAUMONT-PORT ARTHUR, TX Amdt. 8, MAR 8, 1990 ELEV **16**
JEFFERSON COUNTY
RADAR—121.3 322.3

	RWY	GS/TCH/RPI	CAT	DH/ MDA-VIS	HAT/ HAA	CEIL-VIS	CAT	DH/ MDA-VIS	HAT/ HAA	CEIL-VIS
ASR	12		ABC	**420**/40	404	(500—¾)	D	**420**/50	404	(500—1)
	34		ABCD	**440**—1¼	424	(500—1¼)				
	16		AB	**560**—1¼	544	(600—1¼)	C	**560**—1½	544	(600—1½)
			D	**560**—1¾	544	(600—1¾)				
	30		AB	**560**—1¼	547	(600—1¼)	C	**560**—1½	547	(600—1½)
			D	**560**—1¾	547	(600—1¾)				
CIRCLING			AB	**560**—1¼	544	(600—1¼)	C	**560**—1½	544	(600—1½)
			D	**580**—2	564	(600—2)				

When control tower not in operation, approach not authorized.
Category D S—12 visibility increased ¼ mile for inoperative MALSR.
▲

DALLAS, TX Amdt. 25, MAY 5, 1988 ELEV **487**
LOVE FIELD
RADAR—123.9 252.9

	RWY	GS/TCH/RPI	CAT	DH/ MDA-VIS	HAT/ HAA	CEIL-VIS	CAT	DH/ MDA-VIS	HAT/ HAA	CEIL-VIS
ASR	13L		A	**960**/24	475	(500—½)	B	**960**/40	475	(500—¾)
			C	**960**/50	475	(500—1)	D	**960**—1½	475	(500—1½)
	13R		A	**960**—1	483	(500—1)	B	**960**—1¼	483	(500—1¼)
			C	**960**—1½	483	(500—1½)	D	**960**—2	483	(500—2)
CIRCLING			A	**960**—1	473	(500—1)	B	**1020**—1¼	535	(600—1¼)
			C	**1020**—1¾	533	(600—1¾)	D	**1180**—2¼	693	(700—2¼)

▼ ▲

EL PASO, TX Amdt. 12, OCT 13, 1983 ELEV **3956**
EL PASO INTL
RADAR—124.15 307.0

	RWY	GS/TCH/RPI	CAT	DH/ MDA-VIS	HAT/ HAA	CEIL-VIS	CAT	DH/ MDA-VIS	HAT/ HAA	CEIL-VIS
ASR	22		ABC	**4320**—½	376	(400—½)	DE	**4320**—1	376	(400—1)
	26L		AB	**4380**—1	424	(500—1)	CD	**4380**—1¼	424	(500—1¼)
			E	**4380**—1½	424	(500—1½)				
	4		AB	**4400**—1	479	(500—1)	C	**4400**—1¼	479	(500—1¼)
			D	**4400**—1½	479	(500—1½)	E	**4400**—1¾	479	(500—1¾)
CIRCLING			A	**4420**—1	464	(500—1)	B	**4460**—1	504	(600—1)
			C	**4460**—1½	504	(600—1½)	DE	**4520**—2	564	(600—2)

Procedure not authorized when control tower closed.
CAUTION: Steeply rising terrain 4.5 NM West of airport.
Categories D and E S—22 visibility increased ¼ mile for inoperative SSALR. Category E circling west of airport not authorized. ▼ ▲

SC-2

RADAR INSTRUMENT APPROACH MINIMUMS

RADAR MINS N2

Fig. 8-7. Continued.

(Continued from page 231.)
and locations, these diagrams are invaluable for pilots operating into or out of unfamiliar airports.

Opposite the inside back cover of the TPP is the airport loran TD correction table. Unless corrected, seasonal variations in loran signals might cause position errors of thousands of feet. These errors can be partially offset by the pilot entering time difference (TD) corrections into the navigation receiver. Seasonal changes are published on this page for those loran approaches contained in the TPP volume. TD values from the table should be transferred to the TD correction box shown in the plan view of the loran approach for the destination airport. Published TD correction values must be entered into the loran airborne receiver prior to beginning the approach.

The inside back cover of the TPP contains a rate of descent table (Fig. 8-8) that provides descent rates in feet per minute for planning and executing precision descents based on glide slope angle, plus the respective known or approximate ground speed. For example, notice in the profile diagram that the glide slope angle for the ILS approach is 3.00°. Plan a 100 knot IAS approach, with reported surface winds straight down the runway at 10 knots. If DME is available with the approach and the aircraft is equipped with a DME receiver, ground speed would be known; however, under the given conditions we would estimate ground speed at 90 knots (100 − 10 = 90). Considering a glide angle of 3°, at a ground speed of 90 knots (Fig. 8-8), the approximate rate of descent required would be 480 feet per minute. ATC normally provides the pilot with approximate rate of descent information on precision radar approaches.

USING APPROACH CHARTS

Approach control informs a pilot which approach to expect when the airport has more than one instrument procedure, perhaps an IAP or CVFP. This information is provided by the controller or broadcast on the ATIS. When the advertised approach cannot be executed, or another procedure is desired, it must be specifically requested by the pilot.

Years ago I accompanied an instrument student in a Mooney on a flight to San Diego's Lindbergh Field. The approach in use was the localizer back-course, which at the time required the use of a marker beacon receiver. The Mooney was only equipped with VOR/localizer and ADF receivers. We planned to execute a front-course localizer approach, utilizing the ADF to establish outer and middle marker locations. ATC suggested and we accepted a radar vector below the overcast, over the ocean due to traffic in the area. Ground control subsequently instructed us to call approach control on the phone; I explained the situation and was relieved when the ATC supervisor replied, "Oh, OK."

Figure 8-9 contains two ILS procedures for the Dallas-Fort Worth International Airport: ILS-1 RWY 18R and CONVERGING ILS-2 RWY 36L. The ILS-1 RWY 18R is a Category II procedure as indicated in the margin and by the special note: "CATEGORY II ILS- SPECIAL AIRCREWS & AIRCRAFT CERTIFICATION REQUIRED." (The ILS-2 RWY 18R is a converging approach and must be a separate procedure.) The converging ILS-2 Runway 36L approach converges at a 50° angle with the converging ILS-2 Runway 31R approach. These approaches meet the criteria of converging angle, missed approach

(Continued on page 242.)

DESCENT TABLE

RATE OF DESCENT TABLE

A rate of descent table is provided for use in planning and executing precision descents under known or approximate ground speed conditions. It will be especially useful for approaches when the localizer only is used for course guidance. A best speed, power, altitude combination can be programmed which will result in a stable glide rate and altitude favorable for executing a landing if minimums exist upon breakout. Care should always be exercised so that the minimum descent altitude and missed approach point are not exceeded.

(ft. per min.)

ANGLE OF DESCENT (degrees and tenths)	GROUND SPEED (knots)										
	30	45	60	75	90	105	120	135	150	165	180
2.0	105	160	210	265	320	370	425	475	530	585	635
2.5	130	200	265	330	395	465	530	595	665	730	795
3.0	160	240	320	395	480	555	635	715	795	875	955
3.5	185	280	370	465	555	650	740	835	925	1020	1110
4.0	210	315	425	530	635	740	845	955	1060	1165	1270
4.5	240	355	475	595	715	835	955	1075	1190	1310	1430
5.0	265	395	530	660	795	925	1060	1190	1325	1455	1590
5.5	290	435	580	730	875	1020	1165	1310	1455	1600	1745
6.0	315	475	635	795	955	1110	1270	1430	1590	1745	1950
6.5	345	515	690	860	1030	1205	1375	1550	1720	1890	2065
7.0	370	555	740	925	1110	1295	1480	1665	1850	2035	2220
7.5	395	595	795	990	1190	1390	1585	1785	1985	2180	2380
8.0	425	635	845	1055	1270	1480	1690	1905	2115	2325	2540
8.5	450	675	900	1120	1345	1570	1795	2020	2245	2470	2695
9.0	475	715	950	1190	1425	1665	1900	2140	2375	2615	2855
9.5	500	750	1005	1255	1505	1755	2005	2255	2510	2760	3010
10.0	530	790	1055	1320	1585	1845	2110	2375	2640	2900	3165
10.5	555	830	1105	1385	1660	1940	2215	2490	2770	3045	3320
11.0	580	870	1160	1450	1740	2030	2320	2610	2900	3190	3480
11.5	605	910	1210	1515	1820	2120	2425	2725	3030	3335	3635
12.0	630	945	1260	1575	1890	2205	2520	2835	3150	3465	3780

DESCENT TABLE

Fig. 8-8. The rate of descent table provides descent rates in feet per minute for precise execution of the maneuver.

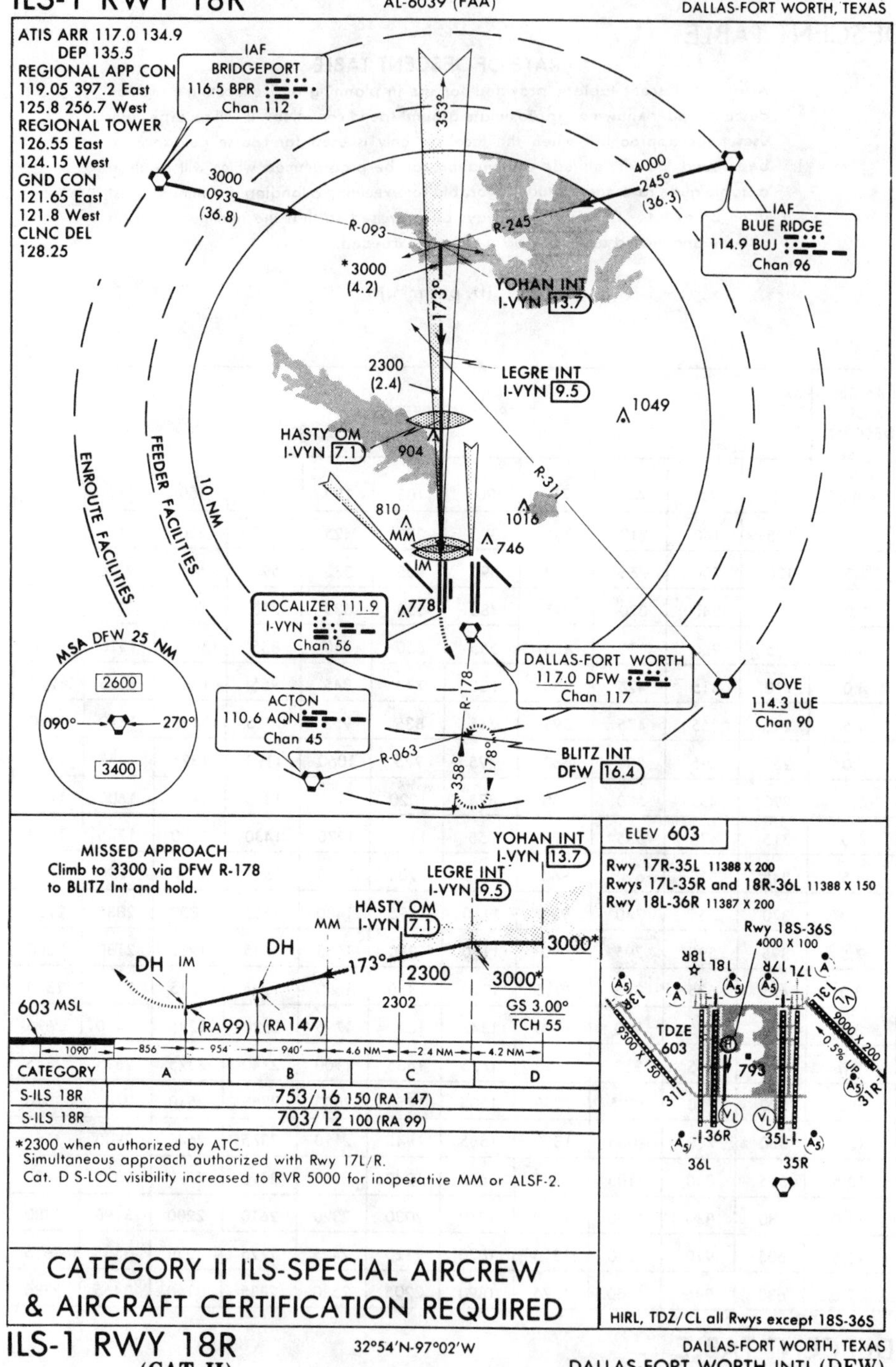

Fig. 8-9. A Category II approach requires special certification for aircrew and aircraft. Converging approaches must meet special criteria.

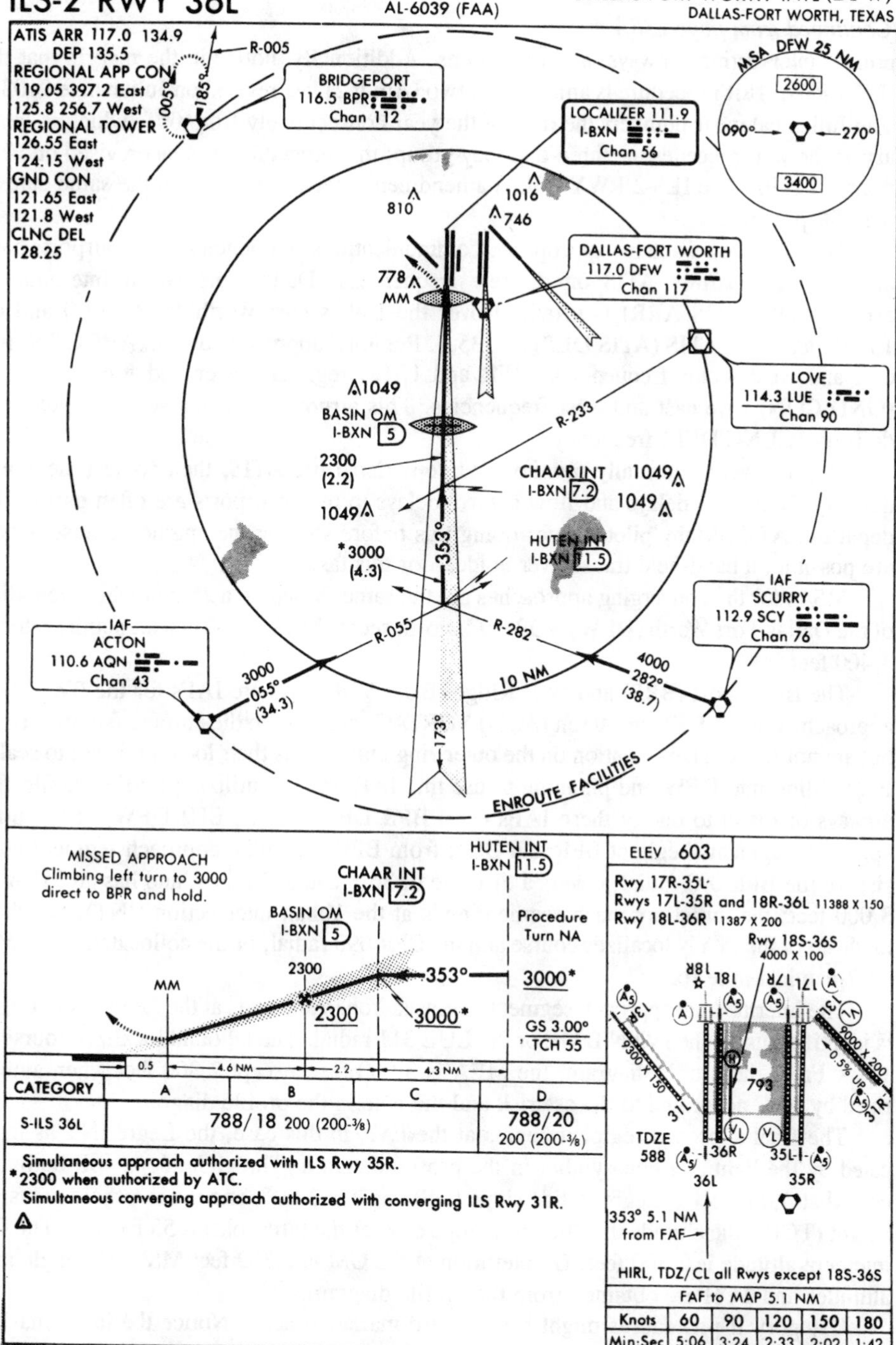

CATEGORY	A	B	C	D
S-ILS 36L	788/18 200 (200-⅜)			788/20 200 (200-⅜)

Simultaneous approach authorized with ILS Rwy 35R.
*2300 when authorized by ATC.
Simultaneous converging approach authorized with converging ILS Rwy 31R.

FAF to MAP 5.1 NM					
Knots	60	90	120	150	180
Min:Sec	5:06	3:24	2:33	2:02	1:42

Fig. 8-9. Continued.

(Continued from page 238.)
points, intersecting runways, and minimums. Additionally, notice in the margin that the ILS-1 RWY 18R procedure is amendment two (Amdt 2) last revised on Julian date 91038. The Julian calendar numbers the days of the year consecutively from 001, which is January 1; the year precedes the three-digit day group; this procedure was last revised on February 7, 1991. The ILS-2 RWY 36L is amendment one, last revised on the same date as the 18R procedure.

The plan view contains appropriate communications frequencies. Busy airports have arrival and departure ATIS on discrete frequencies. Dallas-Fort Worth International arrival ATIS (ATIS ARR) is provided over the Dallas-Fort Worth VOR 117.0 and on 134.9: departure ATIS (ATIS DEP) on 135.5. Regional approach control (APP CON) has east and west sector frequencies, VFR and UHF; regional tower and ground control (GND CON) have east and west frequencies. This airport also has a separate clearance delivery (CLNC DEL) frequency.

(A pilot would normally monitor and copy departure ATIS, then contact clearance delivery. Departure delays and flow control delays to major airports are often part of the departure ATIS. Many pilots prefer doing this before starting the engine because delays are possible; a hand-held transceiver is ideal for this task.)

MSAs for the converging approaches are the same, based on a 25-nautical mile radius of the Dallas-Fort Worth (DFW) VORTAC: north sector MSA is 2,600 feet, south sector is 3,400 feet.

The Bridgeport (BRP) and Blue Ridge (BUJ) VORTACs are IAFs for the RWY 18R approach. Love (LUE) and Acton (AQN) VORTACs are also on the enroute facilities ring, but are not IAFs. Their location on the outer ring only means their location is not to scale. Pilots filing into DFW and planning to use this IAP, and not utilizing a SID, should file airways or direct to one of these IAFs (. . . BPR DFW or . . . BUJ DFW). The initial approach segments begin at BPR and BUJ; from BPR the initial approach segment consists of the BPR 093 radial, covers a distance of 36.8 nautical miles, and has an MEA of 3,000 feet; the initial approach segment ends at the Yohan intersection (INT), which is made up of the IVYN localizer course and the BPR 093 radial, or the collocated ILS DME 13.7-nautical mile fix.

The intermediate approach segment begins at Yohan and ends at the Legre INT. Legre is identified using the IVYN DME or the LUE 311 radial. The inbound localizer course is 173°. From the profile diagram, the MEA is 3,000 feet, except 2,300 feet when authorized by ATC according to the asterisk and the note in the profile diagram.

The final approach segment begins at the FAF, in this case, the Legre INT as indicated by the lightning bolt symbol in the profile diagram. The glide slope will be intercepted at 3,000 feet at Legre. Glide slope (GS) angle is 3.00°, and the threshold crossing height (TCH) (the altitude that the glide slope crosses the threshold) is 55 feet agl. The GS intercept altitude is 2,300 feet; GS elevation at the OM is 2,302 feet MSL. GS angle and altitudes and TCH are obtained from the profile diagram.

Category II procedures might have a third marker beacon. Notice the inner marker (IM) located between the middle marker and the runway on the ILS-1 RWY 18R procedure.

The final approach segment ends at the missed approach point: on this approach, DH

753 (radar altimeter RA 147 feet) or DH 703 (RA 99 feet). The DH 703 is collocated with the inner marker. The different DHs are authorized based upon aircraft equipment and crew certification. Distances between fixes on the final approach segment are provided below the profile diagram.

From the landing minima data section we see that minima apply only to a straight-in landing on Runway 18R (S-ILS 18R). Minima is either DH 753, RVR 1600 or DH 703, RVR 1200 (150 feet agl or 100 feet agl respectively) for all aircraft categories. The RA heights are slightly different due to terrain over the DH points. Additional notes advise that simultaneous approaches are authorized with Runways 17L and 17R. Also, category D straight-in minimum visibility must be increased to RVR 5000 for inoperative middle marker or approach lighting system. This is greater than the standard required by the inoperative components or visual aids table in Fig. 8-5.

From the profile diagram, the MAP for the RWY 18R approach consists of a climb to 3,300 feet via the DFW 178 radial to the Blitz INT and hold. Note that the holding pattern is depicted with a broken line, indicating it is part of the missed approach procedure. Upon reaching DH and not sighting the runway environment, the pilot would immediately initiate a climb to 3,300 feet, and as indicated in the profile diagram, make a slight left turn to intercept the DFW 178 radial. Because the required turn is fewer than 15°, the MAP is considered straight. Blitz INT is designated by the DFW 170 and AQN 063 radials. Blitz can also be identified by 16.4 DME from the DFW VORTAC. Unless otherwise instructed by ATC, the pilot would enter the depicted holding pattern at Blitz, maintaining 3,300 feet.

The airport sketch contains standard airport symbols and information. Field elevation, runway lengths and widths, and lighting are provided. Runway 18R has runway centerline lights, sequence flashing lights are indicated by the dot above the approach lighting symbol, which also indicates the type of system, in this case an ALSF-2 decoded from Fig. 8-4.

The TPP index in Fig. 8-5 indicates that an airport diagram is available. The airport diagram provides much greater detail; therefore, it can be determined that Runway 18S-36S is only available for prop and short takeoff and landing (STOL) aircraft, 12,500 pounds and below. This information could also be obtained from the A/FD.

Most information on the RWY 36L approach is similar to that on the RWY 18R procedure. The Scurry (SCY) and AQN VORTACs are IAFs. A procedure turn is not authorized. The missed approach procedure directs a climbing left turn to 3,000 feet direct to BPR and hold. As soon as a climb has been established, a climbing right turn is executed as soon as safety permits.

Notes below the landing minima data indicate that the approach has nonstandard alternate minimums. Alternate minimums for this approach (Fig. 8-6) are ceiling 900 feet and visibility 2½ miles. The airport sketch provides the distance from the FAF to the runway, 5.1 nautical miles. The airport sketch shows approach light code A5 for this runway, which decodes (Fig. 8-4) as a medium intensity approach lighting system with runway alignment indicator lights (MALSR).

Figure 8-10 is two approaches: Mesquite/Phil L. Hudson Municipal, Mesquite, Texas (HQZ), LOC/DME BC RWY 35 and Van Nuys, California (VNY), LDA C.

(Continued on page 246.)

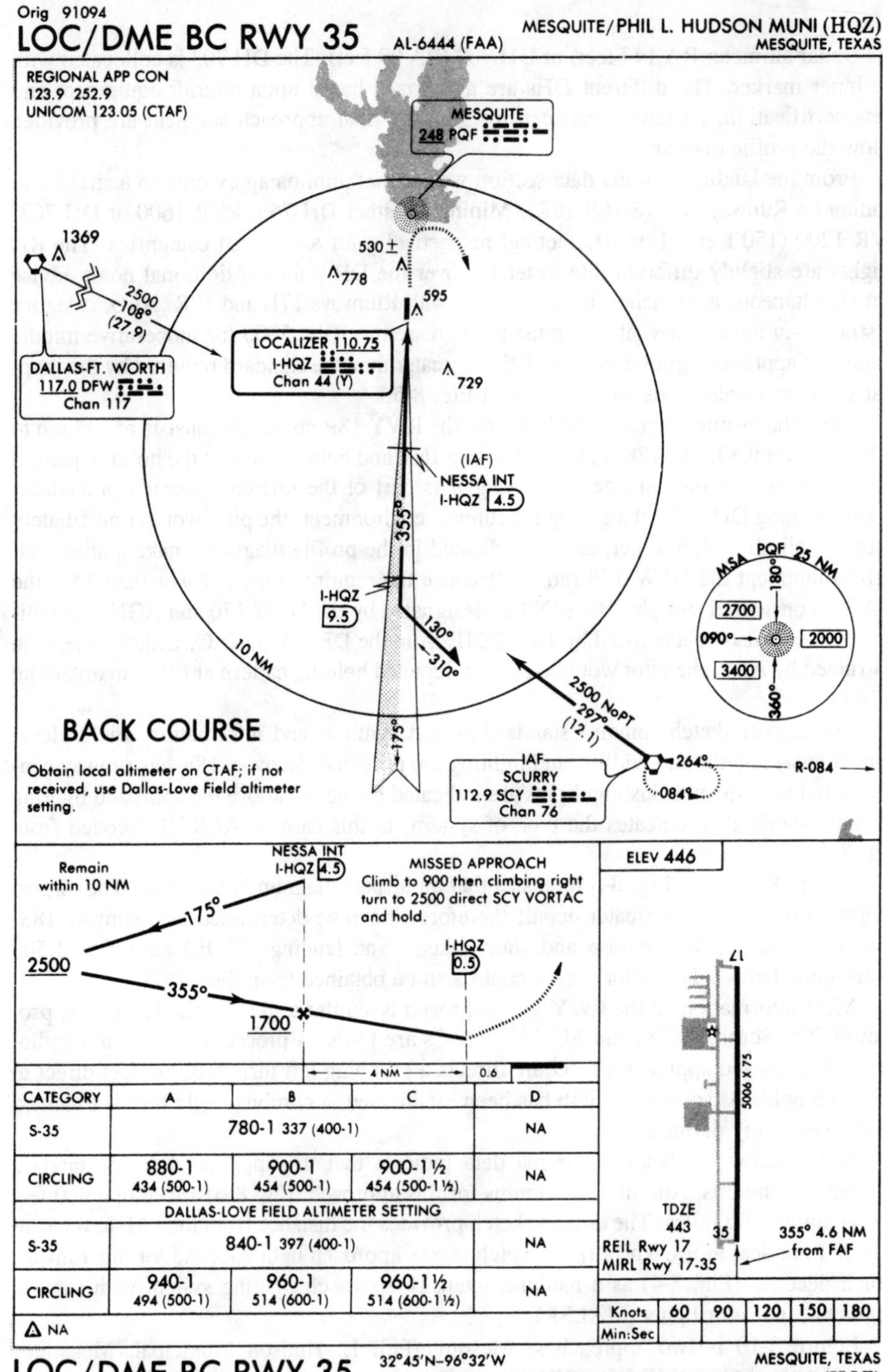

CATEGORY	A	B	C	D
S-35	780-1 337 (400-1)			NA
CIRCLING	880-1 434 (500-1)	900-1 454 (500-1)	900-1½ 454 (500-1½)	NA
DALLAS-LOVE FIELD ALTIMETER SETTING				
S-35	840-1 397 (400-1)			NA
CIRCLING	940-1 494 (500-1)	960-1 514 (600-1)	960-1½ 514 (600-1½)	NA

△ NA

Fig. 8-10. Localizer and LDA approaches are nonprecision procedures based on localizer course guidance.

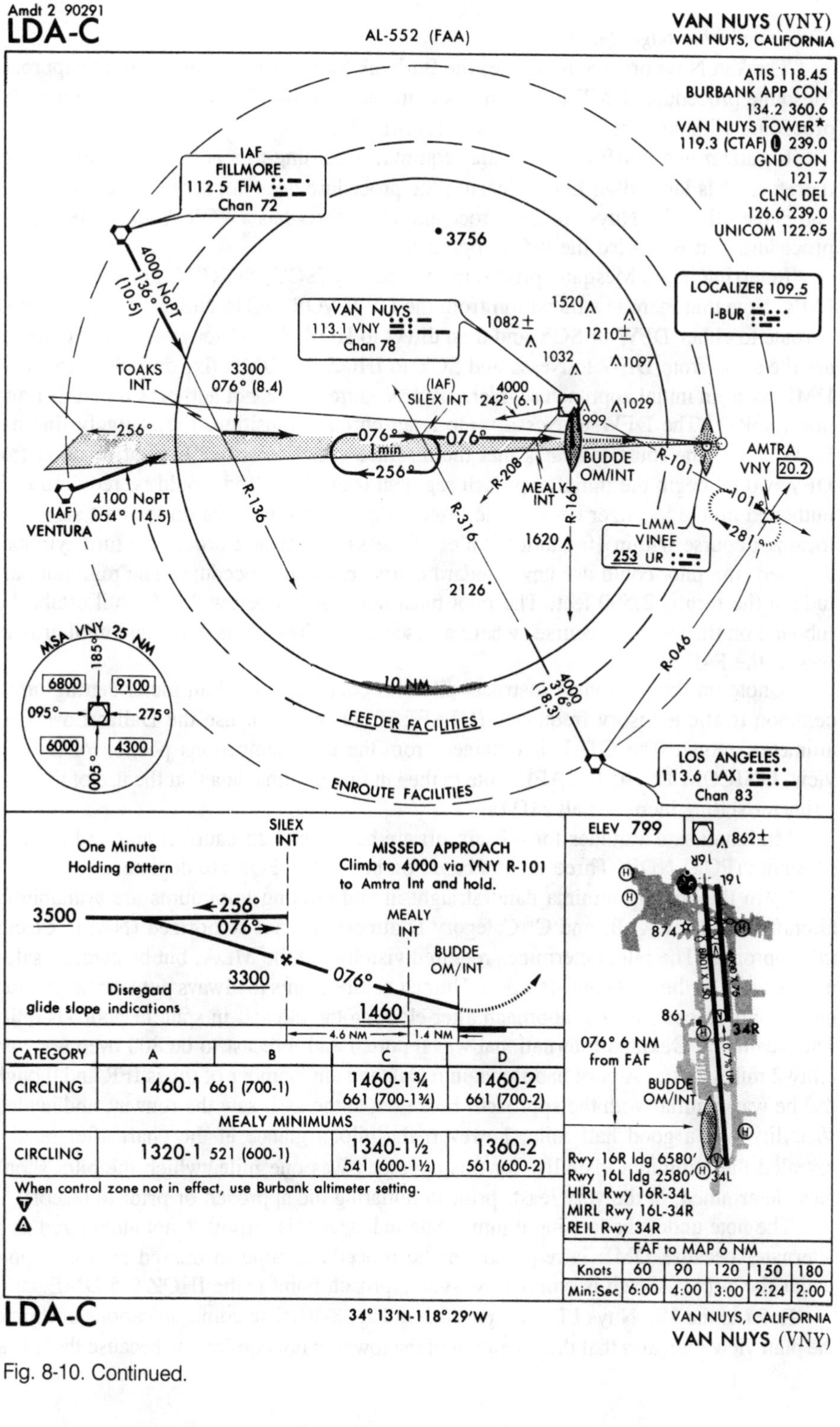

Amdt 2 90291

LDA-C

AL-552 (FAA)

VAN NUYS (VNY)
VAN NUYS, CALIFORNIA

ATIS 118.45
BURBANK APP CON
134.2 360.6
VAN NUYS TOWER*
119.3 (CTAF) 239.0
GND CON
121.7
CLNC DEL
126.6 239.0
UNICOM 122.95

IAF
FILLMORE
112.5 FIM
Chan 72

4000 NoPT
136°
(10.5)

3756

LOCALIZER 109.5
I-BUR

VAN NUYS
113.1 VNY
Chan 78

1520
1082±
1210±
1032
1097
1029
999

TOAKS
INT

3300
076° (8.4)

(IAF)
SILEX INT

4000
242° (6.1)

256°
076°
1 min
256°
076°

BUDDE
OM/INT

MEALY
INT

R-208
R-101
R-164
R-316
R-136

AMTRA
VNY 20.2

101°
281°

LMM
VINEE
253 UR

4100 NoPT
054° (14.5)

(IAF)
VENTURA

1620
2126

MSA VNY 25 NM
185°
6800
9100
095°
275°
6000
4300
005°

10 NM

R-046

4000
316°
(18.3)

FEEDER FACILITIES

ENROUTE FACILITIES

LOS ANGELES
113.6 LAX
Chan 83

One Minute
Holding Pattern

SILEX
INT

MISSED APPROACH
Climb to 4000 via VNY R-101
to Amtra Int and hold.

3500
256°
076°

MEALY
INT

BUDDE
OM/INT

076°

3300

Disregard
glide slope indications

1460

4.6 NM
1.4 NM

ELEV 799
862±
16R
16L
19L
874
861
34R
8001 X 150
4000 X 75
0.7% UP

076° 6 NM
from FAF

BUDDE
OM/INT

Rwy 16R ldg 6580'
Rwy 16L ldg 2580'
34L
HIRL Rwy 16R-34L
MIRL Rwy 16L-34R
REIL Rwy 34R

CATEGORY	A	B	C	D
CIRCLING	1460-1	661 (700-1)	1460-1¾ 661 (700-1¾)	1460-2 661 (700-2)
MEALY MINIMUMS				
CIRCLING	1320-1	521 (600-1)	1340-1½ 541 (600-1½)	1360-2 561 (600-2)

When control zone not in effect, use Burbank altimeter setting.

FAF to MAP 6 NM					
Knots	60	90	120	150	180
Min:Sec	6:00	4:00	3:00	2:24	2:00

LDA-C

34° 13'N-118° 29'W

VAN NUYS, CALIFORNIA
VAN NUYS (VNY)

Fig. 8-10. Continued.

(Continued from page 243.)

The Van Nuys procedure utilizes the Burbank localizer course to a circling approach; Mesquite procedure. DME is the only way to establish the FAF of the back course (BC) procedure with straight-in minimums to Runway 35.

Localizer and DME receivers are required, according to a note in the margin of the therefore, it is labeled an LDA. Because the procedure does not serve a specific runway, and is the third Van Nuys circling procedure (Van Nuys has a VOR-A and VOR/DME-B procedures), it is labeled the "-C" approach.

Referring to the Mesquite procedure, the Scurry (SCY) VORTAC and Nessa INT are IAFs. Note that there is a transition from the DFW VORTAC to Nessa. Pilots should file enroute to either DFW or SCY and then direct to HQZ. Even though minimum altitudes are the same from DFW to Nessa and SCY to IHQZ 9.5 DME fix, only SCY to the 9.5 DME fix is an initial approach segment. A procedure turn is not authorized on this transition (NoPT). The DFW to Nessa route is an enroute transition; the squiggly line near DFW across the route indicates that the distance is not to scale. A pilot inbound from DFW would begin the initial approach segment over Nessa and would be required to fly outbound on the localizer backcourse, executing a course reversal on the east side of the localizer course within 10 nautical miles of Nessa. Because a procedure turn symbol is depicted, the pilot could use any standard course reversal procedure. The minimum altitude in the turn is 2,500 feet. The pilot must not descend below 2,500 until established inbound on the localizer course, where a descent to 1,700 feet is permitted until crossing Nessa, the FAF.

A note on the plan view instructs pilots to obtain the local altimeter setting on the common traffic advisory frequency (CTAF); if not received, use the Dallas-Love Field altimeter setting. The CTAF is obtained from the communications portion of the plan view, UNICOM 123.05 (CTAF). Note in the landing minima data that the use of the Love altimeter setting increases all MDAs.

Minimum safe altitudes for this airport are based on a 25 nautical mile radius of the Mesquite (PQF) NDB. Three sectors based on bearing to PQF are depicted.

From the landing minima data, straight-in and circling minimums are available for aircraft categories A, B, and C. Category D aircraft are not authorized (NA) to execute this approach. The pilot determines required visibility at the MDA, but be careful, safety must always be the first consideration. Human nature seems to always put terrific pressure on the pilot to continue the approach after clearing the clouds, in spite of poor visibility. The weather at Oakland International was reported and forecast to be 800 overcast, visibility 2 miles in fog. A pilot had flown into Oakland any number of times IFR and thought that he was familiar with the approach, broke out at the DH, saw the runway, and landed. Visibility was a good half mile; however, the pilot's glance at the chart after landing revealed the minimum visibility for this approach was one mile, which the pilot should have determined, at the very least, prior to initiating the approach or prior to takeoff.

The note under the landing minima data indicates this airport is not authorized as an alternate; because DME is required for the procedure, time to missed approach point under the airport sketch is omitted. Missed approach point is the IHQZ 0.5 DME.

Refer to the Van Nuys LDA-C approach in Fig. 8-10. The communications portion of the plan view indicates that the operation of the tower is not continuous because there is an asterisk. Tower hours could be obtained from the A/FD. When the tower is closed, the CTAF is the tower frequency and runway lights are activated by the pilot, as indicated by the L following the frequency. Also note in the airport sketch that the approach lighting system has a minus sign. This indicates that the approach lights are also pilot controlled.

Enroute transitions are available from the Los Angeles (LAX) VORTAC and the Van Nuys (VNY) VOR/DME. This procedure has four IAFs: Fillmore (FIM), Ventura (VTU), and the Silex and Toaks INT. Pilots utilizing the LAX or VNY transitions would be required to execute a holding pattern course reversal because a holding pattern is depicted. NoPT is indicated for the other transitions.

The profile diagram note instructs pilots to disregard glide slope indications. That's because the glide slope serves the Burbank airport and is not part of this procedure. This approach has a step-down fix, Mealy INT (VNY R-208), inside the FAF. Note from the landing minima data: "MEALY MINIMUMS." The missed approach point is the Budde OM/INT. Budde can be established from the marker beacon or the VNY 164 radial. Other notes instruct the pilot to use the Burbank altimeter setting when the control zone is not in effect—tower closed—and that VNY has nonstandard takeoff and alternate minimums. Finally, the airport sketch provided time to missed approach point data; a pilot could use time, as well as the Budde OM/INT to establish the missed approach point.

Figure 8-11 is the Austin/Lakeway Airpark VOR/DME-C and the Austin/Robert Mueller Municipal VOR/DME or TACAN RWY 17 approaches. DME is required for both procedures to establish the FAF. Lakeway Airport has the symbol -C, which means circling minima only. The Robert Mueller procedure has straight-in minima to Runway 17. This approach is also designed for military TACAN equipped aircraft.

On the Lakeway Airport procedure, the 17 nautical mile DME arc between the 222 and 008 radials serves as an IAF. Any radial or airway that intersects the arc constitutes an IAF. The pilot would then proceed along the arc, turning inbound at the lead-in radial (LR-272 or LR-282), executing the approach with no procedure turn. Note that the Donho INT is also an IAF. A pilot proceeding from the Austin VORTAC would be required to use a holding pattern course reversal, as depicted. From the landing minima data, category D aircraft are not authorized and the procedure cannot be used at night, the Austin Robert Mueller altimeter setting is to be used, nonstandard takeoff minimums apply, and the airport cannot be used for an alternate. The airport sketch shows that both runways have displaced thresholds symbols, as defined in Fig. 8-3. Total runway length is 4,000 feet; runway available for landing is 3,770 for Runway 16 and 3,665 for Runway 34.

The communications portion of the plan view for the Robert Mueller Municipal indicates that ASR radar is available. Gales INT is the only IAF. Note the enroute transitions from the Austin VORTAC to Gales: MEA 3,000 feet, on the 340 radial, at six nautical miles. The pilot would then execute a course reversal on the east side of the radial within 10 nautical miles of Gales. VASI symbols on the airport sketch apply to Runways 17-35, 13R, and 31R; gradients are also shown for runways 17-35 and 13R-31L.

The plan view of the Robert Mueller procedure shows another airport in the vicinity, Bergstrom AFB. Misidentifying a destination airport is discussed in chapter 7, in the section on visual approach procedures. This also occurs when the pilot breaks out of the clouds and fails to continue to navigate the aircraft in accordance with the procedure.

(Continued on page 252.)

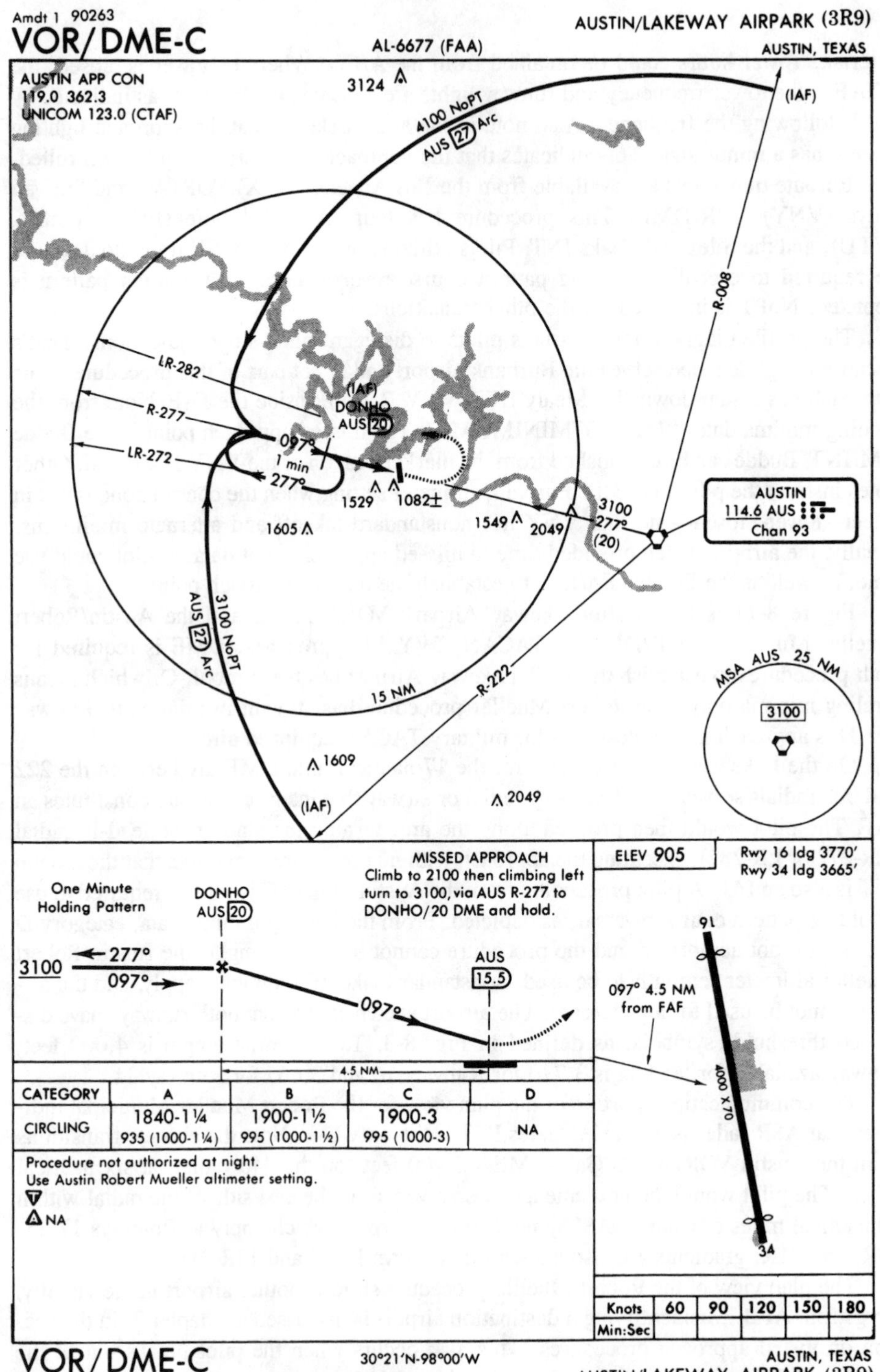

Fig. 8-11. When DME appears in the procedure name it is a required component for the approach.

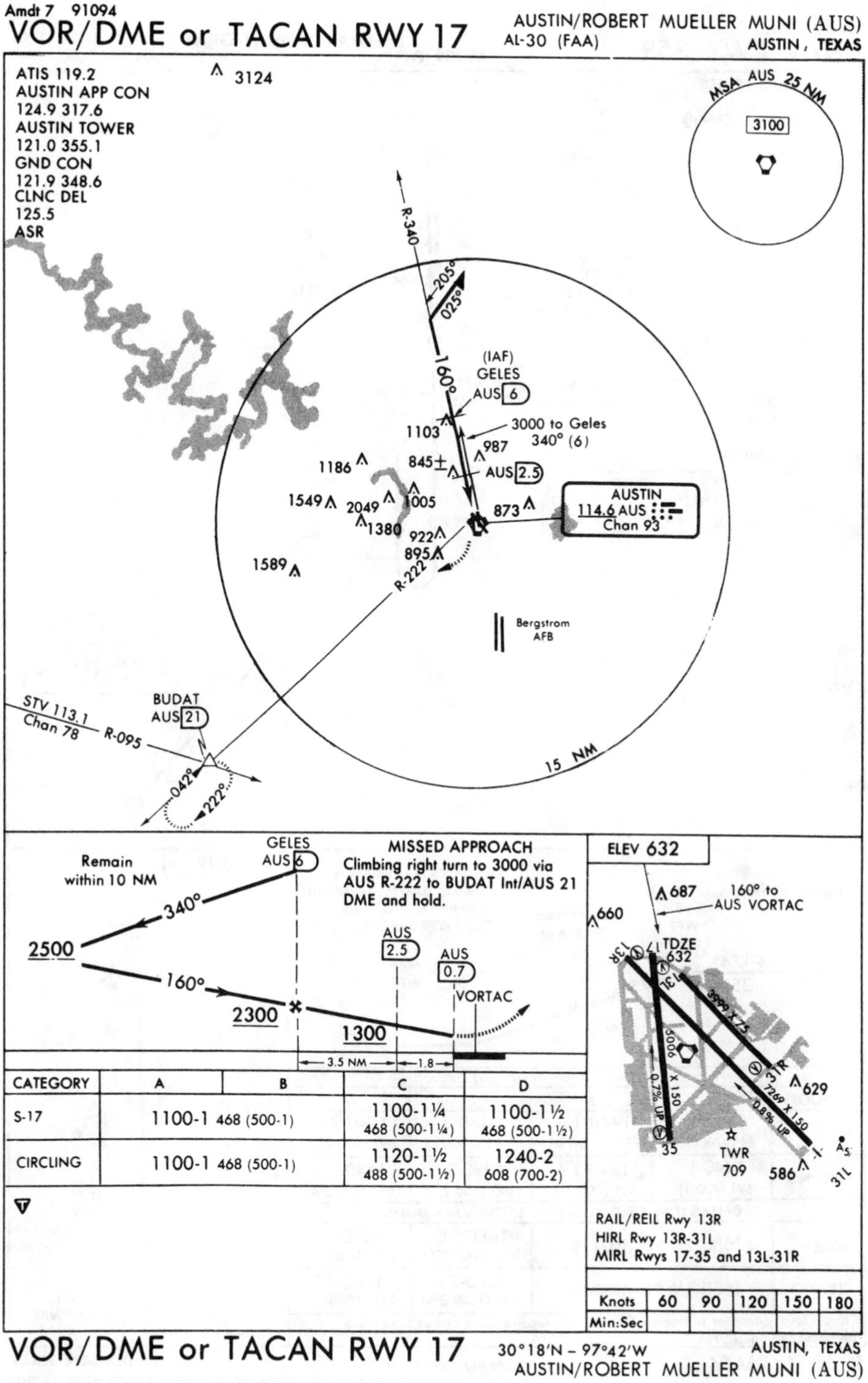

Amdt 7 91094

VOR/DME or TACAN RWY 17

AUSTIN/ROBERT MUELLER MUNI (AUS)
AL-30 (FAA)
AUSTIN, TEXAS

ATIS 119.2
AUSTIN APP CON
124.9 317.6
AUSTIN TOWER
121.0 355.1
GND CON
121.9 348.6
CLNC DEL
125.5
ASR

3124

MSA AUS 25 NM
3100

R-340
205°
025°
160°
(IAF)
GELES
AUS 6
1103
3000 to Geles
340° (6)
987
845
AUS 2.5
1186
1005
1549
2049
1380
873
AUSTIN
114.6 AUS
Chan 93
922
895
1589
R-222
Bergstrom
AFB
BUDAT
AUS 21
STV 113.1
Chan 78
R-095
042°
222°
15 NM

Remain
within 10 NM

GELES
AUS 6

MISSED APPROACH
Climbing right turn to 3000 via AUS R-222 to BUDAT Int/AUS 21 DME and hold.

340°
2500
160°
2300
AUS 2.5
AUS 0.7
VORTAC
1300
3.5 NM
1.8

ELEV 632

687
160° to
AUS VORTAC
660
TDZE
632
13R
17
13L
3999 X 75
5006 X 150
0.7% UP
31R
7269 X 150
0.8% UP
629
35
TWR
709
586
31L

CATEGORY	A	B	C	D
S-17	1100-1 468 (500-1)		1100-1¼ 468 (500-1¼)	1100-1½ 468 (500-1½)
CIRCLING	1100-1 468 (500-1)		1120-1½ 488 (500-1½)	1240-2 608 (700-2)

RAIL/REIL Rwy 13R
HIRL Rwy 13R-31L
MIRL Rwys 17-35 and 13L-31R

Knots	60	90	120	150	180
Min:Sec					

VOR/DME or TACAN RWY 17

30°18′N – 97°42′W

AUSTIN, TEXAS
AUSTIN/ROBERT MUELLER MUNI (AUS)

Fig. 8-11. Continued.

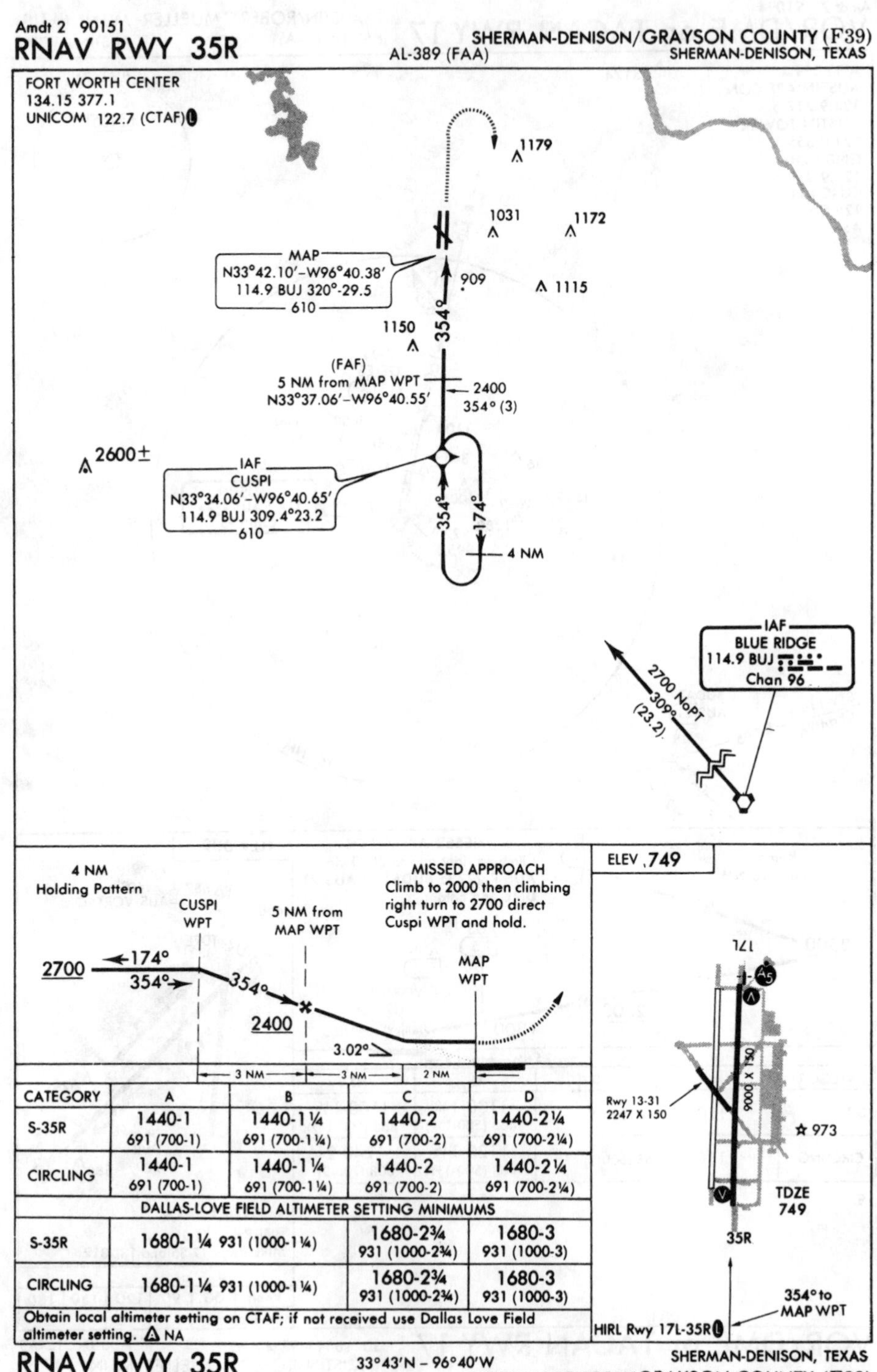

CATEGORY	A	B	C	D
S-35R	1440-1 691 (700-1)	1440-1¼ 691 (700-1¼)	1440-2 691 (700-2)	1440-2¼ 691 (700-2¼)
CIRCLING	1440-1 691 (700-1)	1440-1¼ 691 (700-1¼)	1440-2 691 (700-2)	1440-2¼ 691 (700-2¼)
DALLAS-LOVE FIELD ALTIMETER SETTING MINIMUMS				
S-35R	1680-1¼ 931 (1000-1¼)		1680-2¾ 931 (1000-2¾)	1680-3 931 (1000-3)
CIRCLING	1680-1¼ 931 (1000-1¼)		1680-2¾ 931 (1000-2¾)	1680-3 931 (1000-3)

Obtain local altimeter setting on CTAF; if not received use Dallas Love Field altimeter setting. △ NA

Fig. 8-12. Loran is not approved for other than loran RNAV procedures.

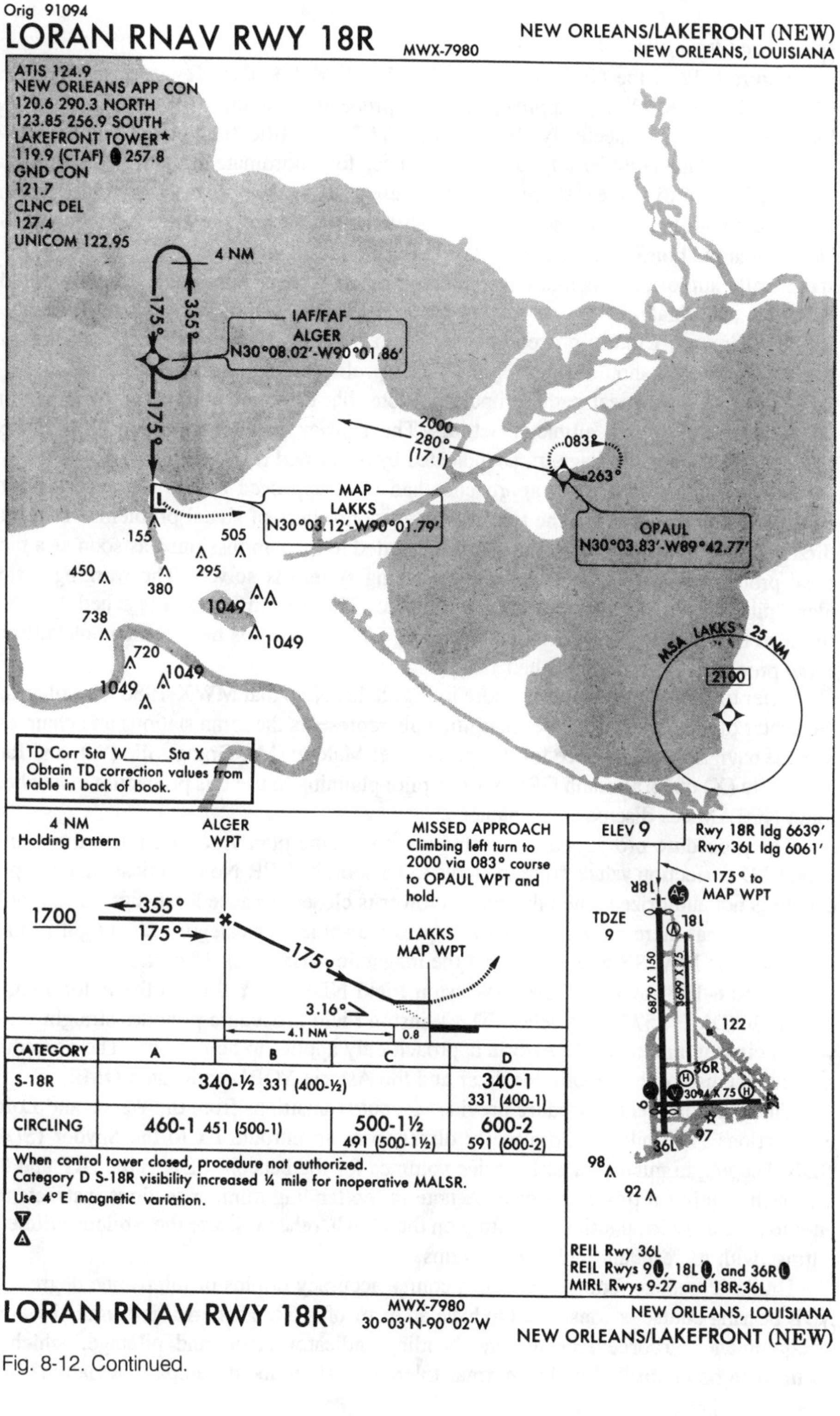

Orig 91094

LORAN RNAV RWY 18R

MWX-7980

NEW ORLEANS/LAKEFRONT (NEW)
NEW ORLEANS, LOUISIANA

ATIS 124.9
NEW ORLEANS APP CON
120.6 290.3 NORTH
123.85 256.9 SOUTH
LAKEFRONT TOWER*
119.9 (CTAF) 257.8
GND CON
121.7
CLNC DEL
127.4
UNICOM 122.95

4 NM
175°
355°
IAF/FAF
ALGER
N30°08.02′-W90°01.86′
175°
2000
280°
(17.1)
083°
263°
MAP
LAKKS
N30°03.12′-W90°01.79′
OPAUL
N30°03.83′-W89°42.77′
155
505
295
450
380
1049
738
1049
720
1049
1049
MSA LAKKS 25 NM
2100

TD Corr Sta W____Sta X____
Obtain TD correction values from table in back of book.

4 NM
Holding Pattern

ALGER
WPT

MISSED APPROACH
Climbing left turn to 2000 via 083° course to OPAUL WPT and hold.

1700
355°
175°
175°
LAKKS
MAP WPT
3.16°
4.1 NM
0.8

ELEV 9
Rwy 18R ldg 6639′
Rwy 36L ldg 6061′
175° to MAP WPT
18R
TDZE
9
18L
6879 X 150
3699 X 75
122
36R
9
3094 X 75
27
97
36L
98
92

REIL Rwy 36L
REIL Rwys 9, 18L, and 36R
MIRL Rwys 9-27 and 18R-36L

CATEGORY	A	B	C	D
S-18R	340-½ 331 (400-½)			340-1 331 (400-1)
CIRCLING	460-1 451 (500-1)		500-1½ 491 (500-1½)	600-2 591 (600-2)

When control tower closed, procedure not authorized.
Category D S-18R visibility increased ¼ mile for inoperative MALSR.
Use 4° E magnetic variation.

LORAN RNAV RWY 18R

MWX-7980
30°03′N-90°02′W

NEW ORLEANS, LOUISIANA
NEW ORLEANS/LAKEFRONT (NEW)

Fig. 8-12. Continued.

(Continued from page 247.)

Figure 8-12 is the Grayson County RNAV RWY 35R and New Orleans/Lakefront LORAN RNAV RWY 18R approaches. Both procedures require IFR certified RNAV or loran equipment, respectively. Way point (WPT) identification boxes on the RNAV approach contain latitude/longitude coordinates, for coordinate RNAV equipment, and VORTAC radial/distance. At least for the present, RNAV approach procedures are not authorized with loran equipment. Additionally, the RNAV box provides reference facility elevation at the bottom of the box (610 feet for the Cuspi way point). The loran RNAV is specifically authorized for loran users. The loran way point boxes only contain latitude/longitude coordinates.

The Grayson County procedure provides a holding pattern course reversal of four nautical mile legs, should it be required. The profile diagram indicates a glide angle of 3.02° for use by vertical path computers. Note that different minima apply depending upon the source of the altimeter setting. The runway parallel to 17L-35R is closed, according to the airport diagram, as indicated by the closed runway symbol from Fig. 8-3.

(Only a half dozen loran approaches had been approved by 1991. The procedures have temporary monitors in the towers to alert controllers of signal problems. Two-hundred and fifty new loran approaches are scheduled to be commissioned as soon as a technical problem with the loran-C cockpit warning system is solved. The warning system alerts pilots when signal reception is unreliable. The warning system is scheduled to be installed on all 28 loran transmitters by spring 1992. Once this has been completed, new loran procedures can be published.)

Refer to the loran RNAV procedure in Fig. 8-12. Note that MWX-7980 is displayed in the center of the upper and lower margin. This represents the loran stations and chain that provide navigational guidance for this procedure: Malone (M), Grangeville (W), and Raymondville (X) on loran chain GRI 7980. A pilot planning to use this procedure must check loran NOTAMs as discussed in chapter 2.

The procedure provides a TD correction box in the plan view and refers the user to obtain TD correction values from the table in back of the TPP. Notes indicate that the procedure is not authorized when the control tower is closed because loran signals cannot be monitored, that there are nonstandard inoperative values for category D straight-in minima when the MALSR is out, and that the magnetic variation is 4° east.

Figure 8-13 shows the Snyder/Winston Field NDB RWY 35 and the Astoria COPTER LOC/DME 257° approaches. The Winston Field procedure provides straight-in, as well as circling minima. The Astoria approach only applies to helicopters. The procedure requires the use of the Astoria localizer and the Astoria VOR's collocated DME.

The Winston Field procedure provides enroute transitions from the Weepe and Loran intersections. The pilot could also fly direct from an enroute fix to the Snyder (SDR) NDB, keeping in mind standard service volumes for the beacon. SDR is the IAF and the approach requires a procedure turn. A note in the landing minima sections instructs the pilot to obtain the local altimeter setting on the CTAF, otherwise use the Abilene altimeter setting, with its resultant higher minimums.

Unlike a localizer course that has a course accuracy of plus or minus one degree, an NDB bearing course is considered to have an error of ±5°. This error is cumulative, due to equipment—airborne receiver and heading indicator error, and pilotage, which is assumed to be controlled within normal tolerance. How about compass deviation? One

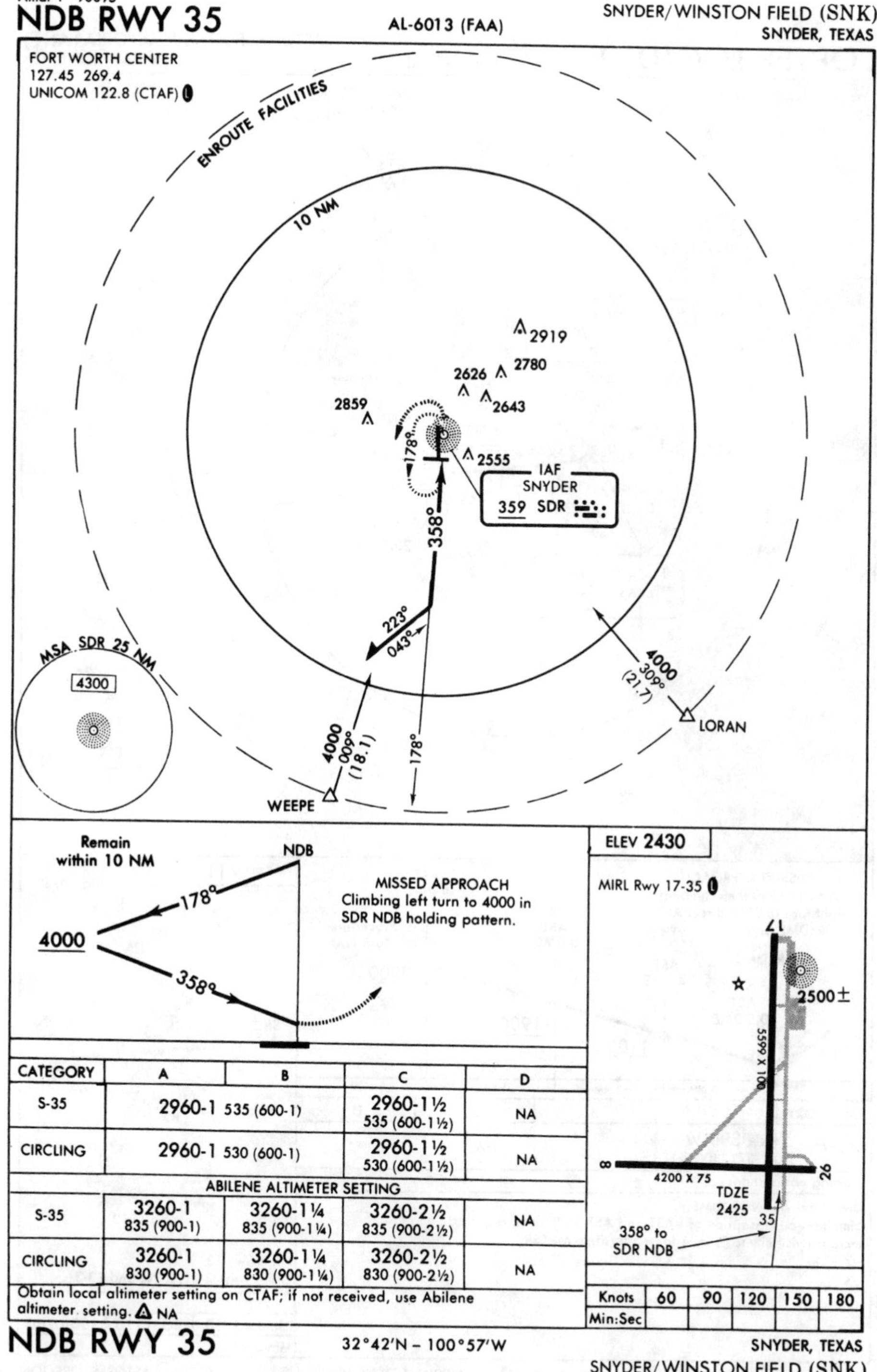

CATEGORY	A	B	C	D
S-35	2960-1 535 (600-1)		2960-1½ 535 (600-1½)	NA
CIRCLING	2960-1 530 (600-1)		2960-1½ 530 (600-1½)	NA
ABILENE ALTIMETER SETTING				
S-35	3260-1 835 (900-1)	3260-1¼ 835 (900-1¼)	3260-2½ 835 (900-2½)	NA
CIRCLING	3260-1 830 (900-1)	3260-1¼ 830 (900-1¼)	3260-2½ 830 (900-2½)	NA

Fig. 8-13. The procedure turn symbol allows the pilot to execute any standard course reversal, its absence mandates the published procedure, such as a DME arc.

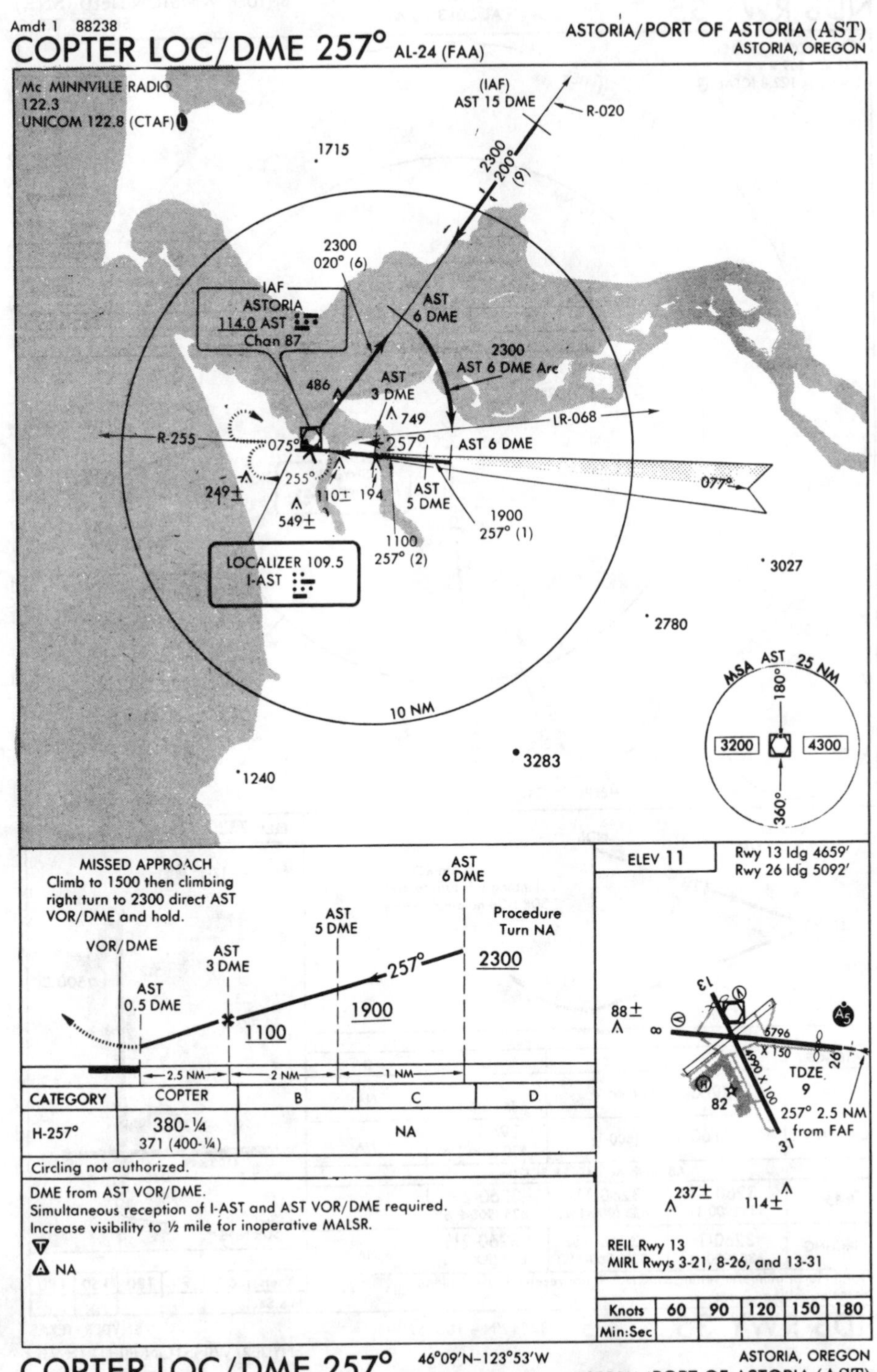

CATEGORY	COPTER	B	C	D
H-257°	380-¼ 371 (400-¼)	NA		

Circling not authorized.

DME from AST VOR/DME.
Simultaneous reception of I-AST and AST VOR/DME required.
Increase visibility to ½ mile for inoperative MALSR.

NA

Knots	60	90	120	150	180
Min:Sec					

COPTER LOC/DME 257° 46°09'N–123°53'W ASTORIA, OREGON
ASTORIA/PORT OF ASTORIA (AST)

Fig. 8-13. Continued.

consequence is that NDB minimums are usually higher than minimums for procedures using other course guidance.

(This is illustrated by a complaint from a large flight school that an NDB was out of tolerance. Pilots at the FAF were complaining of being a mile off course. The NDB was collocated at the ILS middle marker, 7.6 nautical miles from the FAF. By calculating course guidance accuracy and adding as little as $2^1/_2$° of compass deviation, one mile error at the FAF is not unreasonable. Pilots need to be aware of these parameters.)

The missed approach for this NDB procedure consists of a climbing left turn to 4,000 feet in the SDR NDB holding pattern. A miss from a straight-in approach would be a relatively simple climbing direct entry into the depicted pattern. If visual reference is lost while circling, the missed approach specified for the procedure must be followed; however, to become established on the prescribed missed approach course, the pilot should make an initial climbing turn toward the landing runway and continue the turn until established on the missed approach course. Because circling maneuvers may be accomplished in more than one direction, different patterns might be required to establish the aircraft on the missed approach course. This will depend upon the aircraft position at the time that visual reference is lost. For example, in the Winston Field procedure, let's assume we are circling in a right downwind to Runway 17 and lose visual reference. In this case we would initiate a climbing right turn over the airport. If we're north of the beacon we would proceed to the NDB, and use a teardrop entry into the holding pattern; if we're south of the beacon a parallel entry would be appropriate.

The Astoria helicopter procedure has two IAPs: the Astoria (AST) VOR/DME and the AST 020 radial at 15 nautical miles. The pilot would proceed from either IAF to the AST 6 nautical mile DME arc. This approach requires the use of an arc, and DME is specified in the approach name. Minimums are quite low, ceiling 400 feet and visibility one-quarter mile, but helicopters can fly quite slow. Circling is not authorized. The pilot would fly inbound, descending to the MDA, and land straight-in. If this could not be accomplished, a missed approach is required. Notes below the landing minima data show that DME information is obtained from the AST VOR/DME; therefore, simultaneous reception of the IAST localizer and AST VOR/DME are required. And, that inoperative MALSR requires visibility be increased to $^1/_2$ mile. This is an instance where the aircraft must be capable of obtaining course guidance from one facility and DME information from another.

Let's assume that Snyder is our destination and we've lost the radios. We would proceed at the last assigned altitude or the MEA, whichever is higher. We've planned ahead and filed a route that will take us to the Loran intersection. From there we fly at our last assigned altitude or 4,000 feet (the MEA) to the NDB. We will begin a descent and approach as close as possible to an expect further clearance time, expect approach clearance time, or flight plan filed estimated time of arrival, in that order. Depending on our altitude we would descend in the holding pattern to an altitude where at a normal rate of descent we would complete the procedure turn at 4,000 feet.

Refer to the Mesquite procedure in Fig. 8-10. We've been cleared to the destination, via the DFW VORTAC. In this case we would proceed to Nessa INT, hold south, right turns, if necessary until the appropriate time, and commence the approach. With the procedure turn indicated, we can use any type of course reversal, including a holding pattern.

The final scenario would be a situation where the clearance limit is not a fix from which an approach begins. We would leave the clearance limit at the expect further clearance (EFC) time if issued, or from over the clearance limit proceed to a fix from which an approach begins and commence descent, or descent and approach, as close as possible to the estimated time of arrival. Refer to the Dallas-Fort Worth ILS-2 RWY 36L procedure in Fig. 9-9. For the sake of argument let's say we've been cleared short to the Love (LUE) VOR. We lose communications. From LUE we would proceed to a fix from which the approach begins. The closest would be Scurry. We would depart LUE at the issued EFC time, or if none were received, plan to arrive over Scurry as close as possible to the estimated time of arrival and commence descent (if above 4,000 feet) and approach. Hopefully, there will be a published route between any holding fix short of a transition route or IAF to provide us with radials, distances, and altitudes.

Normally, ATC will issue EFC times and route when holding aircraft short of the destination. For example, in this case we might receive a clearance to "hold, expect further clearance via direct Scurry at zero five one five, maintain 6,000." At zero five one five we would depart LUE for Scurry, cross Scurry at 6,000, and commence descent and approach. One final thought, if communications are lost don't forget to monitor VOR voice communication because ATC will attempt to communicate on these frequencies.

JEPPESEN APPROACH CHARTS

E.B. Jeppesen first published his *Airway Manual* in 1934. Jeppesen services and products have become worldwide since then. Jeppesen charts are printed on $5^1/2 \times 8^1/2$ inch sheets, oversized charts are folded to the standard size. Chart dimensions provide a balance between bulk and utility. Other Jeppesen advantages include individual sheets and the reason for the revision. Individual sheets allow the pilot to place charts in clear plastic pockets for easier use. The most often heard disadvantage of the individual Jeppesen charts, especially the approach plates, is the constant requirement to update the *Airway Manual*. (Some pilots prefer NOS procedures because they are simply thrown away and entirely replaced.) To help alleviate the problem of sheet-by-sheet revisions, Jeppesen has Q Service; subscribers receive a standard revision envelop, every two weeks, with a special update index; all terminal charts are replaced at the end of the 16-week Q Service cycle and the update procedure begins again.

Jeppesen provides IAP chart coverage for the world. Fifteen coverage options are available for the U.S., ranging from the entire U.S. to sections similar to individual volumes of the TPP. Canada, Alaska, and the Aleutians have three coverage areas; therefore, pilots are able to subscribe to only those areas where they do most of their flying. Jeppesen provides special trip coverage on a one-time sale basis.

Currency differences between NOS and Jeppesen charts are discussed in chapter 2. Terminology and symbols used on Jeppesen products have the same definitions as those on NOS and DMA charts, but charts are presented in a slightly different format (Jeppesen IAP chart features):

- Looseleaf format for updating
- Entire plan view drawn to scale
- Highest point within the plan view depicted
- Procedure turn fully depicted

- Touchdown zone elevation
- Inoperative component minimums shown in landing minima data
- RVR converted to visibility in miles and fractions in landing minima data
- Descent rates in feet per minute
- Changes made since previous chart

Unlike the TPP, Jeppesen IAPs contain inoperative component minimums on the chart. The minima section also provides RVR visibility converted to miles and fractions. And, descent rates in feet per minute for various airspeeds from 70 to 160 knots for ILS procedures appear on the chart. This saves the pilot from having to look up data in separate sections and calculate inoperative component minimums. These Jeppesen features are seen in Fig. 8-14 (compared to NOS in Fig. 8-9) and Fig. 8-15 (compared to NOS in Fig. 8-10).

The other major difference between NOS and Jeppesen IAPs is the airport sketch. Jeppesen provides a larger, more detailed airport sketch on the reverse side of the first IAP chart for each location. A separate scaled airport diagram chart is published when the airport is large and complex. Features include:

- Bearing, distance, frequency, and identifier from the nearest NAVAID to the airport
- Airport magnetic variation
- Airport notes including traffic pattern information, when other than standard
- Latitude/longitude grid
- Airport reference point location
- Runway threshold elevations
- Taxiway designators
- Distance beyond displaced threshold
- Runway restrictions
- Takeoff minimums
- Alternate minimums
- IFR departure procedures

Jeppesen airport diagrams include general information, additional runway information, takeoff minimums, departure procedure, and alternate minimums. This information is normally located below the airport diagram, except when a large diagram is required, then this data is printed on the reverse side of the airport layout. General information consists of traffic patterns and other miscellaneous data. General notes, such as airport restrictions, bird activity, and the availability of special services are provided. Additional runway information provides details on runway surface, width, length available for takeoff and landings, lighting information, and remarks. The availability and type of approach lighting aids are provided for each runway.

IFR departure procedures, and takeoff and alternate minimums are provided on the chart, unlike the TPP that has the information elsewhere. Additionally, the airport diagram contains a latitude/longitude grid, airport reference point locations, runway threshold elevations, taxiway designators, and any runway restrictions. Jeppesen also includes a detailed air carrier ramp diagram, and coordinates for gate locations, when necessary.

(Continued on page 264.)

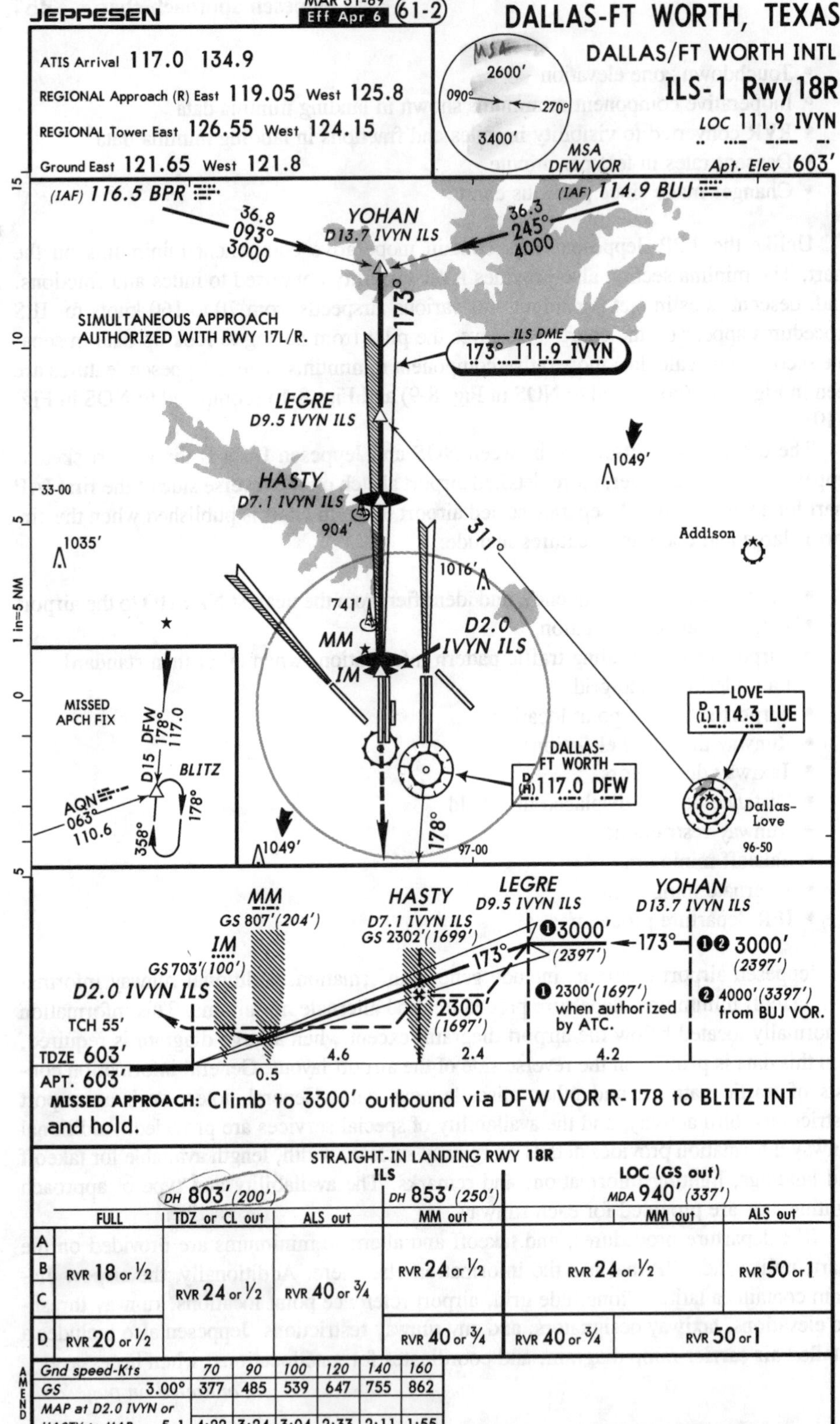

JEPPESEN
MAR 31-89
Eff Apr 6
61-2
DALLAS-FT WORTH, TEXAS
DALLAS/FT WORTH INTL
ILS-1 Rwy18R
LOC 111.9 IVYN
ATIS Arrival 117.0 134.9
REGIONAL Approach (R) East 119.05 West 125.8
REGIONAL Tower East 126.55 West 124.15
Ground East 121.65 West 121.8
2600'
090°
270°
3400'
MSA
DFW VOR
Apt. Elev 603'
(IAF) 116.5 BPR
(IAF) 114.9 BUJ
36.8
093°
3000
36.3
245°
4000
YOHAN
D13.7 IVYN ILS
173°
SIMULTANEOUS APPROACH AUTHORIZED WITH RWY 17L/R.
ILS DME
173° 111.9 IVYN
LEGRE
D9.5 IVYN ILS
1049'
HASTY
D7.1 IVYN ILS
904'
311°
Addison
1035'
1016'
741'
MM
D2.0 IVYN ILS
IM
MISSED APCH FIX
D15 DFW
178°
117.0
BLITZ
AQN
063°
110.6
358°
178°
LOVE
114.3 LUE
DALLAS-FT WORTH
117.0 DFW
Dallas-Love
1049'
97-00
96-50
33-00
1 in=5 NM
MM
GS 807'(204')
HASTY
D7.1 IVYN ILS
GS 2302'(1699')
LEGRE
D9.5 IVYN ILS
YOHAN
D13.7 IVYN ILS
IM
GS 703'(100')
3000'
(2397')
173°
3000'
(2397')
D2.0 IVYN ILS
2300'
(1697')
2300'(1697') when authorized by ATC.
4000'(3397') from BUJ VOR.
TCH 55'
TDZE 603'
APT. 603'
4.6
2.4
4.2
0.5
MISSED APPROACH: Climb to 3300' outbound via DFW VOR R-178 to BLITZ INT and hold.
STRAIGHT-IN LANDING RWY 18R
ILS
DH 803' (200')
DH 853' (250')
LOC (GS out)
MDA 940' (337')
FULL
TDZ or CL out
ALS out
MM out
MM out
ALS out
A
B
C
D
RVR 18 or ½
RVR 24 or ½
RVR 40 or ¾
RVR 24 or ½
RVR 24 or ½
RVR 50 or 1
RVR 20 or ½
RVR 40 or ¾
RVR 40 or ¾
RVR 50 or 1
Gnd speed-Kts 70 90 100 120 140 160
GS 3.00° 377 485 539 647 755 862
MAP at D2.0 IVYN or HASTY to MAP 5.1 4:22 3:24 3:04 2:33 2:11 1:55
AMEND 2
CHANGES: Procedure.
© JEPPESEN SANDERSON, INC., 1986, 1989. ALL RIGHTS RESERVED

Fig. 8-14.

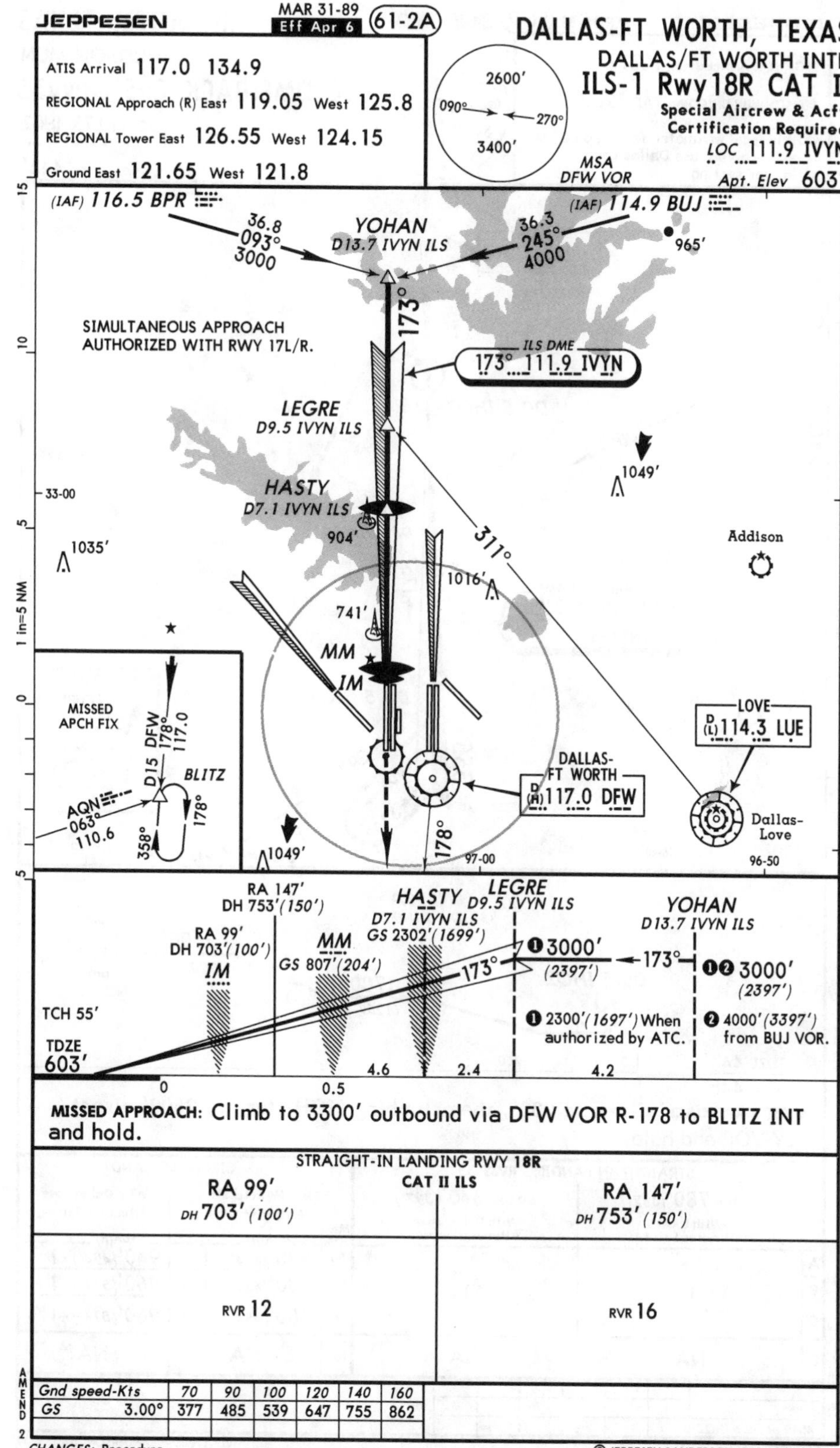

Fig. 8-14. Continued.

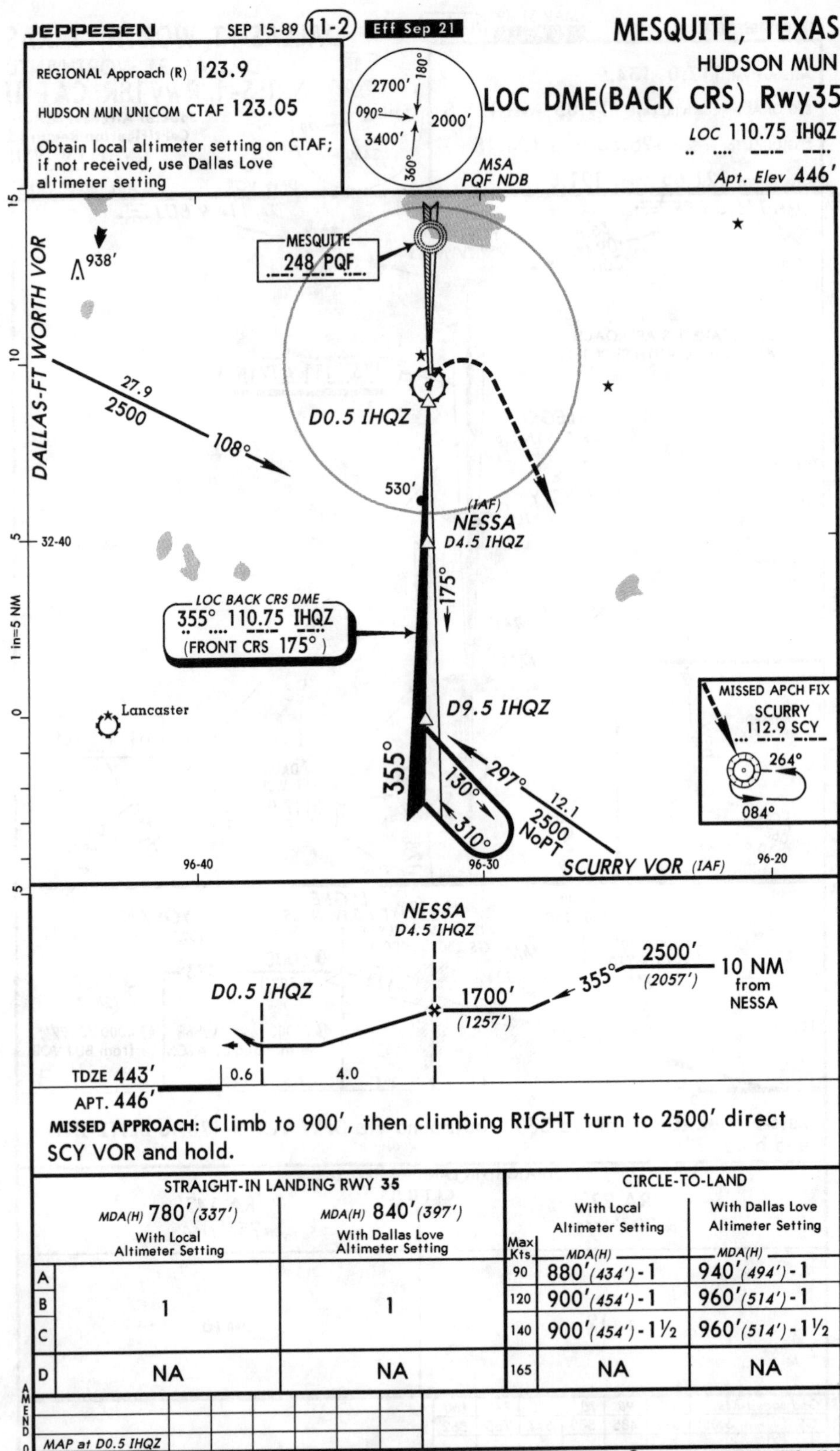

	STRAIGHT-IN LANDING RWY 35			CIRCLE-TO-LAND	
	MDA(H) 780' (337') With Local Altimeter Setting	MDA(H) 840' (397') With Dallas Love Altimeter Setting	Max Kts	With Local Altimeter Setting MDA(H)	With Dallas Love Altimeter Setting MDA(H)
A	1	1	90	880' (434') - 1	940' (494') - 1
B	1	1	120	900' (454') - 1	960' (514') - 1
C	1	1	140	900' (454') - 1½	960' (514') - 1½
D	NA	NA	165	NA	NA

Fig. 8-15.

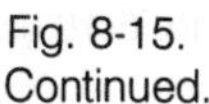

Fig. 8-15. Continued.

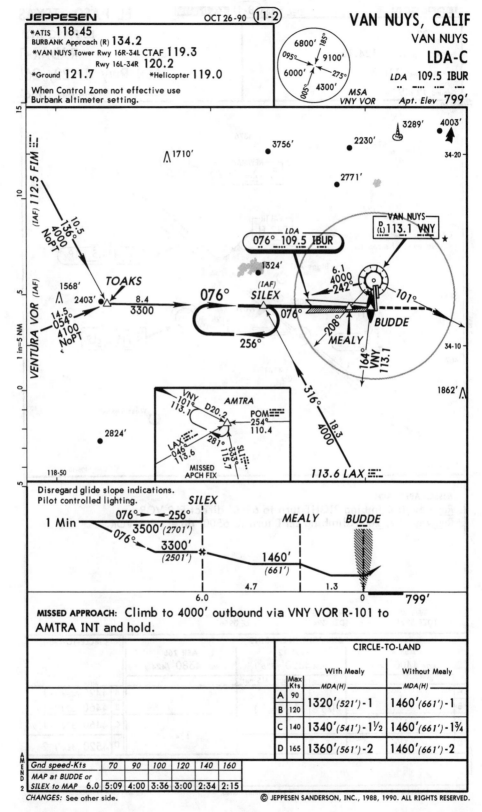

JEPPESEN OCT 26-90 (11-2)

VAN NUYS, CALIF

VAN NUYS

LDA-C

LDA 109.5 IBUR

Apt. Elev 799′

*ATIS 118.45
BURBANK Approach (R) 134.2
*VAN NUYS Tower Rwy 16R-34L CTAF 119.3
Rwy 16L-34R 120.2
*Ground 121.7 *Helicopter 119.0

When Control Zone not effective use Burbank altimeter setting.

MSA VNY VOR: 6800′ (095°–185°), 9100′ (185°–275°), 4300′ (275°–005°), 6000′ (005°–095°)

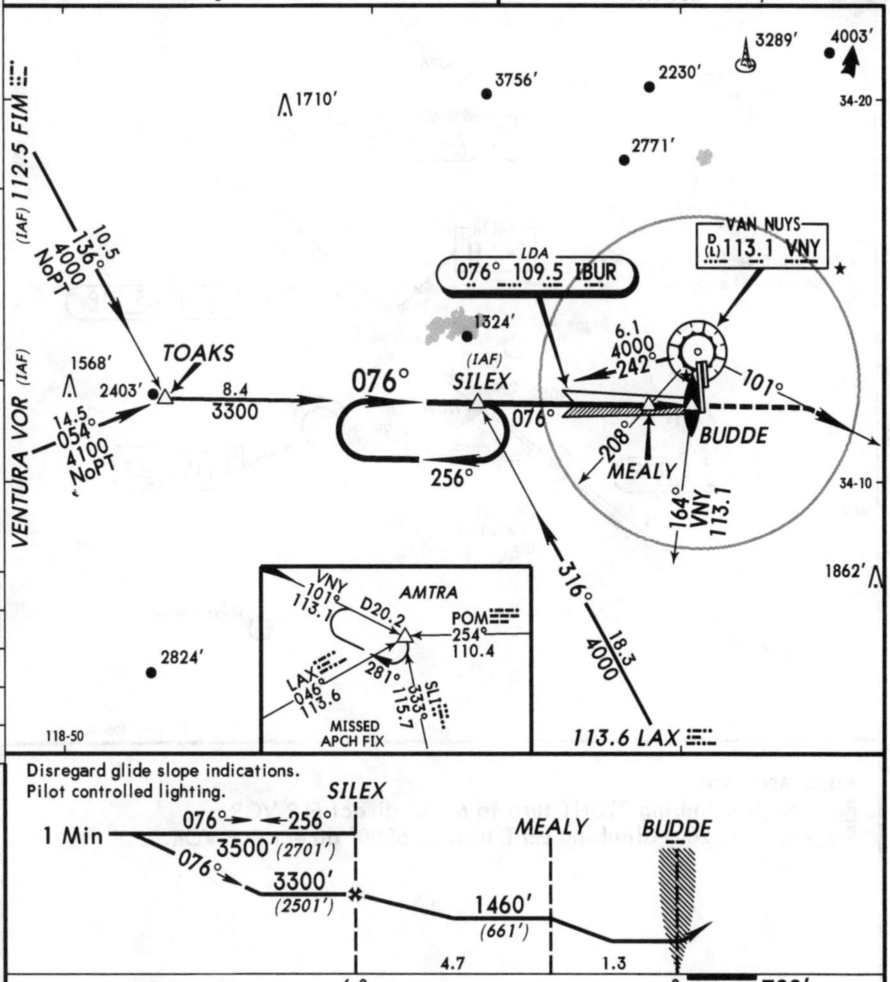

Disregard glide slope indications.
Pilot controlled lighting.

MISSED APPROACH: Climb to 4000′ outbound via VNY VOR R-101 to AMTRA INT and hold.

CIRCLE-TO-LAND

	Max Kts	With Mealy MDA(H)	Without Mealy MDA(H)
A	90	1320′(521′)-1	1460′(661′)-1
B	120	1320′(521′)-1	1460′(661′)-1
C	140	1340′(541′)-1½	1460′(661′)-1¾
D	165	1360′(561′)-2	1460′(661′)-2

Gnd speed-Kts	70	90	100	120	140	160
MAP at BUDDE or SILEX to MAP 6.0	5:09	4:00	3:36	3:00	2:34	2:15

AMEND 2

CHANGES: See other side.

JEPPESEN DEC 9-88 (18-1) Eff Dec 15

EL PASO, TEXAS
EL PASO INTL
RADAR-1
ASR Rwy 4, 22, 26L

Fig. 8-16.

ATIS 120.0
EL PASO Approach (R) 124.15
EL PASO Tower 118.3
Ground 121.9

Apt. Elev 3956'

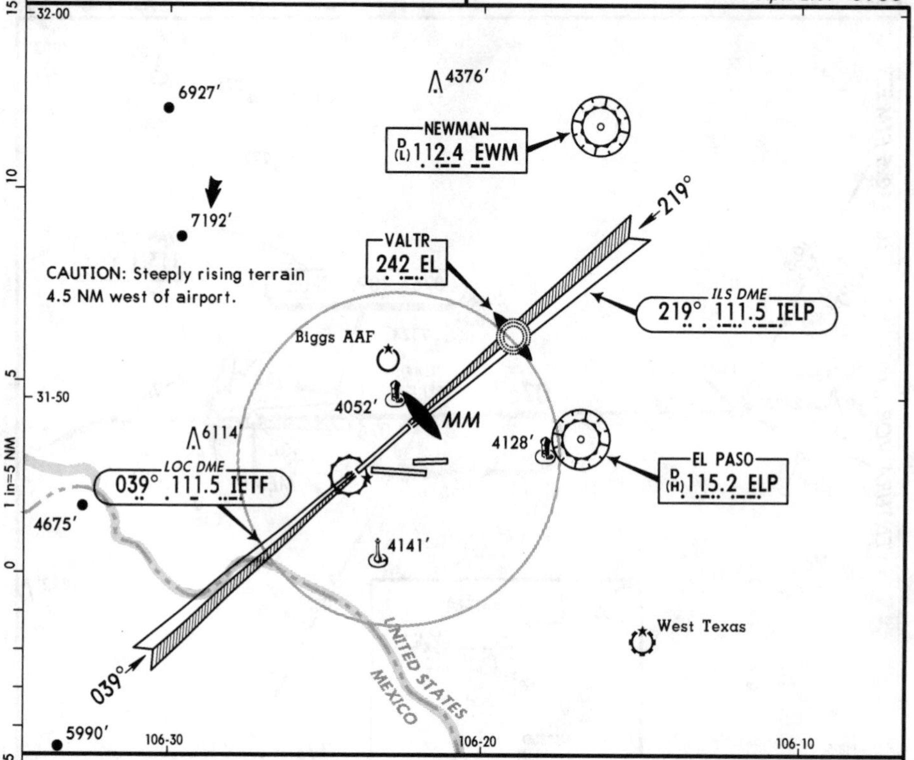

NEWMAN D (L) 112.4 EWM
VALTR 242 EL
ILS DME 219° 111.5 IELP
LOC DME 039° 111.5 IETF
EL PASO D (H) 115.2 ELP
CAUTION: Steeply rising terrain 4.5 NM west of airport.
Biggs AAF
West Texas
UNITED STATES
MEXICO
MM
219°
039°
6927' 7192' 4376' 4052' 4128' 6114' 4675' 4141' 5990'
32-00 31-50 106-30 106-20 106-10
1 in=5 NM

MISSED APPROACH:
Runway 4: Climbing RIGHT turn to 6500' direct ELP VOR.
Runways 22, 26L: Climbing LEFT turn to 6500' direct ELP VOR.

RWY 4 TDZE 3921' | RWY 22 TDZE 3944' | RWY 26L TDZE 3956'

	STRAIGHT-IN LANDING					CIRCLE-TO-LAND
	ASR 4 MDA 4400' (479')	ASR 22 MDA 4320' (376')			ASR 26L MDA 4380' (424')	MDA
			RAIL out	ALS out		
A	1	1/2	3/4	1	1	4420' (464')-1
B	1	1/2	3/4	1	1	4460' (504')-1
C	1 1/4	1/2	3/4	1	1 1/4	4460' (504')-1 1/2
D	1 1/2	1	1 1/4	1 1/4	1 1/4	4520' (564')-2

AMEND 12

CHANGES: IPOX LOC Back Crs decommissioned, IETF LOC commissioned.

Fig. 8-16. Continued.

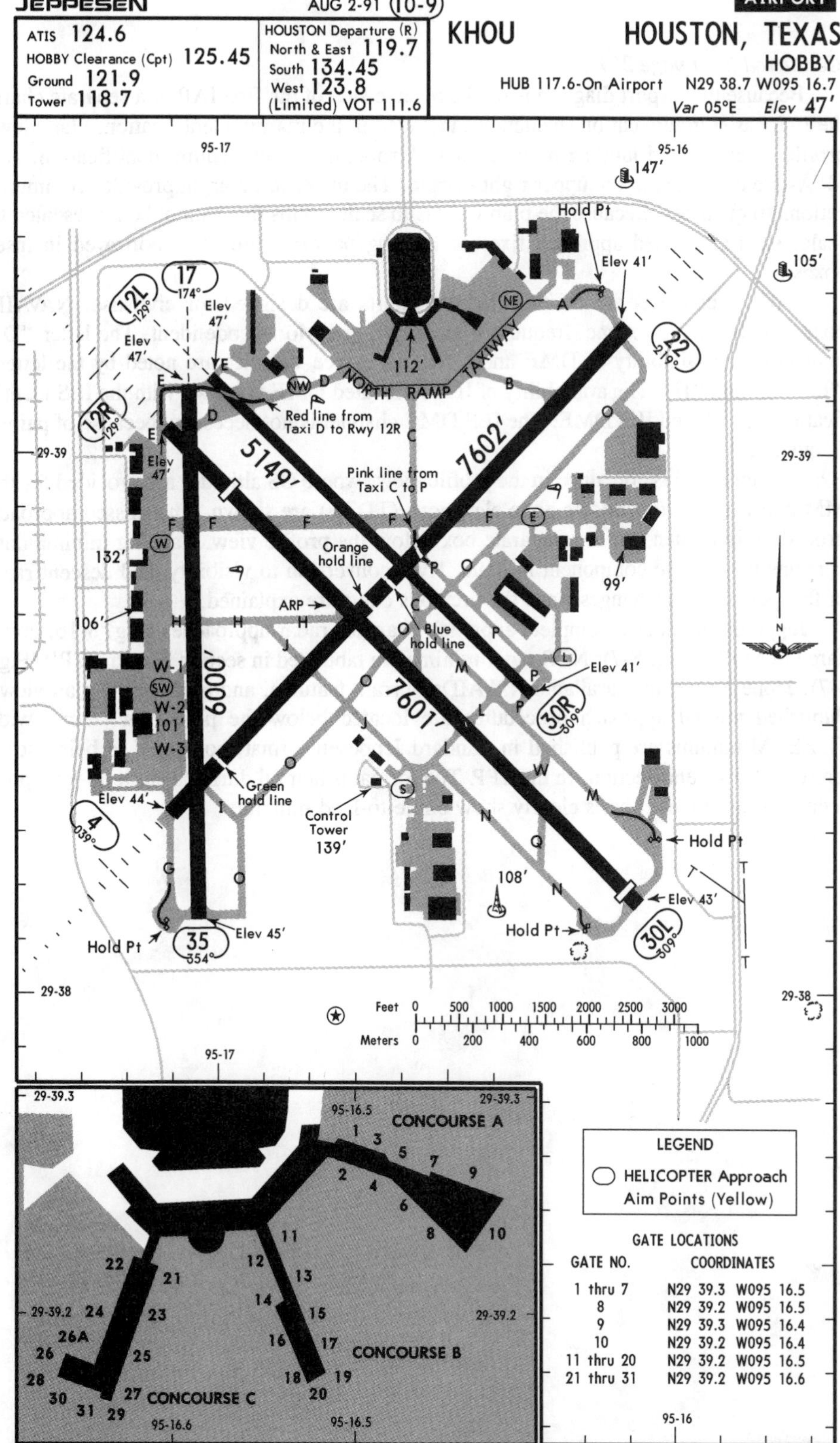

(Continued from page 257.)

Because the airport diagram is on the reverse side of the first IAP or a separate chart, the IAP has communication frequencies included in the margin identification, plan view, profile diagram, and landing minima data. Airport name, procedure identification, and MSAs are contained in the upper right margin. The upper left margin provides communications frequencies. Because the plan view is to scale, items that cannot be represented to scale, such as missed approach fixes or holding patterns, are often portrayed in inset boxes.

TACAN channels are omitted because charts are designed for civil use. NAVAID boxes contain facility name, frequency, identifier, and Morse code ident. The letter "D" indicates the availability of DME and standard service volumes are noted by the letters "T," "L," and "H." The availability of ILS collocated DME is shown with the ILS identification box labeled ILS DME. The ILS DME channel is not necessary because of paired frequencies.

Greater detail is contained in the profile view, where GS altitudes are provided for the MM and IM, and touchdown zone elevations (TDZE) are shown. The missed approach procedure is written out in a separate box below the profile view. Landing minima data contains inoperative component minima, RVR conversion to visibility, and descent rates in feet per minute. Changes since the previous chart are explained.

Jeppesen provides a complete approach chart for radar approaches (Fig. 8-16, compared to NOS in Fig. 8-7). NOS radar minima are tabulated in section N of the TPP (Fig. 8-7). Jeppesen includes available NAVAIDs, terrain features, and notes in the plan view. Published missed approach procedures are located below the plan view, along with TDZE. Minimums are published in standard Jeppesen format, compared to being scattered among several sections in the TPP. The format is helpful during a circle-to-land procedure because these charts clearly show circle-to-land minima.

9
Publications

EFFECTIVE APPLICATION OF CHARTS DURING PREFLIGHT PLANNING AND while navigating enroute is dependent upon an understanding of supplemental publications. Simple charts published in the early days of aviation (*see* chapter 1) naturally evolved hand in hand with the complexities of flying and before too long charts could no longer reasonably depict all the data. Charted information changed more rapidly than it was possible to update, reprint, and distribute charts in a cost efficient manner (*see* chart currency in chapter 2). Most aeronautical publications are merely extensions of aeronautical charts.

Aeronautical publications are most often produced and supplemented by the same agency that publishes the respective charts; NOS charts are supplemented by the *Airport/ Facility Directory* (A/FD) published by NOS, and the NOTAM publication and NOTAM system are administered by the FAA. DMA supports its charts through flight information publications (FLIP) and the aeronautical chart updating manual (CHUM). Canada and other countries that produce charts have similar supplementary publications, such as the Canada flight supplement.

This chapter focuses on publications that supplement NOS visual and instrument aeronautical charts: the A/FD, Alaska supplement, Pacific chart supplement (Fig. 9-1), and the NOTAM publication. Chapter 10 discusses the FLIP, CHUM, and other publications.

I subscribe to the volume of the A/FD where I do most of my flying. It's in my flight case and I have found it to be of immense value. Although charts provide essential data, the directory provides the details. For long trips, out of the area of coverage of my directory, I visit the FSS and use the directories that cover the route. Directories are also available on a one time sale basis from many chart suppliers. Pilots planning long trips would be well advised to obtain the directories that cover their route. By reviewing the directory all pertinent data can be obtained and noted or logged on a flight planning form.

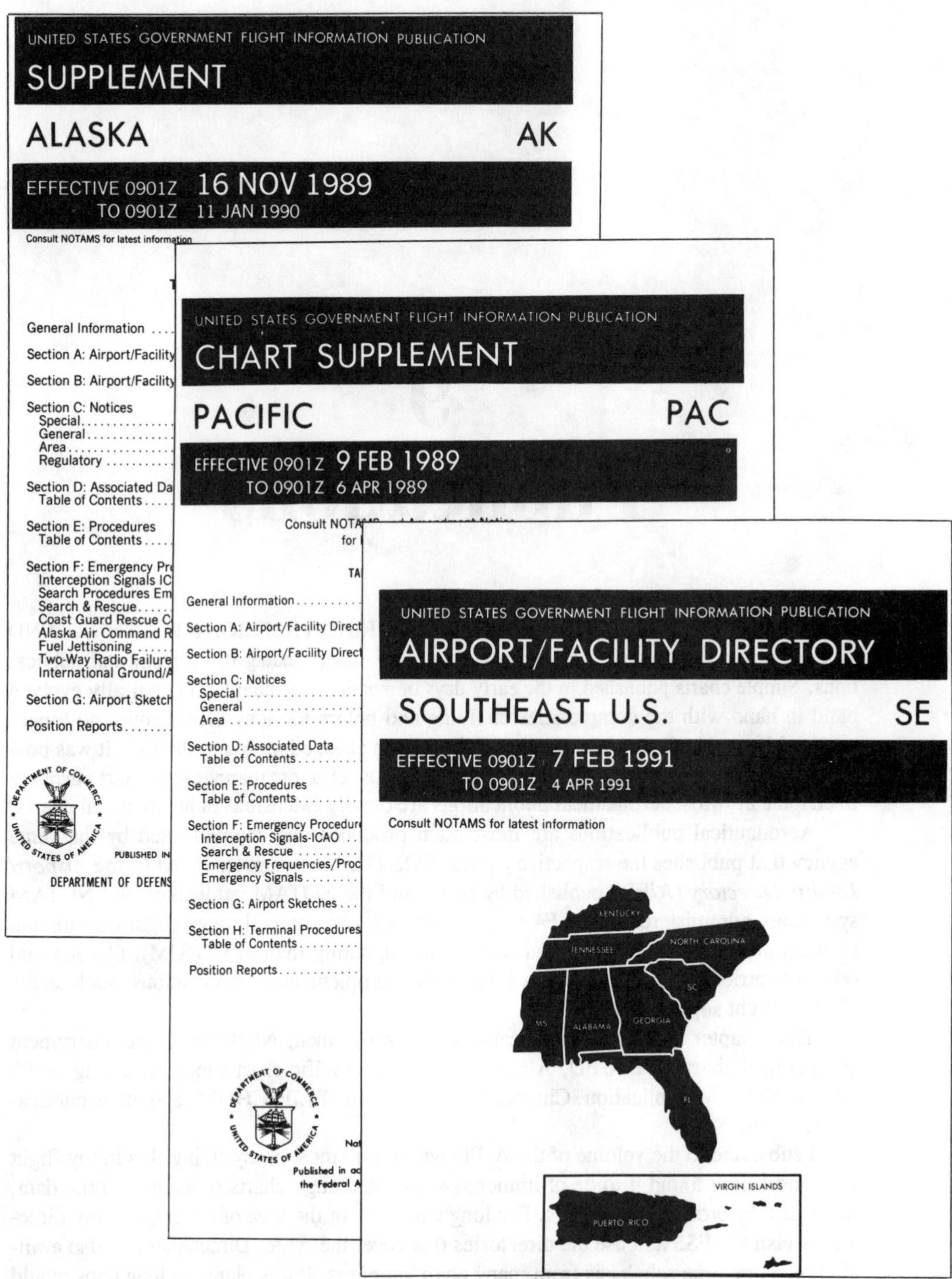

Fig. 9-1. The *Airport/Facility Directory*, Alaska supplement, and Pacific chart supplement support NOS visual and instrument charts for the respective regions.

AIRPORT/FACILITY DIRECTORY

The A/FD is published in paperback books $5^3/_8 \times 8^1/_4$ inches, on the standard 56-day revision cycle. The directory is an alphabetical listing of data on record with the FAA for all airports that are open to the public, associated terminal control facilities, air route traffic control centers (ARTCC), and radio aids to navigation within the contiguous U.S., Puerto Rico, and the Virgin Islands. Radio aids and airports are listed alphabetically. Airports and associated cities are cross-referenced when necessary. The directory directly supports visual charts through the airport listing and aeronautical chart bulletin. Instrument charts are supported through NAVAID restrictions, and detailed airport services and information not available on charts. Features include:

- NOTAM service
- Location identifier
- Airport location
- Time conversion
- Geographic position of airport
- Charts on which the facility is located
- Elevation
- Rotating light beacon
- Servicing available
- Fuel available
- Oxygen available
- Traffic pattern altitude
- Airport of entry and landing rights airports
- FAR 139 crash, fire, rescue availability
- FAA inspection data
- Runway data
- Airport remarks
- Weather data sources
- Communications
- Radio aids to navigation
- Bearing and distance from nearest usable VORTAC
- Detailed directory legend

The A/FD is divided into seven booklets, coverages are depicted in Fig. 9-2. Each directory contains the following:

- General information
- Abbreviations
- Legend, A/FD
- A/FD
- Heliports
- Seaplane bases
- Notices

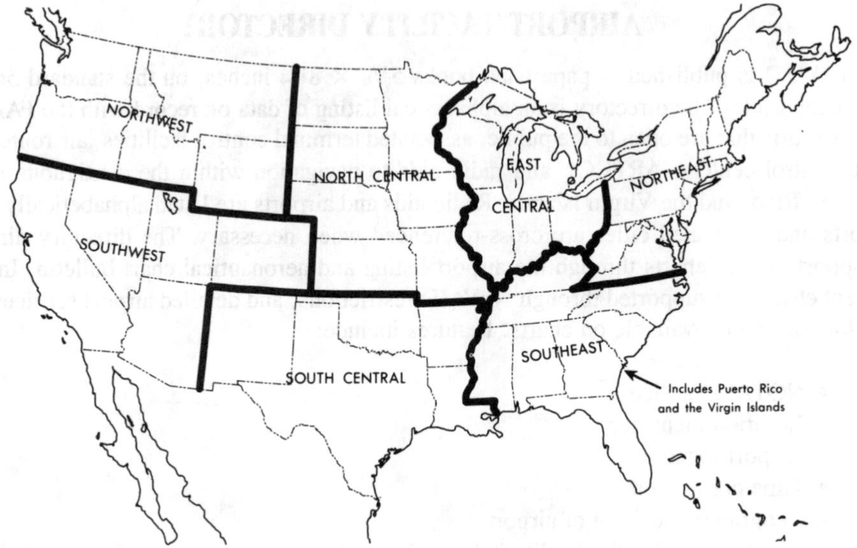

Fig. 9-2. The seven booklets of the *Airport/Facility Directory* cover the United States, including Puerto Rico and the Virgin Islands.

- FAA and National Weather Service telephone numbers
- Air route traffic control centers
- GADO and FSDO addresses and telephone numbers
- Preferred IFR routes
- VOR receiver checkpoints
- Parachute jumping areas
- Aeronautical chart bulletin
- National Weather Service upper air observing stations
- Enroute flight advisory service outlets and frequencies

The inside front cover of the directory provides general information. This consists of information about corrections, comments, and procurement of the directory, with addresses and telephone numbers. General information contains the directory publishing schedule, along with airport and airspace information publication cutoff dates.

Next is a table of contents and list of abbreviations commonly used in the directory. Other abbreviations are contained in the legend and not duplicated in the list. This section is followed by the directory legend.

The A/FD lists public use airports and NAVAIDs that are part of the National Airspace System (NAS). These are listed alphabetically by city or facility name, within the state, and cross-referenced when necessary. This section is followed by a listing of public use heliports and seaplane bases.

Special notices contain additional information within the coverage of the directory. For example, a listing of simultaneous operations on intersecting runways is provided. Other data in this section consists of advance flight plan filing requirements, special flight procedures, temporary closure of facilities, and other general information.

Telephone numbers are provided for FAA and NWS facilities within the directory area of coverage, listed alphabetically within the state. The availability of special services is noted. For example, recorded aviation weather and fast file flight plan filing services. Flight standards district office (FSDO) addresses and telephone numbers are listed.

Air route traffic control center sector frequencies are provided. These are listed alphabetically by location and altitude stratum (low or high), within the individual ARTCC, for the area of coverage of the directory.

Preferred IFR routes are listed. This system has been established to help pilots in planning their routes, to minimize route changes during flight, and aid in the efficient, orderly management of air traffic.

Approved VOR receiver checkpoints and VOR test facilities (VOT) are listed alphabetically, within the states. Type of check, ground or airborne, and checkpoint description are provided. VOT facilities are listed separately.

Parachute jumping areas, depicted by a small parachute symbol on sectional and TAC charts, are tabulated alphabetically, within the states. This is where a pilot would obtain details on charted parachute jumping areas. Unless otherwise indicated, all activities are conducted during daylight hours and in VFR weather conditions. This section also outlines procedures for parachute jumping areas to qualify for inclusion on charts.

The aeronautical chart bulletin provides major changes in aeronautical information that have occurred since the last publication date of each sectional, terminal area, and helicopter chart. Additionally, users of world aeronautical and U.S. Gulf coast VFR aeronautical charts should make appropriate revisions to their charts from this bulletin.

Figure 9-3 is an excerpt from an aeronautical chart bulletin. A number of airports are closed or no longer available for public use on the San Antonio sectional. The MEF has been amended for a quadrant of the U.S. Gulf coast helicopter route chart. On the Wichita sectional several obstructions, under construction, have been added. Additional entries in the aeronautical charts bulletin consist of revisions to transition and control areas, control zones, NAVAID frequencies and identifications, changes in SUA, and special notices.

The page opposite the inside back cover of the directory contains a graphic of National Weather Service (NWS) upper air observing stations and weather radars. These are scheduled balloon releases (1100 UTC—coordinated universal time—and 2300 UTC); therefore, pilots cannot expect to be notified by NOTAM, unless the release is delayed beyond 1130 UTC or 2330 UTC. NOTAMs are issued for unscheduled balloon releases.

The inside back cover of the directory provides the location and frequency of flight watch outlets within the area served by the volume. All low altitude flight watch outlets are on the common frequency of 122.0; therefore, only individual flight watch control station high altitude frequencies are listed. Each center's airspace is served by a discrete high altitude flight watch frequency.

Listings in the airport/facility portion of each directory merely indicate the airport operator's willingness to accommodate transient aircraft, and does not represent that the facility conforms with any federal or local standards, or that it has been approved for use on the part of the general public. Information on obstructions is taken from reports submitted to the FAA. This information has not been verified in all cases. Pilots are cautioned that obstructions not indicated in this tabulation, or on charts, might exist that can create a hazard to flight operations. Detailed specifics concerning services and facilities in the directory are contained in the *Airman's Information Manual, Basic Flight Information and ATC Procedures*.

270 **AERONAUTICAL CHART BULLETIN**

SAN ANTONIO SECTIONAL
46th Edition, December 13, 1990

Delete MONTEITH RANCH APRT 30°58'00"N, 097°38'00"W Delete WEST TEXAS BOYS RANCH 31°20'00"N, 100°37'03"W Delete GEORGE McENTIRE RANCH 31°53'19"N, 101°08'31"W Delete PHILLIPS PLANTATION 32°20'33"N, 097°15'02"W Delete BATESVILLE ARPT 28°57'12"N, 099°37'55"W Delete RIVER RIDERS RANCH 31°59'15"N, 098°02'30"W Add obst 1998'MSL (318'AGL); 29°30'50"N, 100°23'46"W

Military Training Routes
No Changes

U.S. GULF COAST HELICOPTER ROUTE CHART
5th Edition, November 15, 1990

Add MEF 400' in quadrant 28°00'00"-29°00'00"N, 88°00'00"-89°00'00"W Add arpt Salaika Aviation (pvt) 29°14'25"N, 95°20'40"W

Military Training Routes
No Changes

WICHITA SECTIONAL
45th Edition, November 15, 1990

Add obst 3875'MSL (466'AGL)UC 39°13'09"N 101°20'35"W Add obst 3002'MSL (302'AGL)UC 36°36'56"N 100°19'27"W Add obst 1542'MSL (345'AGL)UC 36°07'06"N 98°15'48"W Add obst 3277'MSL (382'AGL)UC 36°37'53"N 100°46'36"W Add obst 1937'MSL (333'AGL)UC 36°15'55"N 98°37'13"W Add obst 2052'MSL (270'AGL)UC 36°48'35"N 98°55'20"W

Add obst 2105'MSL (380'AGL) 36°31'34"N 99°01'38"W Add obst 1662'MSL (340'AGL)UC 36°49'44"N 98°35'51"W

Military Training Routes
IR 501 Revised
IR 524 Added

Fig. 9-3. The aeronautical chart bulletin revises visual charts as needed between publication cycles.

Directory legend

Figure 9-4 is an example entry in the directory. The following discussion is keyed to the circled numbers on the sample.

1. City/airport name. Airports and facilities are listed alphabetically by associated state and city. Where a city name is different from the airport name, the city name appears on the line above the airport name, as in Fig. 9-4. Airports with the same associated city name are listed alphabetically by airport name and will be separated by a dashed line. All others are separated by a solid line.

2. NOTAM service. The NOTAM service symbols (the section symbol) have been removed from the directory. Recent changes in the NOTAM system have made them redundant because all airports now receive NOTAM-D distribution. Airport NOTAM file identifier is shown following the associated FAA data for individual airports (NOTAM FILE ORL, the NOTAM file for this airport is Orlando [ORL]). This is important because

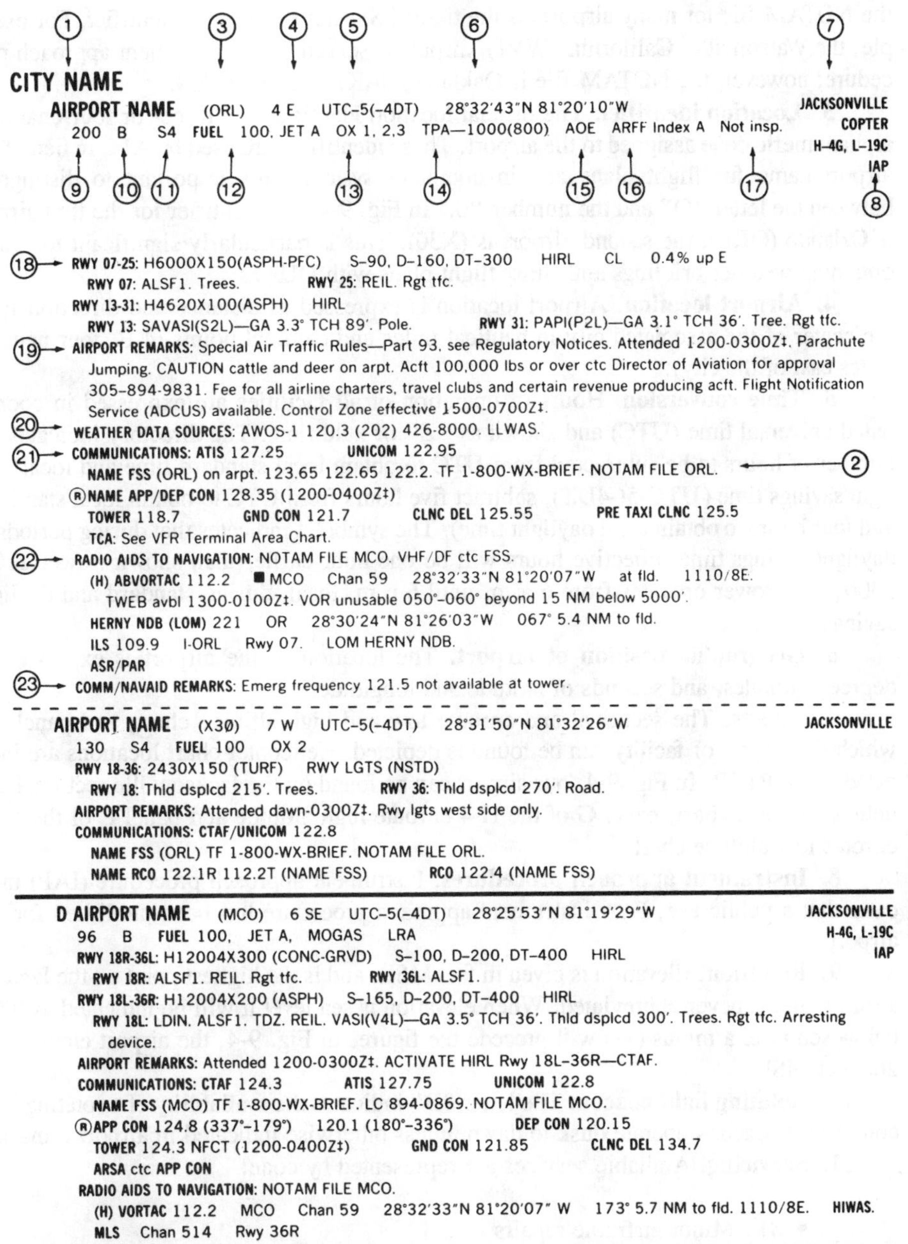

2 **DIRECTORY LEGEND**

SAMPLE

(1) (3) (4) (5) (6) (7)

CITY NAME

AIRPORT NAME (ORL) 4 E UTC–5(–4DT) 28°32′43″N 81°20′10″W **JACKSONVILLE**

200 B S4 **FUEL** 100, JET A OX 1, 2,3 TPA—1000(800) AOE ARFF Index A Not insp. **COPTER**

H-4G, L-19C

IAP

(9) (10) (11) (12) (13) (14) (15) (16) (17) (8)

(18) **RWY 07-25:** H6000X150(ASPH-PFC) S–90, D–160, DT–300 HIRL CL 0.4% up E

RWY 07: ALSF1. Trees. **RWY 25:** REIL. Rgt tfc.

RWY 13-31: H4620X100(ASPH) HIRL

RWY 13: SAVASI(S2L)—GA 3.3° TCH 89′. Pole. **RWY 31:** PAPI(P2L)—GA 3.1° TCH 36′. Tree. Rgt tfc.

(19) **AIRPORT REMARKS:** Special Air Traffic Rules—Part 93, see Regulatory Notices. Attended 1200-0300Z‡. Parachute Jumping. CAUTION cattle and deer on arpt. Acft 100,000 lbs or over ctc Director of Aviation for approval 305-894-9831. Fee for all airline charters, travel clubs and certain revenue producing acft. Flight Notification Service (ADCUS) available. Control Zone effective 1500-0700Z‡.

(20) **WEATHER DATA SOURCES:** AWOS-1 120.3 (202) 426-8000. LLWAS.

(21) **COMMUNICATIONS: ATIS** 127.25 **UNICOM** 122.95

NAME FSS (ORL) on arpt. 123.65 122.65 122.2. TF 1-800-WX-BRIEF. NOTAM FILE ORL. (2)

Ⓡ**NAME APP/DEP CON** 128.35 (1200-0400Z‡)

TOWER 118.7 **GND CON** 121.7 **CLNC DEL** 125.55 **PRE TAXI CLNC** 125.5

TCA: See VFR Terminal Area Chart.

(22) **RADIO AIDS TO NAVIGATION:** NOTAM FILE MCO. VHF/DF ctc FSS.

(H) ABVORTAC 112.2 ■ MCO Chan 59 28°32′33″N 81°20′07″W at fld. 1110/8E.

TWEB avbl 1300-0100Z‡. VOR unusable 050°-060° beyond 15 NM below 5000′.

HERNY NDB (LOM) 221 OR 28°30′24″N 81°26′03″W 067° 5.4 NM to fld.

ILS 109.9 I-ORL Rwy 07. LOM HERNY NDB.

ASR/PAR

(23) **COMM/NAVAID REMARKS:** Emerg frequency 121.5 not available at tower.

AIRPORT NAME (X3Ø) 7 W UTC–5(–4DT) 28°31′50″N 81°32′26″W **JACKSONVILLE**

130 S4 **FUEL** 100 OX 2

RWY 18-36: 2430X150 (TURF) RWY LGTS (NSTD)

RWY 18: Thld dsplcd 215′. Trees. **RWY 36:** Thld dsplcd 270′. Road.

AIRPORT REMARKS: Attended dawn-0300Z‡. Rwy lgts west side only.

COMMUNICATIONS: CTAF/UNICOM 122.8

NAME FSS (ORL) TF 1-800-WX-BRIEF. NOTAM FILE ORL.

NAME RCO 122.1R 112.2T (NAME FSS) **RCO** 122.4 (NAME FSS)

D AIRPORT NAME (MCO) 6 SE UTC–5(–4DT) 28°25′53″N 81°19′29″W **JACKSONVILLE**

96 B **FUEL** 100, JET A, MOGAS LRA **H-4G, L-19C**

RWY 18R-36L: H12004X300 (CONC-GRVD) S–100, D–200, DT–400 HIRL **IAP**

RWY 18R: ALSF1. REIL. Rgt tfc. **RWY 36L:** ALSF1.

RWY 18L-36R: H12004X200 (ASPH) S–165, D–200, DT–400 HIRL

RWY 18L: LDIN. ALSF1. TDZ. REIL. VASI(V4L)—GA 3.5° TCH 36′. Thld dsplcd 300′. Trees. Rgt tfc. Arresting device.

AIRPORT REMARKS: Attended 1200-0300Z‡. ACTIVATE HIRL Rwy 18L-36R—CTAF.

COMMUNICATIONS: CTAF 124.3 **ATIS** 127.75 **UNICOM** 122.8

NAME FSS (MCO) TF 1-800-WX-BRIEF. LC 894-0869. NOTAM FILE MCO.

Ⓡ**APP CON** 124.8 (337°-179°) 120.1 (180°-336°) **DEP CON** 120.15

TOWER 124.3 NFCT (1200-0400Z‡) **GND CON** 121.85 **CLNC DEL** 134.7

ARSA ctc **APP CON**

RADIO AIDS TO NAVIGATION: NOTAM FILE MCO.

(H) VORTAC 112.2 MCO Chan 59 28°32′33″N 81°20′07″ W 173° 5.7 NM to fld. 1110/8E. **HIWAS.**

MLS Chan 514 Rwy 36R

E AIRPORT NAME (See PLYMOUTH)

All Bearings and Radials are Magnetic unless otherwise specified.

All mileages are nautical unless otherwise noted.

All times are UTC except as noted.

HORIZONTAL DATUM: Alaska, Canada and Conterminous United States based on 1927 North American Datum.

All other areas based on local datum.

Fig. 9-4. The *Airport/Facility Directory* provides detailed information that cannot be included on charts.

the NOTAM file for many airports is the tie-in FSS, not the airport identifier. For example, the Watsonville, California (WVI), airport is served by an instrument approach procedure; however, the NOTAM file is Oakland (OAK), the tie-in FSS.

3. Location identifier. The official location identifier is a three- or four-character alphanumeric code assigned to the airport. These identifiers are used by ATC in lieu of the airport name for flight plans and in computer systems. It's important to distinguish between the letter "O" and the number "0." In Fig. 9-4, the identifier for the first airport is Orlando (ORL), the second airport is (X30). This is particularly significant for pilots obtaining weather briefings and filing flight plans with DUATs.

4. Airport location. Airport location is expressed as distance and direction from the center of the associated city in nautical miles and cardinal points (4 E, four nautical miles east of the city).

5. Time conversion. Hours of operation of all facilities are expressed in coordinated universal time (UTC) and shown as "Z" or zulu time. The directory indicates the number of hours to be subtracted from UTC to obtain local standard time and local daylight savings time (UTC-5(-4DT), subtract five hours from UTC to obtain local standard, and four hours to obtain local daylight time). The symbol ‡ indicates that during periods of daylight savings time, effective hours will be one hour earlier than shown (tower 1100-2300‡, this tower operates from 6 a.m. until 6 p.m. local, during standard and daylight savings time).

6. Geographic position of airport. The location of the airport is expressed in degrees, minutes, and seconds of latitude and longitude.

7. Charts. The sectional and enroute low and high altitude chart, and panel, on which the airport or facility can be found is depicted. Helicopter chart locations are indicated as COPTER. In Fig. 9-4, our airport can be found on the Jacksonville sectional and helicopter route chart, panel G of the H-4 enroute high altitude and panel C of the L-19 enroute low altitude chart.

8. Instrument approach procedures. Instrument approach procedure (IAP) indicates that a public use, FAA instrument approach procedure has been published for the airport.

9. Elevation. Elevation is given in feet MSL, and is the highest point on the landing surface, and is never abbreviated. When elevation is sea level it will be indicated as (00), below sea level a minus (−) will precede the figure. In Fig. 9-4, the airport elevation is 200 feet MSL.

10. Rotating light beacon. The letter "B" indicates the availability of a rotating beacon. These beacons operate dusk to dawn unless otherwise indicated in airport remarks.

11. Servicing. Available services are represented by code:

- S1 Minor airframe repairs
- S2 Minor airframe and minor powerplant repairs
- S3 Major airframe and minor powerplant repairs
- S4 Major airframe and major powerplant repairs

12. Fuel. Availability and grade of fuel are also coded:

- 80 Grade 80 gasoline (red)

- 100 Grade 100 gasoline (green)
- 100LL 100LL gasoline, low lead (blue)
- A Jet A kerosene freeze point −40 °C
- A1 Jet A-1 kerosene freeze point −50 °C
- A1+ Kerosene with icing inhibitor, freeze point −50 °C
- B Jet B wide-cut turbine fuel freeze point −50 °C
- B+ Jet B wide-cut turbine fuel with icing inhibitor freeze point −50 °C
- MOGAS Automobile gasoline used as an aircraft fuel (Automobile gasoline may be used in specific aircraft engines that are FAA certified. MOGAS indicates automobile gasoline, but grade, type, and octane rating are not published. Due to a variety of factors, the fuel listed might not always be obtainable to transient pilots. Confirmation of availability should be made directly with fuel vendors at planned refueling locations.)

13. Oxygen. The availability of oxygen is indicated by one of the following:

- OX 1 High pressure
- OX 2 Low pressure
- OX 3 High pressure—replacement bottles
- OX 4 Low pressure—replacement bottles

14. Traffic pattern altitude. The first figure shown is traffic pattern altitude (TPA) above MSL, the second figure, in parentheses is TPA above airport elevation. In Fig. 9-4, TPA is 1,000 feet MSL, 800 feet agl. This is consistent with an airport elevation of 200 feet.

15. Airport of entry and landing rights airports. Airport of entry (AOE)—A customs airport of entry where permission from U.S. Customs is not required, but at least one hour advance notice of arrival must be furnished. Landing rights airport (LRA)—Application for permission to land must be submitted in advance to U.S. Customs, and at least one hour advance notice of arrival must be furnished. Advance notice of arrival at AOE and LRA airports may be included in the flight plan when filed in Canada or Mexico, where flight notification service (ADCUS) is available. Airport remarks will indicate this service. This notice will also be treated as an application for permission to land in the case of an LRA. Although advance notice of arrival may be relayed to Customs through Mexico, Canada, and U.S. communications facilities by flight plan, the aircraft operator is solely responsible for ensuring that customs receives the notification.

16. Certificated airport (FAR 139). Airports serving Department of Transportation certified carriers and certified under FAR Part 139 are indicated by the ARFF index, which relates to the availability of crash, fire, and rescue equipment. Index definitions are listed in the directory and FAR 139, "Certification and Operations: Land Airports Serving Certain Air Carriers." When the ARFF index changes, due to temporary equipment failure or other reasons, a NOTAM D will be issued advertising the condition.

17. FAA inspection. All airports not inspected by the FAA will be identified by the note: Not insp. This indicates that airport information has been provided by the owner or operator of the field.

18. Runway data. Runway information is shown on two lines. Information common to the entire runway is shown on the first line while information concerning the runway ends is shown on the second or following line. Lengthy information will be placed in airport remarks. Runway directions, surface, length, width, weight bearing capacity, lighting, gradient, and remarks are shown for each runway. Direction, length, width, lighting, and remarks are shown for seaplanes. The full dimensions of helipads are shown. Runway lengths prefixed by the letter "H" indicate that the runways are hard surface concrete or asphalt. If the runway length is not prefixed, the surface is sod, clay, and the like. Runway surface composition is indicated in parentheses after runway length:

- AFSC Aggregate friction seal coat
- ASPH Asphalt
- CONC Concrete
- DIRT Dirt
- GRVD Grooved
- GRVL Gravel, or cinders
- PFC Porous friction courses
- RFSC Rubberized friction seal coat
- TURF Turf
- TRTD Treated
- WC Wire combed

Runway strength data is derived from available information and is a realistic estimate of capability at an average level of activity. It is not intended as a maximum allowable weight or as an operating limitation. Many airport pavements are capable of supporting limited operations with gross weights of 25 to 50 percent in excess of the published figures. Permissible operating weights, insofar as runway strengths are concerned, are a matter of agreement between the owner and user. When desiring to operate into any airport at weights in excess of those published, users should contact the airport management for permission. Runway weight bearing capacity is indicated by code:

- S Single-wheel type landing gear (DC-3)
- D Dual-wheel type landing gear (DC-6)
- DT Dual-tandem type landing gear (B707)
- DDT Double dual-tandem type landing gear (B747)

Quadricycle and dual-tandem are considered virtually equal for runway weight bearing consideration, as are single-tandem and dual-wheel. The omission of weight bearing capacity indicates information is unknown. Three zeros are added to the figures for gross weight capacity. For example, S-90 single-wheel type landing gear, weight 90,000 pounds.

Lighting available by prior arrangement only or operating part of the night only, or pilot controlled, and with specific operating hours are indicated under airport remarks. Because obstructions are usually lighted, obstruction lighting is not included in the lighting code. Unlighted obstructions on or surrounding an airport will be noted in airport remarks. Runway light nonstandard (NSTD) are systems for which the light fixtures are not FAA approved—color, intensity, or spacing does not meet FAA standards. Nonstandard lighting will be shown in airport remarks. Types of lighting are shown with the runway or runway end they serve. Lighting contractions can be decoded by referring to Fig. 8-1 or Fig. 8-4.

The type of visual approach slope indicator (VASI) and its location are described by a three-digit alphanumeric code. The code begins with the letter V, indicating a VASI system. The next figure indicates the number of boxes utilized (2, 2-box; 4, 4-box; 6, 6-box). The last letter indicates which side of the runway has the unit when it is a single side installation (L, left; R, right). For example, V6R would be a 6-box VASI on the right side of the runway; V16 is a 16-box VASI on both sides of the runway.

Runway gradient will be shown only when it is 0.3 percent or more. When available, the direction of upward slope will be indicated. Lighting systems, obstructions, and displaced thresholds will be shown on the specific runway end. Right-hand traffic patterns for specific runways are indicated by "Rgt tfc."

In Fig. 9-4, RWY 07-25 is hard surface, 6,000 × 150 feet. The surface is asphalt with porous friction courses. Wheel bearing capacity is single-wheel 90,000, dual-wheel 160,000, and dual-tandem 300,000 pounds. High intensity runway lights and runway centerline lights are available. Runway gradient is 0.4 percent up towards the east. Runway 07 has ALSF-1 approach lighting system, with trees on the approach. Runway 25 is equipped with runway end identification lights, with a right traffic pattern.

19. Airport remarks. Airport remarks provide supplemental information on data already shown, or additional airport information. Data is confined to operational items affecting the status and usability of the airport.

20. Weather data sources. This section indicates the availability of weather data or an automated weather observing system (AWOS). AWOS is available in one of four systems:

- AWOS-A: reports altimeter setting only
- AWOS-1: reports altimeter setting, wind, and usually temperature, dewpoint, and density altitude
- AWOS-2: data in AWOS-1, plus visibility
- AWOS-3: data in AWOS-1, plus visibility and cloud/ceiling information

Other types of weather information are denoted by one of the following contractions:

- SAWRS Supplemental aviation weather reporting station for current weather information
- LAWRS Limited aviation weather reporting station for current weather information
- LLWAS Low level wind shear alert system
- HIWAS Under radio aids to navigation—hazardous in-flight weather advisory service, a continuous broadcast of SIGMETs and AIRMETs, and urgent pilot reports

21. Communications. Communications are listed in the following order along with the frequency:

• CTAF	Common traffic advisory frequency
• ATIS	Automatic terminal information service
• UNICOM	Aeronautical advisory station
• FSS	Flight service station
• APP CON	Approach control, "R" indicates the availability of radar
• Tower	Control tower
• GND CON	Ground control
• DEP CON	Departure control
• CLNC DEL	Clearance delivery
• PRE TAXI CLNC	Pretaxi clearance

Pretaxi clearance procedures have been established at certain airports to allow pilots of departing IFR aircraft to receive the IFR clearance before taxiing for takeoff.

22. Radio aids to navigation. The directory lists all NAVAIDs, except military TACANs, that appear on NOS visual or IFR charts, and those upon which the FAA has approved instrument approach procedures. NAVAIDs within the National Airspace System have an automatic monitoring and shutdown feature in the event of malfunction. Unmonitored (UNMON), means that an ATC facility cannot observe the malfunction or shutdown if the facility fails. NAVAID NOTAM files are listed on the radio aids to navigation line. At times this NOTAM file will be different from the airport NOTAM file. For example, the Watsonville localizer and NDB, and Salinas (SNS) VOR are listed in its radio aids to navigation section. The localizer and NDB NOTAM file is OAK. The Salinas VOR NOTAM file is SNS.

Radio class designators and standard service volume (SSV) classifications, discussed in chapter 6, are listed: (T) terminal, (L) low altitude, and (H) high altitude. In addition to SSVs, restrictions within the normal altitude or range of a NAVAID are published. For example, "VOR unusable 030–090 beyond 30 nautical miles below 5,000′" indicates that the VOR cannot be relied upon for navigation between the 030 and 090 radials beyond 30 nautical miles, below an altitude of 5,000 feet.

The availability of TWEB is indicated by a black box preceding the facility ident, as shown in Fig. 9-4. Additionally, latitude and longitude coordinates, relation of the NAVAID to the airport, facility elevation, and magnetic variation are provided. In Fig. 9-4, the facility is at the field, has an elevation of 1,110 feet with a magnetic variation of 8°E (1110/8E).

ASR/PAR indicates that surveillance (ASR) or precision (PAR) radar instrument approach minimums are published. The availability of automatic weather broadcast (AB), direction finding service (DF), and HIWAS are also listed.

23. Communication/NAVAID remarks. Pertinent remarks concerning communication and NAVAIDs are included in this section. For example, possible interference to approach aids due to aircraft taxiing in the vicinity of the antenna, nonavailability of the emergency frequency at the tower, or tower local control sectorization will be listed in this section.

Refer to the "D AIRPORT NAME" portion of Fig. 9-4. Let's translate some of the airport data beginning with information for Runway 18L. This runway has a lead-in lighting system (LDIN) with the ALSF-1 approach lighting system. Additional lighting aids are touchdown zone (TDZ) and runway end identification lights (REIL). The runway is served with a visual approach slope indicator (VASI), a 4-box VASI on the left side of the runway (V4L), the glide angle is 3.5° that crosses the threshold at 36 feet (TCH 36′). The threshold is displaced 300 feet. Trees are close to the final approach, right traffic pattern, and an arresting gear device serves the runway. Airport remarks indicate that the airport is attended between 1200Z and 0300Z, and zulu time changes one hour during daylight savings. The HIRL are activated on the CTAF frequency. The FSS has toll free (TF) and local (LC) telephone numbers. The radar approach control uses frequency 124.8 for northwest through south arrivals (337°–179°), and 120.1 for south through northwest arrivals (180°–336°). Finally, the "E AIRPORT NAME" is an example of cross-referencing. The pilot is directed to "See PLYMOUTH."

ALASKA SUPPLEMENT

The Alaska supplement (AK) provides an A/FD for the state of Alaska, a joint civil/military flight information publication: also, a FLIP-A/FD for Alaskan civil and military visual and instrument charts. The Alaska supplement contains the following sections:

- General information
- A/FD legend
- A/FD
- Notices
- Associated data
- Procedures
- Emergency procedures
- Airport sketches
- Position reports

General information, legend, and directory contain generally the same information as the A/FD, except that they include military and private airports. The legend and directory also contain information on jet aircraft starting units, military specifications for aviation fuels and oils, military oxygen specifications, and arresting gear.

Facilities covered by the FAA and DOD NOTAM system are indicated by a diamond symbol for FAA/DOD NOTAMs or the section symbol for civil NOTAMs only. Pilots flying to airports not covered by the NOTAM system should contact the nearest flight service station or the airport operator for applicable NOTAM information.

Airports in the supplement are classified into two categories: military/federal government and civil airports open to the general public, plus some selected private airports. Airports are identified using an abbreviation:

- A U.S. Army
- AF U.S. Air Force
- ANG U.S. Air National Guard

- AR U.S. Army Reserve
- CG U.S. Coast Guard
- DND Canadian Department of National Defense
- FAA Federal Aviation Administration
- MC U.S. Marine Corps
- MOT Canadian Ministry of Transport
- N U.S. Navy
- NG U.S. Army National Guard
- PVT Private use only, closed to the public
- NMFS National Marine Fisheries Service
- USFS U.S. Forest Service

No classification indicates an airport open to the general public.

Airport lighting is indicated by number:

1. Portable lights—electrical
2. Boundary lights
3. Runway floods
4. Runway or strip
5. Approach lights
6. High intensity runway lights
7. High intensity approach lights
8. Sequenced flashing lights (SFL)
9. Visual approach slope indicator system (VASI)
10. Runway end identifier lights (REIL)
11. Runway centerline lights (RCL)

An "L" by itself indicates temporary lighting such as flares, smudge pots, or lanterns. An asterisk preceding an element indicates that it operates on request only, by phone, telegram, radio, or letter. Otherwise, lights operate sunset to sunrise, except where pilot controlled lighting (PCL) is indicated.

NAVAIDs providing scheduled weather broadcasts are indicated by radio class code B. FAA flight service stations broadcast at 15 minutes past the hour and Canadian stations at 20 and 50 minutes past the hour. These broadcasts contain weather reports, weather advisories, pilot reports, and NOTAMs for locations within 150 miles of the broadcast location.

In addition to the information contained in the A/FD, the Alaska supplement provides unique data for the area it serves. In Alaska, some FAA flight service stations provide long distance communications: air/ground and a weather broadcast via VOLMET on HF (high frequency). These are published in the supplement. The supplement also provides military air refueling and military training route data.

The procedures section of the supplement contains weather/NOTAM procedures, ARTCC communications, military and civilian air defense identification zone (ADIZ) information, and other general data. A separate section provides emergency procedures. This section contains air intercept signals, air/ground emergency signals, and search and rescue procedures. The final section provides airport sketches, similar to those found on instrument approach procedures charts. And, the back cover contains position report, flight plans, and change of flight plan sequences for in-flight operations.

PACIFIC CHART SUPPLEMENT

The Pacific chart supplement (PAC) is a civil flight information publication. It serves as an A/FD for the state of Hawaii and those areas of the Pacific served by U.S. facilities (American Samoa, Kiribati—Christmas Island; Tern, Kure, and Wake Island; and, the Caroline, Mariana, and Marshall Islands). The supplement contains ATC procedures for operating in the Pacific, including the same information found in domestic terminal procedures publication, for its area of coverage. The Pacific chart supplement contains the following:

- General information
- A/FD legend
- Airport/FD
- Notices
- Associated data
- Procedures
- Emergency procedures
- Airport sketches
- Terminal procedures
- Position reports

General information, legend, and directory are similar in content and format to the domestic directory and the Alaskan supplement. Notices are divided in special, general, and area categories. Special notices include information of a permanent or temporary nature, and sectional chart corrections. General notices include navigational warning areas, preferred routes, and general information on flying to Hawaii. Area notices provide general information for operations in the covered areas, including terminal area graphics and Hawaiian island reporting service. Associated data contains NAVAID and communications information, VOR receiver checkpoints, parachute jumping areas, special use airspace, visual navigation chart bulletin, and military training routes.

Procedures provide information on oceanic navigation and communications requirement, oceanic position reports, routes to the U.S. mainland, and Scatana (security control of air traffic and air navigation aids) and ADIZ procedures. Emergency procedures and airport sketches contain the same information as the Alaska supplement. The final section of the supplement contains terminal procedures. This section contains the same information in the same format, as the domestic terminal procedures publication. SIDs, STARs, and instrument approach procedures are contained in this section.

NOTAM PUBLICATION

The Notices to Airmen, Class II booklet is published every 14 days. The international term Class II refers to the fact that the data appears in printed form for mail distribution, rather than distributed on the FAA's telecommunications systems (FDC NOTAMs and

NOTAM Ds). The booklet is divided into two sections. Section one contains information of a general nature, such as airways, flight restrictions, airports, facilities, and procedural NOTAMs. Section two contains special notices too long for section one that concern a wide or unspecified geographical area, or items that do not meet the criteria for section one. Information in section two varies widely, but is included because of its impact on flight safety, such as airport radar service areas, terminal area graphics, loran-C status information, fly-ins, and the like.

The NOTAM publication provides NOTAM information current to approximately three weeks prior to its issuance date. Because the publication must be assembled and mailed, it cannot be more timely. Information that is not known far enough in advance for publication, or of a temporary nature, is distributed as NOTAM Ds and FDC NOTAMs on the FAA's telecommunications systems. The publication advises users of the last FDC NOTAM number contained in the booklet: FDC NOTAMs listed thru 1/1901 . . .—the most recent FDC NOTAM is 1/1901. FDC NOTAMs issued subsequent to this number are obtained from the FSS and through a DUAT vendor. FDC NOTAMs for temporary flight restrictions are not published in the NOTAM publication; however, they are almost always issued for a short duration.

NOTAMs are issued for the opening, closing, or any change in the operational status of the following facilities or services:

- Airports
- Runway data
- Airport operating restrictions (ARFF)
- Approach lighting systems
- Control zones
- Displaced thresholds
- Runway lighting
- Navigational facilities
- Airport traffic control towers
- Flight service stations
- Weather (AWOS)

OBTAINING CHARTS AND PUBLICATIONS

Additional sources of useful information for pilots are available from NOS and other agencies. NOS suggests two publications for basic reference and supplementary data: NOAA catalog of aeronautical charts and related products plus the NOAA subscription order brochure for aeronautical charts and related products.

The catalog of aeronautical charts and related products describes each IFR and VFR aeronautical chart and chart-related publication, including digital products. Selected related products available from the FAA are also described. Lists of chart agents, prices, chart coverage, and other information needed to select and order charts and publications are provided.

The subscription order brochure for aeronautical charts and related products contains complete ordering information and order forms for all NOS aeronautical products available on a subscription basis. Both publications are available free, upon request from:

NOAA Distribution Branch, N/CG33
National Ocean Service
Riverdale, MD 20737-1199

10 Supplemental publications

ADDITIONAL PUBLICATIONS THAT SUPPORT CHARTS ARE PUBLISHED BY the public and private sector; DMA's flight information publications, Canada flight supplement, and NOS supplemental documents (Fig. 10-1). Private vendor publications are most often airport directories that offer supplemental airport information, such as an airport sketch and names, types and telephone numbers of airport, restaurant, lodging, and transportation services.

FLIGHT INFORMATION PUBLICATIONS

The DMA's equivalent to the *Airport/Facility Directory* are flight information publications (FLIP) planning documents intended primarily for use in ground planning at base, squadron, and unit operations offices. They are revised between publication dates by issuing replacement pages or a planning change notice (PCN) on a schedule or as required basis: separate documents are general planning and area planning.

General planning contains general information on FLIPs, divisions of airspace, aviation weather codes, aircraft categories and codes, loran/omega chart coverage, and information on operations and firings over the high seas. They also include information on flight plans and pilot procedures that have common worldwide applications, and information on International Civil Aviation Organization (ICAO) procedures.

Area planning documents contain planning and procedural data for specific areas of the world. They include those theater, regional, and national procedures that differ from the standard procedures. Additionally, area planning military training routes are available.

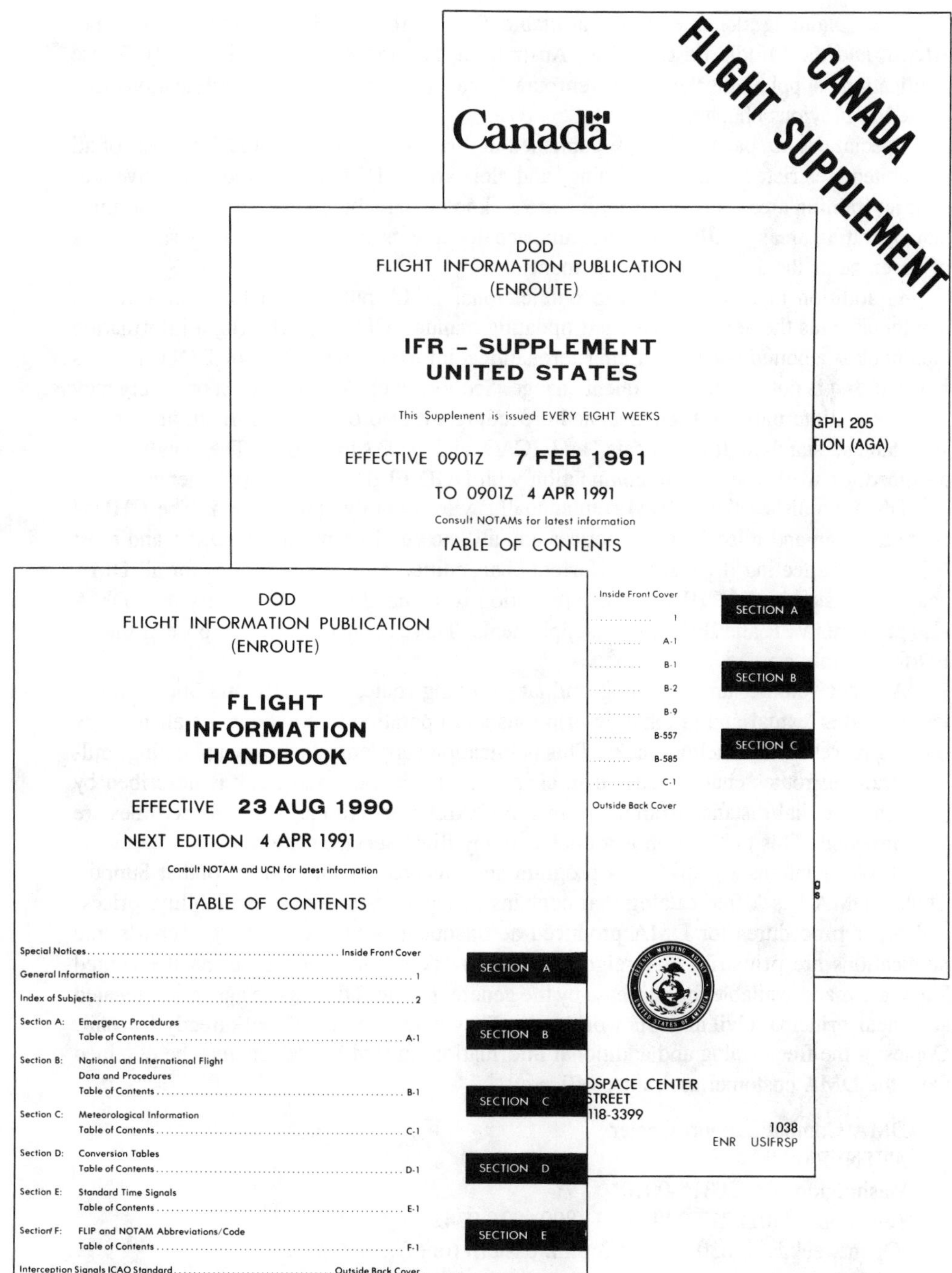

Fig. 10-1. Various chart publishers also issue flight information booklets to supplement the charts.

Area planning documents are available for North and South America; Europe, Africa, and the Middle East; Pacific, Australasia, and the Antarctic (Fig. 10-2). These publications supplement the visual, enroute, area, and terminal DMA publications discussed in previous chapters.

Special use airspace FLIPs are published in three books that contain tabulations of all prohibited, restricted, danger, warning, and alert areas. They also include intensive student jet training areas, military training areas, known parachute jumping areas, and military operating areas. FLIP special use airspace documents are available for the same areas of coverage as the area planning documents.

In addition to FLIP charts and publications, DMA publishes a flight information handbook, plus the aeronautical chart updating manual (CHUM). The flight information handbook is a bound book containing aeronautical information required by DOD aircrews in flight, but is not subject to frequent change. Sections include information on emergency procedures, international flight data and procedures, meteorological information, conversion tables, standard time signals, and ICAO and NOTAM codes. The handbook is designed for worldwide use in conjunction with DOD FLIP enroute supplements.

DMA publishes the CHUM semiannually with monthly supplements. The CHUM contains loran and miscellaneous notices and all known discrepancies to DMA and most NOS charts affecting flight safety. Current chart edition numbers and dates for all DMA charts are listed in the CHUM. The publication is intended for U.S. military use. DMA also produces VFR and IFR enroute supplements. These serve the same purpose as the A/FD for the military.

Another volume, area planning, military training routes for North and South America, provides textual and graphic descriptions and operating instructions for all military training routes, and refueling tracks. This publication supplements the area planning, military training routes chart discussed in chapter 5. Each route and track is described by location, in radial/distance from the nearest NAVAID, and altitude. Normal use times are also provided. This publication is available at any flight service station.

DMA maintains a public sales program administered by the DMA Combat Support Center. DMA has a free catalog that contains product descriptions, availability, prices, and order procedures for DMA produced aeronautical products. Although, charts and publications are primarily of foreign areas, many domestic charts covering the United States are made available for purchase by the general public. DMA sales agents are located at or near principal civil airports worldwide. They may also be ordered directly by mail. Copies of the free catalog and additional information on DMA products may be obtained from the DMA customer assistance office at:

DMA Combat Support Center
ATTN: PMA
Washington, DC 20315-0010
Telephone: (301) 277-2495 or 1 (800) 826-0342
(Open weekdays 6:30 a.m. to 3 p.m. eastern time.)

Use of any obsolete charts or publications for navigation is dangerous. Aeronautical information changes rapidly. It is critical that pilots have current charts and publications.

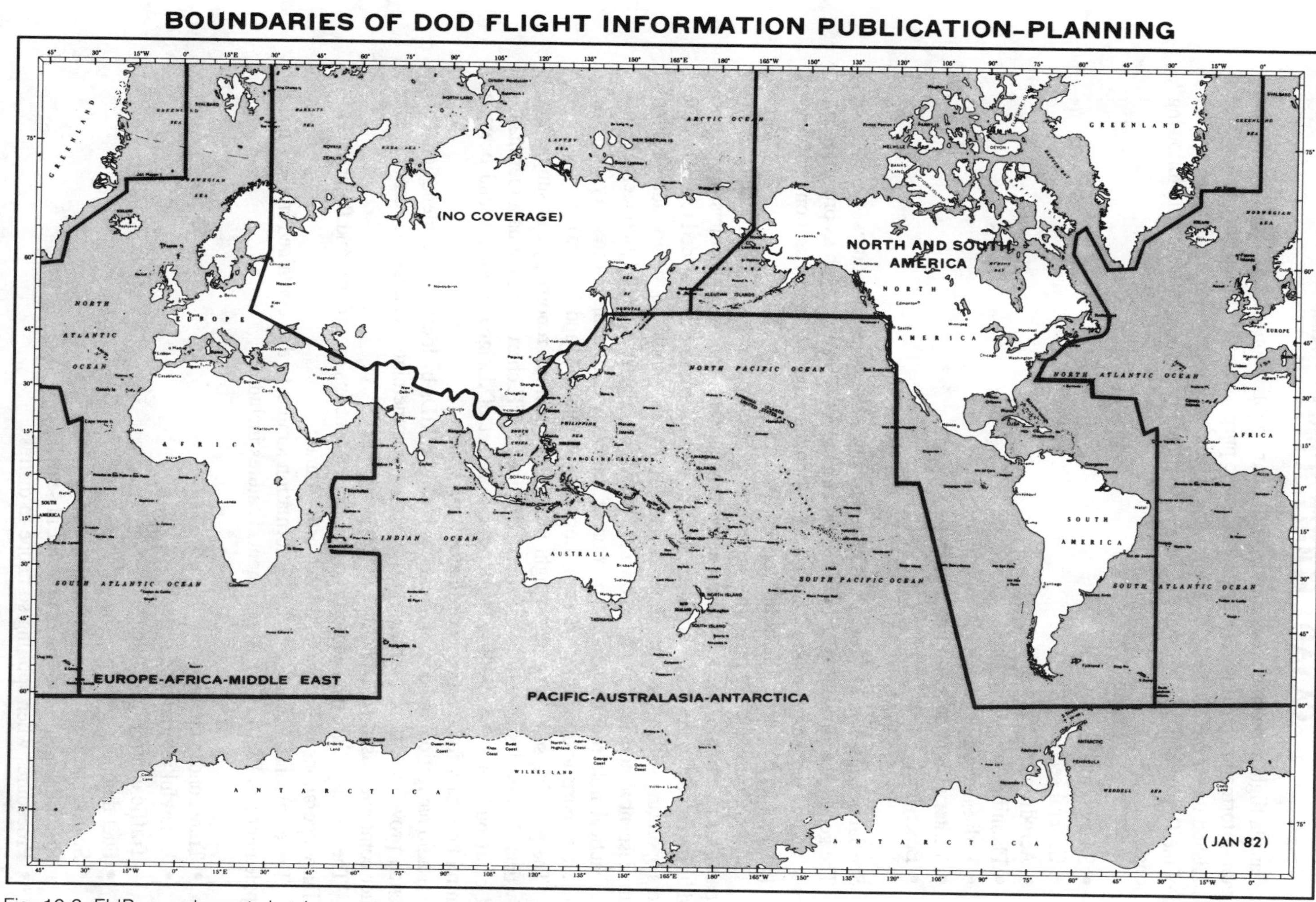

Fig. 10-2. FLIPs supplement visual, enroute, area, and terminal DMA charts.

CANADA FLIGHT SUPPLEMENT

The Canada flight supplement is a joint civil/military publication issued every 56 days that contains information on airports, serving the same purpose as the United States' A/FD. It is published under the authority of Transport Canada, Aeronautical Information Services and of the Department of National Defense. The flight supplement contains the following sections:

- Special notices
- General section
- Aerodrome/facility directory
- Planning
- Radio navigation and communications
- Military flight data and procedures
- Emergency

Special notices direct the pilot's attention to new or revised procedures.

The general section contains information of a universal nature, such as procurement, abbreviations, and a cross reference airport location identifier and name section. This section provides an airport and facilities legend for the supplement.

The aerodrome/facility directory serves the same purpose as its U.S. directory counterpart. This section might include an aerodrome sketch. The sketch depicts the airport and its immediate surrounding area. An obstacle clearance circle is provided to assist VFR pilots operating within close proximity to the airport, but should not be considered minimum descent altitudes. Altitudes shown are the highest obstacle, plus 1,000 feet within a five nautical mile radius. The last portion of this section contains a directory of North Atlantic airports and facilities (Azores, Bermuda, Greenland, and Iceland).

The planning section contains flight plan filing, position reporting, and supplementary data. It provides definitions of airspace classes, VFR chart updating data (aeronautical chart bulletin), and preferred IFR routes. A list of designated airway and oceanic control boundary intersection coordinates is provided.

Radio navigation and communications lists NAVAIDs by location and identifier. This section provides lists of marine radio beacons, commercial broadcasting stations, and other commonly used frequencies and services.

The military flight data and procedures section contains procedures and flight data for military operations in Canada and the North Atlantic.

The emergency section provides information similar to that contained in the Alaska supplement and Pacific chart supplement. Features include:

- Transponder operation
- Unlawful interference (hijack)
- Traffic control light signals
- Fuel dumping
- Search and rescue
- Recommended procedures to assist in search
- Procedures when spotting someone in distress

- Small craft distress signals
- Avoidance of search and rescue areas
- Emergency radar assistance
- Emergency communication procedures
- Two way communications failure
- Information signals
- Military visual signals
- Interception of civil aircraft
- Interception signals
- Signals for use in the event of interception

Canada produces a water aerodrome supplement that provides tabulated and textual information to supplement Canadian VFR charts. This bound booklet contains an aerodrome/facilities directory of all water landing areas shown on Canadian VFR charts. Communications, radio aids, and associated data are also listed.

Pilots planning flights into Canada may obtain a pamphlet, *Air Tourist Information Canada*, free of charge. This pamphlet describes procedures when entering Canada and also lists pertinent aeronautical information publications. This publication is available from:

Transport Canada
AANDHD
Ottawa, Ontario, Canada K1A 0N8

Canadian charts and publications and a free Canadian aeronautical chart catalog are available from:

Canada Map Office
Energy, Mines and Resources Canada
615 Booth Street
Ottawa, Ontario, Canada K1A 0E9
Telephone: (613) 952-7000

Pilots planning flights to Canada would be well advised to obtain the tourist pamphlet and chart and publications catalog. Publications are available on a one time sale basis. Once in flight is no time to realize how helpful a particular chart or publication would be.

AIRMAN'S INFORMATION MANUAL

The *Airman's Information Manual* (AIM) contains basic information needed for safe flight in the U.S. National Airspace System. It includes chapters describing navigation aids, airspace, the air traffic control system, and provides information on flight safety and safe operating practices. It also includes a pilot-controller glossary. The AIM is designed to provide pilots with basic flight information and ATC procedures. As well as fundamentals required to fly in the system, it contains items of interest to pilots concerning health and medical factors. The AIM is revised on a 112 day cycle, approximately three times a year.

AIM is the FAA's resource for pilots to understand operating in today's air traffic control system. With the exception of most regulations, the AIM is the pilot's handbook to operating in the airspace. All of the navigational aids, lighting aids, and procedural descriptions, not within the scope of this book, are defined and explained in the AIM. I recommend that any serious pilot subscribe to, or purchase from one of the many private sources, a copy of this document. It is invaluable to the flight instructor, as well as the student pilot. Features include:

- Navigation aids
- Aeronautical lighting and other airport visual aids
- Airspace
- Air traffic control
- Air traffic procedures
- Emergency procedures
- Safety of flight
- Medical facts for pilots
- Pilot controller glossary

Obtain AIM from the Government Printing Office (GPO). GPO will also provide, free of charge, subject bibliographies on a variety of aviation related topics, which will guide the reader to government publications available through the Superintendent of Documents. Related subject bibliographies include: aircraft, airports, and airways; aviation information and training; and, weather. These bibliographies and a free catalog are available from:

Superintendent of Documents
U.S. Government Printing Office
Washington, D.C. 20402
Telephone: (202) 783-3238

OTHER SUPPLEMENTAL PRODUCTS

Various additional supplemental products are available. Among these are a digital aeronautical chart supplement and a digital obstacle file.

IFR and VFR Pilot *Exam-O-Grams* contain brief explanations of important aeronautical knowledge. These items are critical to aviation safety. *Exam-O-Grams* are developed from the most often missed questions on FAA written exams.

The *Pilot's Handbook of Aeronautical Knowledge* contains essential information used in training. Subjects include the principles of flight, airplane performance, flight instruments, basic weather, navigation and charts, and excerpts from other flight information publications. This is a basic text, and ideal for the person getting started in aviation or interested in obtaining a general overview of aviation related topics.

A Guide to Federal Aviation Administration Publications is a 60-page document that contains information on the wide range of FAA documents and publications, and how they can be obtained. It lists available publications by category and gives the various sources. Listed also are civil aviation related publications issued by other federal agencies. Obtain a free copy by ordering FAA-APA-PG-9 from:

U.S. Department of Transportation
M-494.3
Washington, D.C. 20590

The *International Flight Information Manual* is designed primarily as a preflight and planning guide for use by U.S. nonscheduled operators, business, and private aviators contemplating flights outside of the U.S. This manual, which is complemented by the international NOTAM publication, contains foreign entry requirements, a directory of airports of entry, including data that is rarely amended, and pertinent regulations and restrictions. Information of a rapidly changing nature such as hours of operation, communications frequencies, and danger area boundaries, including restricted and prohibited areas, are not included and the pilot assumes the responsibility for securing that information from other sources: charts, NOTAMs, and enroute supplements. The basic manual is revised every April, with changes issued in July, October, and January. This manual is available at most flight service stations.

The international NOTAM publication provides NOTAM service on a worldwide basis. It covers temporary hazardous conditions, changes in facility operational data and foreign entry procedures and regulations. It supplements the international flight information manual, and is also available at most flight service stations. The respective international NOTAM publication and manual are available on a one year subscription service.

The U.S. *Aeronautical Information Publication* (AIP) is issued every two years with changes every 16 weeks. This is a comprehensive aeronautical publication containing regulations and data for safe operations in the National Airspace System. It is produced in accordance with the recommended standard of the International Civil Aviation Organization (ICAO).

The *Location Identifiers Handbook* 7350.5 lists the location identifiers authorized by the Federal Aviation Administration, Department of the Navy, and Transport Canada. It lists U.S. airspace fixes with latitude, longitude, Rho Theta descriptions, and procedure codes. The handbook also includes guidelines for requesting identifiers and procedures for making assignments.

The *Contractions Handbook* 7340.1 lists contractions for general aeronautical, National Weather Service, air traffic control, and aeronautical weather usage. The handbook provides encode and decode sections, plus air carrier, air taxi, and nationality identifiers: N (USA), C (Canadian), G (United Kingdom), and the like. The last sections provide a list of aircraft type contractions: C150, BE35, PA28, and the like.

Digital supplement

The digital aeronautical chart supplement (DACS) is designed to provide digital airspace data that is not otherwise readily available to the public. This publication was originally used only by air traffic controllers, but is now available to pilots for use in flight planning, and may soon be available on diskettes and tape. The following sections of the DACS are available separately or as a set:

- Section 1: High altitude airways—contiguous U.S.
- Section 2: Low altitude airways—contiguous U.S.

- Section 3: Reserved
- Section 4: Military training routes
- Section 5: Alaska, Hawaii, Puerto Rico, Bahama, and selected oceanic routes
- Section 6: STARs and profile descent procedures
- Section 7: SIDs
- Section 8: Preferred IFR routes
- Section 9: Air route and airport surveillance radar facilities

Features include:

- Routes listed numerically by official designation
- NAVAIDs and fixes listed by official location identifier
- Fixes without official location identifiers (airway intersections, ARTCC boundary crossing points) listed by five digit FAA computer code
- Latitude and longitude for each fix listed to tenths of seconds
- Magnetic variations at NAVAIDs
- Controlling ARTCC
- Military training route descriptions (scheduling activity, altitude data, and route width)
- Preferred IFR route (include departure or arrival airport name, and effective times)
- Radar facilities (ground elevation, radar tower height and type of radar facility)
- Data that is new or deleted since the last edition is clearly marked or listed

The NAVAID digital data file contains the geographic position, type, and unique identifier for every navigational aid in the United States, Puerto Rico, and the Virgin Islands. These data are chart-independent and can be applied to an NOS chart for which the data are required. Loran and RNAV avionics can use these data without modification. The data is government certified, and is compatible with the ARTCC system. This information is made available to the public, including avionics manufactures, software developers, flight planning services, pilots, navigators, and other chart producers. Features include:

- NAVAID identifier
- Type NAVAID
- NAVAID status (commissioned or not commissioned)
- Latitude and longitude to 10ths of a second
- Name of NAVAID
- NAVAID service volume category
- Frequency of NAVAID
- NAVAID elevation
- Magnetic variation
- ARTCC code where NAVAID is located
- State or country where NAVAID is located

These documents are used by chart producers and programmers of navigation systems. This information allows loran equipment to alert pilots of military and restricted areas and assist pilots in navigation.

Digital obstacle file

This quarterly file contains a complete listing of verified obstacles for the United States, Puerto Rico, and the Virgin Islands with limited coverage of the Pacific, Caribbean, Canada, and Mexico. Each obstacle is assigned a unique NOS numerical identifier. The obstacles are listed in ascending order of latitude within each state. A monthly revision file contains all changes made to verified obstacles during the previous four week period. The old record, as it appeared before the change, and the new record are shown. Features include:

- Unique NOS obstacle identifier
- Verification status
- State
- Associated city
- Latitude and longitude
- Obstacle type
- Number of obstacles
- Height agl
- Height MSL
- Lighting
- Horizontal and vertical accuracy code
- Marking, if known
- FAA study number
- Julian date of last change

COMMERCIAL PRODUCTS

In addition to supplementary products from government sources, private vendors produce a variety of publications. These cover the entire spectrum from the copier quality reproductions of local ATC and UNICOM frequencies to high quality airport sketch and data publications.

Jeppesen supplements its chart services with Federal Aviation Regulations and an airport and information directory, known as the J-AID. Features include:

- Radio aids to navigation, including coordinates, variation, and elevation
- Weather information sources
- Sunrise and twilight tables, and other common conversions
- A pilot controller glossary, airport and NAVAID lighting and markings, and services available to the pilot
- Federal Aviation Regulations
- International entry requirements

- Emergency procedures
- Airport diagrams and other airport information

A free catalog is available:

Jeppesen Sanderson
55 Inverness Drive East
Englewood, CO 80112-5498
Telephone: (303) 799-9090

The Aircraft Owners and Pilots Association (AOPA) publishes *Aviation USA*, available for sale to members and nonmembers, that provides a guide to AOPA member services, as well as an airport directory. It also has a section containing much of the material in the AIM. Publications of this type provide information on airport operators, transportation, lodging, and food services.

There are a number of commercially available publications similar to *Aviation USA*, but with subscription update service. These manuals provide a comprehensive listing of airports, with airport sketches, communication and NAVAID frequencies, phone numbers for services, restaurants, and hotels and motels. The Oakland FSS subscribes to a publication that covers the state of California; telephone numbers for FBOs, hotels, and restaurants have been tremendous time savers when locating an overdue aircraft and a pilot that has inadvertently forgotten to close a flight plan.

Pilots should not overlook the value of these publications. Most airport pilot shops carry these products or they may be ordered from an aviation catalog.

Glossary

above ground level (agl) Height, usually in feet, above the surface of the earth.

above sea level (ASL) *See* mean sea level.

Air Defense Identification Zone (ADIZ) The area of airspace over land or water, within which the ready identification, location, and the control of aircraft are required in the interest of national security.

aircraft approach category Landing minima established for aircraft, based on approach speed, and divided into A, B, C, D, E, and COPTER.

area navigation (RNAV) A method of navigation that permits aircraft operation on any desired course within the coverage of station-referenced navigation signals or within the limits of a self-contained system capability. Types include VORTAC, OMEGA/VLF, inertial, and loran.

ARFF *See* certificated airport.

automated weather observing system (AWOS) A computerized system that measures some or all of the following: sky conditions, visibility, precipitation, temperature, dewpoint, wind, and altimeter setting.

bernoulli effect The venturi effect of terrain that causes a decrease in air pressure, resulting in altimeter error.

bogs Areas of moist, soggy ground, usually over deposits of peat.

category *See* aircraft approach category; ILS category.

catenary As depicted cn aeronautical charts, a cable, power line, cable car, or similar structure suspended between peaks, a peak and valley below, or across a canyon or pass.

certificated airport (FAR 139) An airport certified for commercial air carriers under FAR Part 139, which relates to the requirement for crash, fire, and rescue equipment (ARFF).

changeover point (COP) The charted point where a pilot changes from one navigational facility to the next for course guidance.

CHUM *Chart Updating Manual* used to supplement Defense Mapping Agency charts.

climb gradient A rate of climb in feet per nautical mile, usually associated with a requirement for a departure procedure.

common traffic advisory frequency (CTAF) A frequency designed for the purpose of airport advisory practices, pilot's position and intentions during takeoff and landing, at uncontrolled airports.

Consol/CONSOLAN A long range radio aid to navigation, the emissions of which, by means of their radio frequency modulation characteristics, enable bearings to be determined.

coordinated universal time (UTC) Formerly Greenwich mean time (GMT), also known as Z or zulu time, UTC is the international time standard.

critical elevation The highest elevation in any group of related and more or less similar relief formations.

culture Features of the terrain that have been constructed by man, including roads, buildings, canals, and boundary lines.

datum Any quantity, or set of quantities, which might serve as a reference or base for other quantities.

decision height (DH) That altitude on a precision approach where the pilot must make the decision to continue the approach or execute a missed approach.

direct user access terminal (DUAT) A computer terminal where pilots can directly access meteorological and aeronautical information without the assistance of an FSS.

drainage patterns Drainage patterns are the overall appearance of features associated with water, such as shorelines, rivers, lakes, and marshes, or any similar feature.

DUAT *See* direct user access terminal.

feeder fix A fix depicted on instrument approach procedure charts that establishes the starting point of the feeder route.

feeder route A depicted course on instrument approach procedure charts to designate a route for aircraft to proceed from the enroute structure to an initial approach fix.

final approach fix (FAF) The fix where the final approach segment begins.

final approach point (FAP) The point on an instrument approach where the final approach segment begins. This occurs when the final approach segment begins at the NAVAID used for final approach guidance, and no final approach fix is designated.

final approach segment That part of an instrument approach that begins at the final approach fix or point and ends at the missed approach point.

fixed base operator (FBO) A private vendor of airport services, such as fuel, repairs, and tiedown facilities.

flight level (FL) Altitude with the altimeter set to standard pressure (29.92 inches or 1013.2 millibars), pressure altitude. Flight level is the altitude reference for high altitude flights, usually above 18,000 feet.

FLIP *Flight Information Publication* supplements Defense Mapping Agency charts.

flumes Water channels used to carry water as a source of power, such as a waterwheel.

great circle A circle on the surface of the earth, the plane of which passes through the center of the earth.

Greenwich meridian The meridian through Greenwich, England, serving as the reference for Greenwich time (now coordinated universal time UTC). It is accepted almost universally as the prime meridian, or the origin of measurements of longitude.

Julian date The date based on the Julian calendar. Days of the year are numbered consecutively from 001 for January 1; the year precedes the three-digit day group (91244 means September 1, 1991).

hachures A method of representing relief upon a map or chart by shading in short disconnected lines drawn in the direction of the slopes.

hazardous inflight weather advisory service (HIWAS) A continuous broadcast of hazardous weather conditions over selected NAVAIDs.

height above airport (HAA) The height of the minimum descent altitude above published airport elevation.

height above touchdown (HAT) The height of the decision height or minimum descent altitude above the highest runway elevation in the touchdown zone.

HIWAS *See* hazardous inflight weather advisory service.

horizontal datum A geodetic reference point that is the basis for horizontal control surveys, where latitude and longitude are known.

hummocks A wooded tract of land that rises above an adjacent marsh or swamp.

hydrography The science that deals with the measurements and description of the physical features of the oceans, seas, lakes, rivers, and their adjoining coastal areas, with particular reference to their use for navigational purposes.

hypsography The science or art of describing elevations of land surfaces with reference to a datum, usually sea level.

hypsometric tints A method of showing relief on maps and charts by coloring, in different shades, those parts that lie between selected levels.

ICAO *See* International Civil Aviation Organization.

ILS category The term category, with respect to the operation of aircraft, refers to a straight-in ILS approach. Category II and Category III operations allow specially trained crews, operating specially equipped aircraft, using specially certificated ILS systems, lower landing minimums than are available with Category I.

initial approach fix (IAF) The point where the initial approach segment begins. IAFs are charted as a military requirement.

initial approach segment That part of an instrument approach that begins at the initial approach fix and terminates at an intermediate fix.

intermediate segment That part of an instrument approach that begins at the intermediate fix and ends at the final approach fix.

International Civil Aviation Organization (ICAO) A specialized agency of the United Nations whose objective is to develop the principles and techniques of international air navigation and foster planning and development of international civil air transport.

isogonic lines Lines of equal magnetic declination for a given time, the difference between true and magnetic north.

latitude A linear or angular distance measured north or south of the equator.

location identifier Consisting of three to five alphanumeric characters, location identifiers are contractions used to identify geographical locations, navigational aids, and airway intersections.

longitude A linear or angular distance measured east or west from a reference meridian, usually the Greenwich meridian.

loran A *lo*ng *ra*nge radio *n*avigation position fixing system using the time difference of reception of pulse type transmissions from two or more fixed stations.

loran TD correction A correction factor, due to seasonal variations in loran signals, that must be set into loran receivers before beginning a loran RNAV approach.

magenta A purplish red color used on aeronautical charts to distinguish different features.

mangrove Any of a number of evergreen shrubs and trees growing in marshy and coastal tropical areas; a nipa is a palm tree indigenous to these areas.

maximum authorized altitude (MAA) The highest altitude, for which an MEA is designated, where adequate NAVAID signal coverage is assured.

maximum elevation figure (MEF) The MEF represents the highest elevation, including terrain or other vertical obstacles bounded by the ticked lines of the latitude/longitude grid on a chart.

mean sea level (MSL) Altitude above mean or average sea level. This is the reference altitude for most charted items. In Canada, it is called above sea level (ASL).

meridian A north-south reference line, particularly a great circle through the geographical poles of the earth, from which lines of longitude are determined.

minimum crossing altitude (MCA) The minimum altitude that a NAVAID or intersection may be crossed.

minimum descent altitude The lowest authorized altitude on a nonprecision approach.

minimum enroute altitude (MEA) The minimum published altitude that assures acceptable navigational signal coverage, meets minimum obstruction clearance requirements, and ensures radio communications.

minimum obstruction clearance altitude (MOCA) The altitude that meets obstruction clearance criteria between fixes, but only assures navigational signal coverage within 22 nautical miles of the NAVAID.

minimum reception altitude (MRA) The lowest altitude required to receive adequate NAVAID signals to determine specific fixes.

missed approach point (MAP) The point on an instrument approach where the missed approach procedure begins.

NAVAID Any type of radio aid to navigation.

nipa *See* mangrove.

nonperennial *See* perennial.

North American datum The horizontal datum for the United States developed in 1927, located at Meades Ranch, Kansas; referred to as the North American Datum 1927 (NAD 27).

NOTAM file As used in a flight supplement, the associated weather or facility identifier (OAK) where notices to airman for the associated facility will be located.

OMEGA A long range hyperbolic navigation system designed to provide worldwide coverage for navigation.

parallel A circle on the surface of the earth, parallel to the plane of the equator and connecting all points of equal latitude.

penstock *See* flume.

perennial A feature, such as a lake or stream that contains water year round, as opposed to nonperennial, a feature that is intermittently dry.

planimetry Planimetry is the depiction of manmade and natural features, such as woods and water, but does not include relief.

prime meridian *See* Greenwich meridian.

projection A systematic drawing of lines on a plan surface to represent the parallels of latitude and the meridians of longitude of the earth.

QNE This is the altitude shown on the altimeter with the altimeter set to 29.92 inches, pressure altitude.

QNH This is altitude above mean sea level displayed on the altimeter when the altimeter setting window is set to the local altimeter setting.

relief Relief is the inequalities of elevation and the configuration of land features on the surface of the earth.

rhumb line A line on the surface of the earth cutting all meridians at the same angle.

runway visual range (RVR) The horizontal distance a pilot will be able to see high intensity lights down the runway from the approach end.

security control of air traffic and navigation aids (SCATANA) A plan for the security control of civil and military air traffic and NAVAIDs under various conditions.

scale The ratio or fraction between the distance on a chart and the corresponding distance on the surface of the earth.

SCATANA *See* security control of air traffic and navigation aids.

scheduled weather broadcast Available only in Alaska, a flight service station broadcast of certain meteorological and aeronautical information over NAVAID voice channels at 15 minutes past the hour.

special use airspace (SUA) Airspace where activities must be confined because of their nature, such as military operations or national security. Restrictions to flight might be imposed, or they might alert pilots to hazards of concentrated, high speed, or acrobatic military flying.

spot elevation A point on a chart where elevation is noted, usually the highest point on a ridge or mountain range.

shaded relief A method of shading areas on a map or chart so that they would appear in shadow if illuminated from the northwest.

standard service volume (SSV) The distances and altitude a particular NAVAID can be relied upon for accurate navigational guidance.

topography The configuration of the surface of the earth, including its relief, the position of its streams, roads, cities, and the like.

touchdown zone The first 3,000 feet of the runway.

touchdown zone elevation (TDZE) The highest elevation in the first 3,000 feet of the landing surface.

tundra A rolling, treeless, often marshy plain, usually associated with arctic regions.

unicom A nongovernment communications facility that may provide airport advisory information.

UTC *See* coordinated universal time.

vignette A gradual reduction in density so that a line appears to fade in one direction, often used to distinguish airspace boundaries.

visual descent point (VDP) The point on a nonprecision straight-in approach where normal descent from the minimum descent altitude to the runway touchdown point may begin.

zulu (Z) *See* coordinated universal time.

Index

A

T

U

V

W

Z

About the Author

Terry Lankford has been working with aeronautical charts almost daily for approximately 25 years. He obtained a private pilot certificate through an Air Force aero club in England in 1967, then through the G.I. Bill progressed to commercial and flight instructor certificates and an instrument rating, subsequently earning a flight instructor Gold Seal from the FAA. The certificate and rating were stepping stones to a position with the Federal Aviation Administration as an air traffic control specialist (station), although Lankford prefers the title flight service station specialist. He has owned two Cessna 150s and has flown across the United States on two occasions in his 150; flight time has also been logged within Hawaii, Canada, and Mexico.

Other Bestsellers of Related Interest

YOUR PILOT'S LICENSE—4th Edition—Joe Christy

Completly revised, this book offers all the information on student training requirements, flight procedures, and air regulations. It tells you what the physical qualifications are, frankly discusses the expense involved, explains the integral role ground study plays in learning to fly, and even supplies a sample written test comparable to the actual Private Pilot's Written Test. Active pilots and flight instructors will find this an excellent refresher reference, too! 176 pages, 73 illustrations. Book No. 2477, $13.95 paperback only

The classic you've been searching for . . .

STICK AND RUDDER: An Explanation of the Art of Flying—Wolfgang Langewiesche

Students, certificated pilots, and instructors alike have praised this book as *"the most useful guide to flying ever written."* The book explains the important phases of the art of flying, in a way the learner can use. It shows precisely what the pilot does when he flies, just how he does, and why. 395 pages, illustrated. Book No. 3820, $19.95 hardcover only

ABCs OF SAFE FLYING—3rd Edition
—David Frazier

This book gives you a wealth of flight safety information in a fun to read, easily digestible format. The author's anecdotal episodes as well as NTSB accident reports lend both humor and sobering reality to the text. Detailed photographs, maps, and illustrations ensure that you'll understand key concepts and techniques. If you want to make sure you have the right skills each time you fly, this book is your one-stop source. 192 pages, illustrated. Book No. 3757, $14.95 paperback only

THE ART OF INSTRUMENT FLYING
—2nd Edition—J. R. Williams

". . . as complete and up-to-date as an instrument book can be." —***Aero*** magazine

Williams has updated his comprehensive guide to include all elements of IFR flight—flight director, Loran-C, and Omega navigational systems. And, en route, area, TCA, and SID/STAR reference charts reflect current designations. The first edition won the 1989 Best Technical Book award of the Western Region of the Aviation/Space Writers Association. 352 pages, 113 illustrations. Book No. 3654, $19.95 paperback only

STANDARD AIRCRAFT HANDBOOK
—5th Edition—Edited by Larry Reithmaier, originally compiled and edited by Stuart Leavell and Stanley Bungay

Now updated to cover the latest in aircraft parts, equipment, and construction techniques, this classic reference provides practical information on FAA-approved metal airplane hardware. Techniques are presented in step-by-step fashion and explained in shop terms without unnecessary theory and background. All data on materials and procedures is derived from current reports by the nation's largest aircraft manufacturers. 240 pages, 213 illustrations. Book No. 3634, $11.95 Vinyl only

GENERAL AVIATION LAW—Jerry A. Eichenberger

Although the regulatory burden that is part of flying sometimes seems overwhelming, it need not take the pleasure out of your flight time. Eichenberger provides an up-to-date survey of many aviation regulations, and gives you a solid understanding of FAA procedures and functions, airman ratings and maintenance certificates, the implications of aircraft ownership, and more. This book allows you to recognize legal problems before they result in FAA investigations and the potentially serious consequences. 240 pages. Book No. 3431, $16.95 paperback only

BECOMING AN AIRLINE PILOT—Jeff Griffin

Discover exactly what it takes to pursue a cockpit career, from the basics of flight school to your probationary year as a commercial pilot. This is a down-to-earth look at what really goes into preparing for and landing a job with a civilian carrier. You'll learn why it's important to start aiming at that career goal as early as your mid-teens. Griffin tells you how to write a resume and cover letter, how and where to send them, plus many more helpful job-hunting tips. 128 pages, 38 illustrations. Book No. 2449, $12.95 paperback only

AVOIDING COMMON PILOT ERRORS: An Air Traffic Controller's View—John Stewart

This essential reference interprets—from the controller's perspective—mistakes pilots frequently make when operating in controlled airspace. It cites examples of situations frequently encountered by controllers that show how improper training, lack of preflight preparation, poor communication skills, and confusing regulations can lead to pilot mistakes. 240 pages, 32 illustrations. Book No. 2434, $16.95 paperback only

EMERGENCY!: Crisis in the Cockpit
—Stanley Stewart

This rare glimpse into fight crew drills and procedures during crises recalls real-life incidents where disaster seemed imminent. Each situation, however, concludes without loss of life thanks to the skill, bravery, and resourcefulness of the pilots, flight engineers, and flight crews involved. Stewart takes you inside the cockpit to observe these dramatic, suspenseful operations as crewmembers prevent emergency situations from becoming fatal accidents. 272 pages, 51 illustrations. Book No. 3499, $14.95 paperback only

COMMERCIAL AVIATION SAFETY
—Alexander Wells, Ed.D.

This ideal on-the-job reference provides aviation professionals with much-needed analysis of current aviation safety policies and programs as they have developed since the Airline Deregulation Act of 1978. It takes a comprehensive look at how existing regulations, procedures, and technologies work to ensure safety in commercial aviation, as well as what both government and industry can do to ease the strain on the nation's increasingly overcrowded airspace. 352 pages, 41 illustrations. Book No. 3754, $32.95 hardcover only